Electronic Surveying and Navigation

Actual laser beam sent to the moon from the University of Hawaii LURE Observatory, Mt. Haleakala (photograph by William Carter).

ELECTRONIC SURVEYING AND NAVIGATION

SIMO H. LAURILA
Professor of Geodesy
University of Hawaii
Honolulu, Hawaii

A WILEY-INTERSCIENCE PUBLICATION

JOHN WILEY & SONS, New York—London—Sydney—Toronto

Copyright © 1976 by John Wiley & Sons, Inc.

All rights reserved. Published simultaneously in Canada.

Library of Congress Cataloging in Publication Data

Laurila, Simo.
  Electronic surveying and navigation.

  "A Wiley-Interscience publication."
  Includes bibliographical references and index.
  1. Electronics in surveying. 2. Electronics in navigation. I. Title.

QB301.L33   526'.028   75-41461
ISBN 0-471-51865-4

Printed in the United States of America

10 9 8 7 6 5 4 3 2 1

To My Family

PREFACE

The phenomenal advancement of electronic techniques has ultimately challenged the surveyor to abandon his measuring tape and the navigator his sextant in everyday use. Economy, accuracy, and flexibility are the key words the surveyors and navigators must keep in mind when they encounter the competition of modern scientific and commercial society. The field of electronic surveying and navigation is literally expanding every day, and new theories, instrument designs, and applications are presented as a continuous stream in the form of journal articles, symposium proceedings, and reports. In writing this book, my ambition has been to stop awhile in that stream, to try to collect between two covers the status quo of developments in electronic surveying and navigation, and to convey this information to surveyors, geodesists, geophysicists, hydrographers, and navigators. At first glance, it seems that the expected users of this book make up a group of professionals with widely varying interests, but one goal is common to all: obtaining precise positioning.

Based on my own experience as a researcher and teacher of this topic at the Helsinki Institute of Technology, the Ohio State University, and the University of Hawaii, my desire in writing this book has been to make it comprehensible to graduate students in the geosciences having a limited background in electronics, and also interesting for engineering students having no background in the geosciences. Therefore, the volume is divided into three parts. Part I consists of the basic principles of electronics and electronic surveying systems intended to be used as a reference for readers who are familiar with electronic concepts but need information on certain new designs presented in the main text. Also, readers who have little or no background in electronics can use Part I as a selected text to introduce the basic components of instrument designs.

Part II, excluding Chapter 20, consists of the mathematical and physical aspects of various systems, data reduction, and special applications; it can best serve as a text for readers who are utilizing these systems and want to dig deeper into the theory. Chapter 20 in Part II, and Part III, offer a comparative reference of design, capability, and accuracy of various instruments available today. All accuracy quotations are referred to one sigma if not otherwise stated, and distances are in metric units, except where the instrument design (modulation frequencies) requires the use of statute or nautical miles.

In writing this combined reference and textbook I wanted to bring the new and exciting field of electronic surveying and navigation to colleges, universities, and offices, and also to private persons interested in knowing how to determine the size and shape of the earth in a modern manner, how to probe satellites and the moon, and how to map oceans and continents. We should not forget the navigator who asks: where am I now, where am I going, and how far is it?

For making the writing of this book possible, I would like to acknowledge the cooperation of many instrument manufacturing companies who provided me with material while I visited their plants or through correspondence. I owe sincere thanks to Dr. *Matti Karras* from Oulu University, Finland, for reading Chapters 2 and 3 and for making many valuable suggestions, and to Miss *Valerie Hanna* for checking the manuscript for correct English usage. I would also like to thank Miss *Betty Matsui* and Mrs. *Carol Koyanagi* for the final typing of the manuscript (by kind cooperation with the Hawaii Institute of Geophysics) and Mrs. *Dolores Senus* for reading Chapters 25 and 26. Finally, I want to thank my wife *Mary* for reading the final manuscript and all proofs and for her patience and encouragement throughout the entire work.

As contributors to the very important chapters in electronic navigation, I owe gratitude to *Walter J. Senus*, U.S. Coast Guard, Washington, D.C., for writing Chapters 25 and 26 on Loran-C and Omega and *Thomas A. Stansell*, Jr., Magnavox, Torrance, California (previously at Johns Hopkins University, Applied Physics Laboratory, Maryland), for writing Chapter 28, "Positioning by Satellite."

Simo H. Laurila

Honolulu, Hawaii
October 1975

CONTENTS

INTRODUCTION TO ELECTRONIC SURVEYING AND NAVIGATION

HISTORY

The era of radio, electronics, and consequently electronic surveying and navigation began in the latter part of the nineteenth century when the famous English physicist, *James Clerk Maxwell* pointed out by mathematical calculations that if voltage and current alternate in an electric circuit, the circuit will radiate electric energy in a way similar to a wave motion. Maxwell called this radiation electromagnetic radiation, and he proved that its propagation velocity is about 300,000 km/sec. The German physicist *Heinrich Hertz* tested Maxwell's hypothesis in practice and in 1887 proved by experiments that radio waves were reflected from solid objects and could be formed into narrow beams by metallic conductors or antennas. The first step toward positioning by radio aids was made by *Guglielmo Marconi*, the Italian inventor and builder of the first wireless, when he presented the first horizontal directional antenna in 1902–1905. In 1924 Marconi strongly recommended the use of short waves for communication purpose and also built the first parabolic antenna applied to a wavelength of less than one meter. He had already ascertained in 1902 that the maximum distances that could be reached were much shorter during daytime than at night, a phenomenon that could only be explained by the existence of reflecting layers. The ionosphere was postulated in 1902 by an Anglo-American physicist *Arthur Kennelly* and an English physicist *Oliver Heaviside*; the Americans *Gregory Breit* and *Merle Tuve* of the Carnegie Institution in Washington, D.C., set the foundation for the pulse-type range measuring technique in 1925 by measuring the height of the Kennelly-Heaviside layer. These investigators transmitted a train of very short pulses of electromagnetic waves and determined the time that elapsed until the reflections were received. In 1928 an American *W. L. Everitt* invented the radioaltimeter, which measures the distance from an aircraft to the ground.

The three great mathematical physicists who revolutionized electron theory in the early twentieth century are the German *Max Planck*, the German-Swiss-American *Albert Einstein*, and the Dane *Niels Bohr*, who all contributed to the development of the *quantum theory*. In 1900 Max Planck introduced the quantum theory and Planck's constant, stating that the bound electron considered as an oscillator has an energy directly proportional to the frequency with Planck's constant as a proportional factor. Albert Einstein developed the quantum theory of light radiation and gave the radiation quantum the name *photon*. Niels Bohr first applied the quantum theory to atomic structure. Based on the new concept of atomic structure, the first breakthrough in modern electronics came years later: *John Bardeen* and *Walter Brattain* at the Bell Telephone Laboratories invented the point-contact transistor in 1948, and *William Shockley*, also of Bell, invented the junction transistor in 1949. Because of fast-developing solid state technology, instruments became more compact, lighter, longer lasting, and less expensive. It is interesting to note, however, that Heinrich Hertz was using semiconductors as early as 1889. Einstein's quantum theory of light radiation led to the invention of the laser (light amplification by stimulated emission of radiation). In the first experiments, stimulating radiation was used at microwave frequencies, and in 1958 *A. L. Schawlow* and *C. H. Townes* proposed the use of optical frequencies. The first operating ruby laser was demonstrated by *T. Maiman* in 1960. The cesium beam standard and the atomic clock, integral parts in modern electronic surveying and navigation equipment, serve to stabilize the transmission frequencies and to synchronize time.

Radar (radio detection and ranging) was developed during World War II to detect the presence of objects by means of reflected radio waves. Significantly, the total expenditure for research, development, and procurement of radar-type equipment in the period 1941–1944 exceeded the amount spent in developing the atom bomb. In addition to radar proper, such navigation systems as *Decca* and *Loran* played important role in the D-Day landing, and systems such as the *Gee-H* and *Oboe* were extensively used in guiding aircrafts to their target areas. After the war the potential use of such equipment in geodetic, geophysical, and navigational applications was immediately realized, and the production of new equipment and methods began to mushroom.

The first postwar attempt to measure the distance to the moon by radar was made on January 22, 1946, at Belmar, New Jersey, by the U.S. Army Corps of Engineers Signal Laboratories. Signals transmitted were reflected back by the moon's surface and were received about 2.5 sec after the transmission, and the reflections were clearly seen on the *cathode ray tube* of the receiver. The Space Age actually started on October 4, 1957, when the U.S.S.R. launched *Sputnik* 1, the first man-made artificial earth satellite. During the past two decades, several satellites have been launched to serve geodetic purposes, and in 1963 a program was initiated by the U.S. Coast and Geodetic Survey to establish a worldwide satellite triangulation network. A striking milestone in the history of using extraterrestrial bodies for investigation of minute movements on the earth's surface was the landing of Apollo 11 on the moon. Two of the astronauts, *Edwin E. Aldrin, Jr.*, and *Neil A. Armstrong*, placed the Lunar Ranging Retroreflector on the lunar surface on July 21, 1969. Similar reflectors were placed in different locations on the moon's surface by the astronauts of Apollo 14 and Apollo 15 on February 5, 1971, and July 31, 1971, respectively. One of the latest attempts of man to explore the earth from space was the launching of the space laboratory *Skylab* on May 14, 1973. In addition to many other scientific instruments, the Skylab carried a radar altimeter, which can measure distances from the spacecraft to the ground. The great significance of this method is that it allows us to measure directly such phenomena as ocean currents, geoidal heights, and sea states.

Today's surveyor or navigator can select the equipment and methods best suited for his needs from a multitude of devices ranging from small, inexpensive instruments capable of measuring distances of a few kilometers, to sophisticated, worldwide satellite navigation systems.

TWO

PRINCIPLES OF ELECTRONICS

To introduce the reader to the terminology and components frequently appearing in the main text, this chapter is confined to an outline of basic principles of electronics and electrical circuitry. The chapter is also intended to serve as a review for those already familiar with the concept of electronics.

2.1. ELECTRICAL UNITS AND BASIC DEFINITIONS

An atom can be subdivided into a positively charged *nucleus* or core and a cloud of negatively charged *electrons* that revolve at a very high speed around the nucleus. For example, the structure of the simplest atom, the hydrogen atom, consists of one electron revolving around the *proton*, which acts as the nucleus. Since the positive charge on the proton equals exactly the negative charge of the electron, the atom is electrically neutral. In solids the outermost electron states form energy bands that continue from atom to atom and through the material. If all the quantum states in a particular band are filled with electrons, there is no conductivity through the band, because no electron can find a free state to move into. Such a band is called the *valence band*, and a material having only filled valence bands is an *insulator*. If, however, the outermost band has only a partial population in its quantum states, electrons can move freely from one atom to another. Such a band is called the *conduction band*, and the material is a *conductor*. There are materials where the energy gap

6

between the valence band and the conduction band is so small that electrons can jump from the filled valence band to an empty conduction band. These materials with limited conductivity are called semiconductors.

The *voltage* in an electric circuit can be compared to pressure, since it is the force that causes the electron flow. The open-circuit voltage of an electric power source is called the *electromotive force* (emf) because it is the force available to cause motion of electrons. Voltage is actually a difference in electric *potential* (E), since the potential energy is able to do work. The unit of voltage or potential is the *volt* (V). The electron flow that results from the application of a voltage is called the current (I), whose fundamental unit is the *ampere* (A). The *resistance* (R) of the circuit is measured in *ohms* (Ω), and one ohm is the resistance through which the voltage of one volt causes the current of one ampere to flow. The resistance is the factor of proportionality between applied voltage and the current flow.

Potential difference between two conductors always creates a field, called the *electric field*. The *field strength* of an electric field, which can be denoted simply as E, is measured in volts per meter (V/m). Electrons moving in the conductor create a field around the conductor. This field is called the *magnetic field*, and its field strength (B) is stated in units of *webers* per square meter (Wb/m^2), which is also called *tesla* (T). According to the *Faraday's law*, an electromotive force is induced into a conductor that is in an alternating magnetic field or moves across a magnetic field.

The combination of the electric and magnetic fields, the *electromagnetic field*, is based on the laws that a moving electric field creates a magnetic field and a moving magnetic field creates an electric field. The electromagnetic field makes possible such things as building an electric motor or creating the radiation of electromagnetic waves or electromagnetic energy.

2.2. ALTERNATING CURRENT SIGNAL

An introduction to AC signal or alternating energy is obtained by reviewing the theory of simple harmonic motion. Let a point move along the perimeter of a circle with radius r counterclockwise at a constant speed (Fig. 2.1). At the time moment zero the point is assumed to be located at A, and t seconds later at P. The distance of the projection P'

from the center O of the circle is then

$$y = r \sin \Theta \qquad (2.1)$$

where Θ is the angle between the x-axis and the direction of the radius vector. If the time needed for one total revolution is T, then

$$\frac{\Theta}{2\pi} = \frac{t}{T} \qquad (2.2)$$

and Eq. 2.1 can be written

$$y = r \sin \frac{2\pi t}{T}$$

or

$$y = r \sin 2\pi f t \qquad (2.3)$$

where f is the frequency.

Application of the angular velocity of the radius vector

$$\omega = \frac{\Theta}{t}$$

into Eq. 2.1 yields

$$y = r \sin \omega t \qquad (2.4)$$

Faraday's law states that the induced emf in the coil of a generator is proportional to the rate of change of magnetic flux included in the coil; thus the general expression for harmonic motion (Eq. 2.4) can be rewritten for alternating sinusoidal voltage as follows:

$$V = V_0 \sin \omega t \qquad (2.5)$$

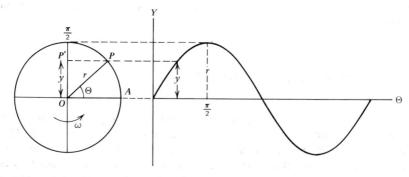

Figure 2.1   Harmonic motion.

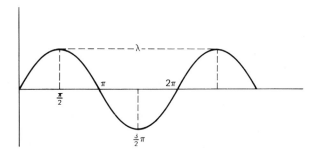

Figure 2.2   Wavelength.

where V_0 is the maximum voltage that can be developed by that particular generator, V is the instantaneous voltage, and ωt represents the *phase angle*. The term "phase angle" refers to time, but since all AC cycles are of the same length where time is concerned, the fraction of the cycle itself can be used as the angular unit. Angular degrees may also be utilized, anticipating that the total cycle is equal to 360°. If, for reasons explained later in the text, the phase of a signal is moved ahead, it is said to *lead* the phase of the original signal; if it is moved behind, we say that it *lags* that signal. If the phase lag is assigned the value of Φ, for instance, Eq. 2.5 should be written as follows:

$$V = V_0 \sin (\omega t - \Phi) \qquad (2.6)$$

The time taken for an alternating current to go through one complete cycle of values is called the period of the wave. One cycle of the wave motion is completed, of course, when one period has been completed, and the number of cycles per unit of time is called the *frequency*. The unit frequency is the *hertz* (Hz), which is one cycle per second. The linear length λ of a wave is the wavelength (Fig. 2.2), which can be determined as a function of the frequency f and the velocity of electromagnetic radiation c, as follows:

$$\lambda = \frac{c}{f} \qquad (2.7)$$

2.3.   RESISTANCE, INDUCTANCE, AND CAPACITANCE

When an alternating voltage E is applied to a *resistance* R, the current I flows exactly in phase with the voltage independent of the frequency, and

Ohm's law can be written as follows

$$R = \frac{E}{I} \qquad (2.8)$$

where R is one ohm when E is one volt and I is one ampere. The in-phase condition is valid only if the resistance is "pure," which means that there are no reactive effects in the circuit. Purity of the resistance is more difficult to achieve in very high frequency and ultrahigh frequency circuits.

Any change of current in a conductor causes a change of the magnetic flux around the conductor. Since a voltage is induced when magnetic flux cuts across a conductor, this change in flux causes the generation of a voltage in the conductor itself as well as in nearby circuits. The ability of a circuit to generate induced electromotive force is known as the *inductance* of the circuit. The symbol for inductance is L, and the unit of inductance is the *henry* (H). A coil of wire, the most common form of inductor, has an inductance of one henry when a voltage of one volt is induced by a uniform rate of current change of one ampere per second. Since the henry is a rather large unit in some components of radio circuits, the millihenry (mH: 10^{-3} H) and the *microhenry* (μH: 10^{-6} H) are frequently used. The induced emf in the coil of the inductor opposes the flow of the alternating current. This opposition, known as *inductive reactance*, is denoted by X_L. The magnitude of the induced emf is proportional to the rate of change of the current through the coil. Thus the voltage across the inductor is a maximum when the rate of change of the current is a maximum. In the sinusoidal wave form the rate of change is largest at zero, hence the phase angle of the voltage is 90° when that of the current is 0°; therefore, it can be said that the current through a pure inductance lags the voltage by 90°. The inductive reactance of a coil is stated in ohms and can be obtained from the following expression:

$$X_L = 2\pi fL \qquad (2.9)$$

where f is the frequency stated in hertz and L is the inductance stated in henrys.

If the medium between two conducting surfaces that are placed parallel and close together is an insulating material or *dielectric*, no electron flow can take place in it, altering the potential difference of the two surfaces. The direct current cannot pass through the space between the surfaces, but if an alternating voltage is applied between the surfaces, the electrons are leaving one surface and arriving at the other during a half-cycle and reversing the direction of flow during the opposite half-cycle. The measure of the ability of two surfaces or electrodes to store energy in an electric field that lies between them is called *capacitance*. A device that is designed to be able to store that energy is called a *capacitor* or a *condenser*. The capacitance of a capacitor increases as the area of the surfaces increases and decreases as the distance between them increases; it also depends on the type of dielectric material used between the surfaces. The symbol for capacitance is C and the unit of capacitance is the *farad* (F). A capacitor has a capacitance of one farad when a current of one ampere flowing into its plates for one second charges the device to one volt. A one-farad capacitor would be extremely large, and therefore microfarad (μF: 10^{-6} F) and picofarad (pF: 10^{-12} F) are commonly used. The amplitude of the alternating current that flows in a capacitor depends on both the size of the capacitor and the frequency, therefore the capacitor offers opposition to the flow of alternating current. This opposition is called *capacitive reactance* and is denoted by X_C. The capacitive reactance of a capacitor is stated in ohms and can be obtained from the following expression:

$$X_C = \frac{1}{2\pi f C} \tag{2.10}$$

where f is the frequency stated in hertz and C is the capacitance stated in farads. A capacitor that is initially uncharged tends to draw a large current when a voltage is first applied. When the charge on the capacitor reaches the applied voltage, no current flows. Thus if a sine-wave voltage is applied to a pure capacitance, the current is a maximum when the voltage begins to rise from zero, and the current is zero when the voltage across the capacitor is a maximum. As a result, the alternating current through a capacitor leads the voltage across it by 90°.

2.4.   RESONANT CIRCUITS

In a circuit that contains resistance, capacitive reactance, and inductive reactance, the total opposition to the flow of alternating current is called the *impedance* and denoted by Z. In this case the basic Ohm's law for a direct current (Eq. 2.8) is no longer valid because there is now a phase angle between the current and the voltage. For instance, the impedance of an inductor in *series* with a resistor does not equal the arithmetic sum of the inductive reactance and the resistance. By defining the above components in terms of voltage and adding them vectorially, the following expression is obtained to determine the impedance:

$$IZ = [(IR)^2 + (IX_L)^2]^{1/2} \tag{2.11}$$

It has been shown that the voltage across a pure inductance leads the current by 90° and the voltage across a pure capacitance lags the current by 90°. Therefore capacitive reactance may be considered negative and inductive reactance positive because their effects in a circuit are 180° out of phase with each other. The phase angle Φ between the applied voltage and the resulting current flow can be somewhere between +90° and −90°, and if the circuit consists of resistance, inductance, and capacitance in *series* (Fig. 2.3), the following general expression following Ohm's law for an AC circuit is obtained from the impedance triangle (Fig. 2.4):

$$Z^2 = (X_L - X_C)^2 + R^2 \tag{2.12}$$

and

$$Z = \sqrt{(X_L - X_C)^2 + R^2} \tag{2.13}$$

where $X_L - X_C = X$ is the net reactance. The phase angle Φ is directly obtained from Fig. 2.4 as follows:

$$\tan \Phi = \frac{X_L - X_C}{R} \tag{2.14}$$

If in a circuit inductive reactance X_L equals capacitive reactance X_C, the circuit is called a *resonant circuit*. The frequency of such a circuit, obtained by solving f simultaneously from Eqs. 2.9 and 2.10, is called the *resonant frequency* (f_0):

$$f_0 = \frac{1}{2\pi\sqrt{LC}} \tag{2.15}$$

If in the case of a series resonant circuit (Fig. 2.3), $X_L - X_C$ equals zero, the impedance will be $Z = R$, according to Eq. 2.13. Also, the phase angle Φ between the applied voltage and the current flow will be zero according to Eq. 2.14. If the frequency of the applied voltage is equal to the

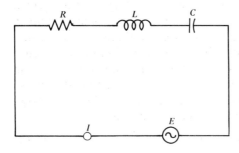

Figure 2.3   Series resonant circuit.

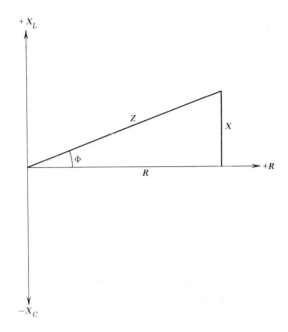

Figure 2.4 Determination of the impedance Z.

resonant frequency, the impedance is at a minimum, consisting only of resistance caused by the wire in the inductor and the wire connecting the circuit components. Therefore, at resonance the current reaches its maximum value. At frequencies below resonance the capacitive reactance becomes larger than the inductive reactance, and the impedance acts capacitively, like a circuit having only capacitor and resistor. The reverse action occurs at frequencies above the resonance and the impedance acts inductively.

A *parallel resonant circuit* consists of a combination of inductance, capacitance, and resistance connected in parallel (Fig. 2.5). The series and parallel resonant circuits are quite similar in some respects. For instance, in the parallel resonant circuit the resonant frequency f_0 equals the one in the series resonant circuit and can be obtained from Eq. 2.15 as a function of the capacitance and inductance. Because of the parallel connection of the circuit components, the net reactance will be

$$X = -\frac{X_L X_C}{X_L - X_C} \qquad (2.16)$$

using the convention given earlier. At the resonant frequency when

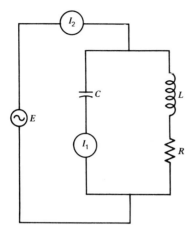

Figure 2.5   Parallel resonant circuit.

$X_L = X_C$, the net reactance, thus the impedance, is *theoretically* infinite and the line current I_l is zero. In effect, there is always some current flowing in the line because the resistance of the inductive reactance causes the phase angle to be slightly less than 90°. The actual impedance at resonance in a parallel circuit is obtained from the following expression:

$$Z = \frac{(2\pi f_0 L)^2}{R}$$
(2.17)

When Eqs. 2.9 and 2.10 are utilized, together with the condition $X_L = X_C$, Eq. 2.17 yields

$$Z = \frac{L}{CR}$$
(2.18)

The circulating current I_2 which flows within the parallel portion of an *LC* circuit, is very large compared with the line current I_1, therefore the parallel resonant circuit is often called the *tank circuit* because it acts as a storage tank when used in certain circuitries.

If a graph of the current is plotted against the frequency to include the resonant frequency f_0, the resultant curve is called the *resonant curve*. The sharper the resonant curve, the more selective the circuit and the more able it is to discriminate against or refuse the frequencies on either side of resonance. The factor indicating circuit selectivity (Q) is expressed as the ratio of the inductive reactance at resonance to the resistance as follows:

$$Q = \frac{X_L}{R} = \frac{2\pi f_0 L}{R}$$
(2.19)

The bandwidth as a function of the selectivity is obtained by dividing the resonant frequency f_0 by the selectivity factor Q as follows:

$$\Delta f = \frac{f_0}{Q} = \frac{R}{2\pi L} \tag{2.20}$$

As an example in computing some of the parameters included in a parallel resonant circuit, let us assume that the circuit inductance is $L = 200\ \mu\text{H}$, the capacitance is $C = 100\ \text{pF}$, and the resistance is $R = 10\ \Omega$. To determine the resonant frequency f_0 and the impedance across the terminals of the circuit, Eqs. 2.15 and 2.18 are applied as follows:

$$f_0 = \frac{1}{2\pi\sqrt{2\times10^{-4}\times10^{-10}}} = 1125.39\ \text{kHz}$$

and

$$Z = \frac{2\times10^{-4}}{10^{-10}\times10} = 2\times10^5\ \Omega$$

2.5. FREQUENCY MIXING

The combination of two frequencies to vary the voltage simultaneously is called *frequency mixing* or *frequency conversion*. Generally speaking, if an audio frequency or any very low information frequency is controlling the radio frequency, the process is termed *modulation* (which can have various forms). If a radio frequency controls another radio frequency, the process is known as a *heterodyne* action. In the modulation process the carrier and modulation frequencies usually differ greatly, and the modulation or pattern signal serves mostly for phase measurements in distance measuring instruments using microwaves or light waves as the carrier. In heterodyning action, two nearby frequencies of similar signals are mixed together, producing a complex combination of basic frequencies, their sums, differences, and higher order harmonics. In applications of heterodyning in distance measuring systems, the sums and differences are of special importance. To achieve more effective amplification of the modulated carrier, the received signal is mixed with a signal generated by a local oscillator, thus introducing a much lower difference frequency, called the *intermediate frequency* (IF), at which the amplification is made. The IF is kept constant by coupling the condensers of the receiving and local oscillating circuits mechanically, thus allowing the utilization of a fixed-tuned amplifier. The amplifier output then contains a wave having the same modulation as the original signal and a carrier frequency that is the difference between the frequency of the received signal and that of the locally superimposed oscillation. The mixed field, including all the different components previously stated, is fed into the tuning tank circuit by inductive coupling, for example. Any desired component, such as the difference of frequencies, may then be introduced by varying the capacitance of the tuning circuit.

Another important use of the heterodyne process in electronic distance measurements is based on a property of the modulation or measuring frequencies in microwave and light wave instruments: that is, such frequencies are usually of the order of tens of megahertz. It is impractical and inaccurate to compare phase angles of the transmitted and received signals at such high frequencies. Before the phase comparison takes place, therefore, the modulation frequencies are heterodyned down to a level of a few kilohertz. A third notable application of the heterodyne principle is in connection with medium long, continuous wave transmission usually employed in land-to-sea hydrographic survey. To distinguish transmissions from two or more ground station transmitters in the survey receiver, *beat frequencies* or *beat notes* are formed between transmissions, and the phase information they carry is conveyed to the survey receiver by various means (described in Chapter 3). The beat frequencies are usually within the low audio frequency band of several hundred hertz and are on either side or on both sides of the principal frequency. The one above the principal frequency, the *upper side frequency*, forms the *sum* or *upper sideband;* the one below the principal frequency, the *lower side frequency,* forms the *difference* or *lower sideband.*

To illustrate the heterodyne principle and the meaning of the sidebands, we can use the example given in the previous section. The selectivity factor Q can be obtained from Eq. 2.19 as follows:

$$Q = 6.283 \times 1125.39 \times 2 \times 10^{-5} = 141.4$$

from which the total bandwidth is obtained by Eq. 2.20:

$$\Delta f = \frac{1125.39}{141.4} = 7.96 \text{ kHz}$$

This means that the receiver tuned to 1125.39 kHz is capable of receiving a beat note up to about 4 kHz on its sidebands.

It is to be noted that heterodyne action can be achieved between two carrier waves as well as between two modulation waves. Also, an important property of the phase measurement should be kept in mind: namely, that *the difference in phase between two heterodyned low frequency signals is the same as that between the original signals.*

2.6. VACUUM TUBES

Metals are conductors because they have partially filled energy bands (conduction bands) that are not restricted to a particular atom but continue through the lattice. For free electrons to escape from metallic surface, sufficient energy must be acquired to enable the electrons to overcome the attracting electrostatic surface forces. If this energy is supplied in the form of heat, the resulting emission of electrons is called *thermionic emission*. The substance from which the electrons escape is the *cathode* (emitter). The high temperatures required to produce satisfactory thermionic emission without melting or sublimation of the material limit the number of substances from which cathodes can be made. The emitting surfaces used in vacuum tubes today are alkali metals and their alloys and oxides of some metals having high melting points and operating temperatures of about 1000°C (indirect heating). The free electrons form an electron cloud around the cathode because other ejected electrons form a shield that hinders their return to the cathode or further advance by the accelerating field to the *anode*. If the anode, another metallic electrode, is present in the vacuum tube and a positive voltage is applied to it relative to the cathode, an electric field is introduced between the two electrodes and negatively charged electrons start flowing from the surface of the electron cloud around the cathode to the anode. This is the basic concept of the *diode*, the simplest type of vacuum tubes.

The diode has several applications in electric circuits. The flow of electrons from cathode to anode is dependent not only on the voltage applied across them but also on the cathode temperature. If the diode is in a DC circuit, the anode current increases with the rising cathode temperature until the current reaches a saturation value that is not surpassed at any higher cathode temperature. By keeping the cathode temperature constant at this point, the diode acts as a current *limiter* because of the action of the electron cloud. Another application is based on recognition that current can flow only in one direction through the diode, namely, from the anode to the cathode. If an alternating voltage is applied in series with the circuit (Fig. 2.6), a current flows through the resistance load only during alternate half-cycles. The flow takes place only when the anode A is positive with respect to the cathode F, and no current flows during the other half-cycle when the anode is negative with respect to the cathode. If two diodes are combined to operate in alternating phasing, the current flows through the resistance load during all half-cycles. The diode in this application is said to act as a *rectifier*, in the former case as a half-wave rectifier and in the latter case as a full-wave

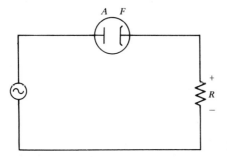

Figure 2.6 Diode as rectifier.

rectifier. The output voltage wave forms are presented in Fig. 2.7 as a pulsating direct current. The principle of this unidirectional characteristic of the diode is also used when the tube serves as a *detector* to resolve the modulating wave form out of a radiofrequency carrier.

When a third electrode, called the *grid* (G), is placed between the cathode and the anode, the vacuum tube is called a *triode*. The grid is designed to permit electrons to get past it to reach the anode; therefore it is an open helix or a mesh of fine wires. A varying grid potential has an important controlling effect on the anode-current output of the tube. If the grid is

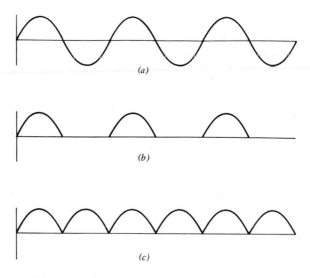

Figure 2.7 Pulsating direct current. (*a*) AC input, (*b*) half-wave rectified output, (*c*) full-wave rectified output.

made sufficiently negative with respect to the cathode, all electrons are repelled by it and are forced back to the cathode. No electrons reach the anode; thus the anode current is zero. The smallest negative voltage between grid and cathode which causes the tube to cease to conduct is called the *cutoff bias*. If the grid is made slightly less negative with respect to the cathode, some electrons get past the grid and move to the anode, producing a small anode current. Further decrease in the negative grid voltage causes further increase in anode current; at zero grid potential no retarding influence is exerted on the electrons, and the action is very similar to that of a diode. The grid thus may be considered to act as a valve to control the anode current, and triodes can be used as *amplifiers* because small changes in input voltage on the grid produce on the anode current effects much larger than those due to changes in anode voltage. When a resistance or impedance load (R_L in Fig. 2.8) is placed in series in the anode circuit, the voltage drop across it, which is a function of the anode current flowing through it, is controlled by the grid voltage. It should be noted, of course, that the output voltage comes from the source of energy (i.e., the battery E_B) in the amplifier itself, not from the source of the input. Amplification is usually made in successive stages, and the amplified output voltage is applied as an input voltage to the grid of the next amplifier stage in the amplification series. Coupling between amplification stages is usually capacitive. This is especially true in IF amplification, where coupling is made through tuned resonant tank circuits. This procedure allows the amplification be made at a constant and common IF value. The important use of the triode as an oscillator is explained in Section 2.8.

By adding more grids in the vacuum tubes, several new desirable characteristics are attained. The most common multielement tubes are *tetrodes*, having two grids, and *pentodes*, which have three grids.

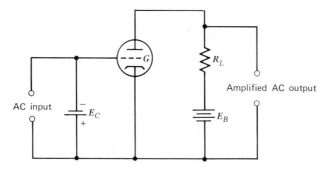

Figure 2.8 Triode as amplifier.

2.7. SEMICONDUCTORS

The use of thermionic or vacuum tubes in electric circuitry has several disadvantages involving size, temperature (which shortens the lifetime of the tubes), and relatively high power consumption. Therefore, the discovery and development of the semiconductors and their applications constitute one of the greatest breakthroughs in electronic technology. The resistance of pieces of solid materials is measured in ohms (Ω). Since conductance is a reverse action to resistance, it is the inverse value of resistance and is measured in mhos ($\mho$). The conductance values between good insulators (e.g., quartz) and good conductors (e.g., copper) vary between about 10^{-12} and $10^{7}\,\mho$/m, respectively, and the semiconductors are in the region of $10-10^{-3}\,\mho$/m. The two most important semiconductors are *germanium* and *silicon*, which have the following common characteristics. In their crystalline structure Ge and Si atoms are bonded together rigidly and symmetrically, but each atom shares one outer-shell or *valence* electron with each of four neighboring atoms, forming *covalent* bonds between the atoms. In conductors, such as metals, each atom has many electrons in the *conduction band*, free to move inside the piece of metal, and the conductivity increases with the number of free *charge carriers*. In insulators the case is reversed, and there are no electrons in the conduction band; in the *valence band*, moreover, all quantum mechanically possible electron states are populated, preventing electrons from moving from atom to atom. If an electron is excited from the valence band to the conduction band, the area the negatively charged electron has left possesses a net positive charge (the vacancy is called the *hole*), and the area in which the electron has settled has a net negative charge. Because the hole represents an empty electron state, an electron from the neighboring atom can move to fill it, and this is how the hole moves, carrying the current as well as the liberated electrons. In the semiconductors and their junctions, conductivity can be controlled with impurities over a wide range, which makes possible the design of *junction diodes, transistors, field effect transistors* (FET), and *photodiodes*, so important in modern electronics.

To comprehend this controlling process we must consider first the characteristics of a pure semiconductor. Relatively few valence electrons are in the conduction band, thus free to move throughout the crystal, the amount of free electrons being a function of temperature, external electrical field, and light exposure. Close to absolute zero temperature every valence electron is "tied" to the covalent bond and there are no current carriers available (electrons and holes). When the temperature rises, the number of conduction electrons increases, but even at room temperature only a small fraction of the valence electrons are in conduction band. In a pure semiconductor the crystal remains electrically neutral, since the number of holes equals the number of free electrons.

The conductivity of a semiconductor can be increased tremendously by *doping* the germanium or silicon with small, accurately measured amounts

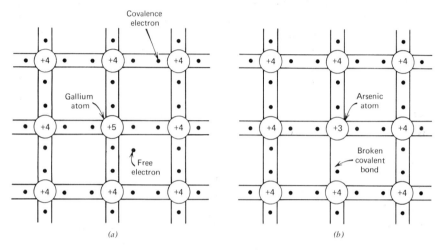

Figure 2.9 Two-dimensional view of atomic structure in germanium crystal. (a) *n*-doping, (b) *p*-doping.

of impurities. Since the germanium (silicon) is a tetravalent element itself, the impurities must be either pentavalent elements such as antimony, phosphorus, and arsenic, which have five valence electrons, or trivalent elements such as aluminum, boron, gallium, and indium, which have three valence electrons. Since each impurity atom from the first group is very easily ionized in the host lattice, it donates one free electron to the semiconductor and is thus called the *donor* atom. In this case free electrons become the *majority carrier*, and the doped semiconductor is called the *n-type* semiconductor with negative charge carriers. Similarly, each impurity atom from the second group accepts one valence electron, thus creating one mobile hole; such atoms are called *acceptor* atoms. In this case the holes become the majority carrier, and the semiconductor is designated a *p-type* semiconductor with positive charge carriers. Figure 2.9 illustrates these two cases of impurity doping.

2.7.1.   The *p-n* Junction Diode

If initially neutral *n*- and *p*-type semiconductors are placed in metallurgical contact with each other, a diffusion of the charge carriers, electrons and holes occurs through the contact plane, the *junction*. During diffusion, the excess electrons in the *n*-type semiconductor cross the junction and settle down in the holes, neutralizing them until an equilibrium is reached. At this point the departing electrons have left behind them positive ions and the *n*-type semiconductor has gained a positive potential over a *transition region*, from which all mobile charge carriers are removed. By absorbing the negative electrons, the *p*-type semiconductor has gained a negative potential, and a *potential barrier* settles in between the two semiconductors to prevent any further diffusion, unless an external voltage is applied across the pair of semiconductors. This potential barrier, also known as contact potential between any materials, is utilized as thermocouple contacts, for example.

When an external DC voltage is applied so that the negative terminal of the voltage supply is connected to the *n*-type material (Fig. 2.10*a*), electrons are attracted across the junction to the positive terminal and holes are attracted across the junction to the negative terminal of the supply. In this case current flows through the *p-n* junction from *p* to *n*. In other words, the potential barrier attracts simultaneously the majority carriers at both sides to pass the junction. When electrons have passed the transition region to the *p*-side, the holes, which are the majority carriers there, are recombining rapidly with the excess electrons, giving rise to hole current. The reverse occurs at the *n*-side. The passed holes recombine with the majority carriers (electrons), causing a local decrease of these particles. This shortage is then made up by the electrons flowing toward the junction at the *n*-side. If the polarity of the voltage is reversed, the attraction of the holes and electrons by the terminals of the

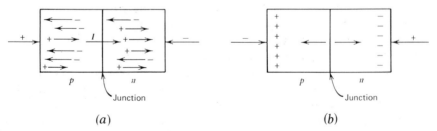

Figure 2.10 Voltage applications across *p-n* junction diode. (*a*) Current flows through junction, (*b*) no current flow.

voltage supply is also reversed, and no current flows across the junction (Fig. 2.10b). Thus if an alternating current is applied across the p-n junction, current will flow only during one-half of each cycle. Because of this characteristic, which is similar to that of a diode vacuum tube, a pair of p- and n-type semiconductors is called a *p-n junction diode,* and it can be used as a rectifier. This type of diode rectifier differs slightly from a diode vacuum tube rectifier in that a small but measurable reverse current is present. This is caused by small amount of *minority carriers,* holes in n-material and electrons in p-material. The p-n junction diodes may also be used as mixers, but in that case, and especially when the microwave technique is being employed, the capacitance between the practically zero-spaced n- and p-plates cannot be neglected. To overcome this difficulty a so-called *point-contact* is established by forming a tiny p-type region under the contact point when n-type material is used for the main body of the diode.

2.7.2. Transistors

Two p-n junction diodes can be connected physically to form one element that is actually composed of the two diodes placed back to back, forming either a *pnp* or an *npn* composition. In reference to the first combination, the common region in the middle is called the *base,* the region p at the left is called the *emitter,* and the region on the right is called the *collector* (Fig. 2.11a). Correspondingly, there are two junctions—namely, the *emitter junction* and the *collector junction.* After physical contact is made between the three semiconductors, diffusion occurs without any external voltage stimulation, and at the point where the state of equilibrium is reached, potential barriers are formed between

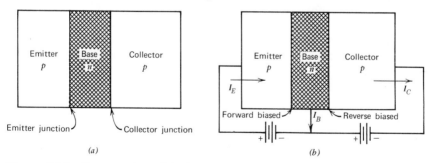

(a)   (b)

Figure 2.11 *Pnp transistor.* (a) Components of *pnp* triode, (b) current flow through *pnp* triode.

p and *n* and *n* and *p*. At this stage there is no external potential difference between the emitter and the collector.

If an external DC voltage is applied across the emitter and base as shown in Fig. 2.11*b*, so that the potential barrier between the emitter and the base is decreased, current will flow through the emitter junction. This means that the holes move from the emitter to the base, and only few electrons move from the base to the emitter because of the very small *n*-impurity concentration in the base. The collector voltage is much larger and has reverse polarity, and normally there would be no current flow in this circuit, but since the base is very thin, however, most emitter-injected holes are able to diffuse through the base without being neutralized by base electrons in the recombination process. The penetrated holes form the collector current because the negative collector attracts them through the transition region. The amplitude of the collector current depends on the amplitude of the emitter current, and for practical purposes $I_C = I_E$, where I_C and I_E are the collector and emitter currents, respectively. The current flow from emitter to base has little internal resistance; thus according to the Ohm's law, $E = IR$, a small input signal voltage in emitter circuit raises the current I_E, which then flows through a large voltage drop E_C of the collector junction. When a resistance load R_L is placed in the collector circuit, I_C is driven through it by the emitter voltage; thus the voltage drop over R_L is a function of I_C. The transistor acts then as an amplifier, where amplification can be controlled by the voltage applied to the base, and the functioning of a transistor resembles to that of the triode vacuum tube in this working mode.

Nowadays semiconductor technology commonly uses *field effect transistors* (FET) as active components. They are simpler to operate and manufacture than junction transistors. An FET contains a piece of semiconductor material forming a current-carrying *channel* between a *source* electrode and a *drain* electrode. The conductivity of this channel is varied by controlling the charge carrier concentration in the channel or by changing the cross-sectional *width* of the channel. The first alternative is facilitated by a *gate* electrode being insulated from the channel by a thin silicone dioxide layer. The electric field of the gate pulls electrons to the conduction band from donor impurities or neutralizes the ionized acceptor, thus reducing the number of holes. The insulated gate construction results in a device called the MOS (metal-oxide-semiconductor) FET. The cross section of the channel is varied in junction FETs, where the channel is separated from the gate by an *n-p* junction. The junction is always reverse biased, and the carrier free transition region grows with increasing bias voltage, thus controlling the width of the channel and its conductivity.

Both modifications of the FETs work with very low power consumption and the gate current is virtually zero in the MOSFET. When the drain-to-source voltage is increased, the channel current increases until a saturation maximum value, depending on a particular gate voltage, is reached. Field effect transistors usually act as amplifiers in the saturation state, and the channel current is controlled by the gate voltage.

In Fig. 2.12 symbols of semiconductor devices are presented, together with the relationship of their elements compared with those of vacuum tubes. Figure 2.12*a* is the symbol of a semiconductor diode, where the

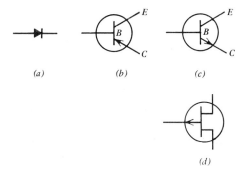

Figure 2.12 Symbols of semiconductor devices. (*a*) Junction diode (*b*) *pnp* transistor, (*c*) *npn* transistor, (*d*) field effect transistor.

arrow represents the anode of a vacuum tube diode and the bar represents the cathode. Figures 2.12*b* and 2.12*c* are the symbols of *pnp*- and *npn*-type transistors, where the emitter *E* represents the cathode, the base *B* represents the grid, and the collector *C* represents the anode of the vacuum tube triode. Fig. 2.12*d* is the FET symbol. The semiconductor diodes and transistors can replace the vacuum tubes in most functions in electronic circuitry.

2.8. OSCILLATORS

Circuitry that converts DC drawn from an energy source (battery) into AC where the electromagnetic energy alternates at the radio frequencies is called an oscillator. Excluding the case of the microwave radiation, an oscillator is composed of the resonant circuit and a triode vacuum tube, for example, or a transistor. There are several ways to connect the triode or transistor into the resonant circuit, but in the following we consider a simple oscillating circuit in which the anode circuit of a triode determines the frequency of the oscillation. In Fig. 2.13 the battery E_A charges the capacitor C instantaneously; the charge will be discharged through the inductor L, causing rapidly damping oscillation. To prevent damping, the instantaneous oscillation is transformed from the inductor by inductive coupling to the grid, thus applying a potential to the grid in addition to the grid bias from battery E_C. Now the triode acts as an amplifier, since the applied grid voltage amplifies the otherwise damping oscillation in the anode circuit. The capacitor charges again and the same cycle of oscillation is repeated undamped, while all the needed energy is drawn from the DC battery. The inductance L and the capacitance C in the tank circuit determine the fundamental frequency at which the oscillator oscillates.

The type of oscillator just described is usually stable enough to be used to generate carrier waves. To generate modulation, or measuring frequencies for which ultimate stability is required, the gradual change in the fundamental frequency caused by temperature changes and the aging of the components must be locked to a fixed, constant value. This is done by placing Q, a piece of quartz crystal (SiO_2), in proper position in the circuit. The use of the crystal is based on its *piezoelectric* characteristics (briefly: the crystal will vibrate both mechanically and electrically in the circuit). If the crystal replaces the LC grid circuit in the tuned-plate–tuned-grid oscillator, the voltage in that circuit causes mechanical changes in the crystal, which starts to vibrate mechanically at a frequency that is dependent on its own shape, size, and especially thickness. The crystal is also affected by temperature and therefore is kept in a thermostatically controlled oven. On the other hand, the mechanical change in the crystal

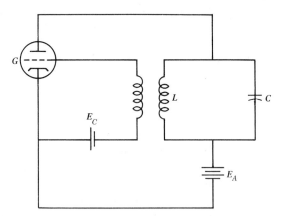

Figure 2.13   Simple oscillating circuit.

causes certain polarization and electrical distribution changes in it, and voltage occurs between the surfaces of the crystal perpendicular to its optical axis. Thus the oscillation in the grid circuit causes the mechanical vibration or oscillation in the crystal, which, in turn, causes electrical oscillation back to the grid circuit; consequently the crystal controls its own oscillation until it is settled to a high degree of stabilization. In instruments utilizing high precision frequency standards, the output of the *voltage-controlled oscillator* (VCO) and the output of the frequency standard are fed to a phase detector. The resulting voltage is equal to the difference in frequency between the two signals. This *error voltage* is then applied to the VCO until it is locked to the standard. The circuitry is called the *phase-locked loop* (PLL).

2.8.1. Velocity-Modulated Tubes; Klystrons

In microwave transmission the carrier wavelength can be of the order of centimeters or even millimeters, and the time needed for electrons to travel from the cathode to the anode can be neglected no longer. Here vacuum tubes such as triodes and pentodes in connection with a conventional resonant circuit must be replaced by so-called *velocity-modulated tubes*. The operation of such a tube depends on changes in the speed of the electrons passing through it. When the electron speed is changed, the tube produces bunches of electrons separated by spaces in which there are few electrons. As a general introduction to this process, let us suppose that a heated cathode emits a cloud of electrons that are attracted by an accelerator grid placed in front of the cathode and having large positive potential with respect to the cathode. The grid has an open wire design that allows all electrons with the same initial high velocity to pass through. The continuous beam of electrons then passes through a pair of closely spaced grids called the *buncher grids*, which are connected to a tuned oscillating circuit. According to the polarity of the alternating voltage across the tuned circuit, at the time when electrons are passing the buncher grids their velocity is either constant, accelerated or retarded. As a result of this velocity difference between electrons, the electrons are no longer propagating uniformly; instead, the faster ones are catching the slower ones to form groups or bunches. The bunches are then allowed to pass through a similar second set of grids called the *catcher grids*, which are coupled to another oscillating circuit. Because the electrons pass through the catcher in bunches rather than in the form of a continuous stream, more energy is delivered to the catcher circuit than is needed for bunching, and the spent electrons are removed from the circuit by a positive collector plate at the end of the tube. If the output from the catcher is fed back into the buncher, and if proper phase relations are maintained between buncher and catcher, the tube operates as an oscillator.

In the devices used in modern microwave operations to generate carrier waves at gigahertz bands, oscillating circuits are not composed of inductors and capacitors connected by open wires; rather, they are in the form of solid cavities. Each of the cavities has a gap through which the electron beam passes. A current caused by the electron beam is induced into the cavity walls and sets up an oscillating field inside the cavity. This oscillation is in resonance with the input microwave signal, and the resonant frequency of the *cavity resonator* depends on the size of cavity space. Also, an electric field is produced across the gap; the field impresses the passing electron beam, and no physical grids exist. This type of velocity-modulated tube, the *klystron*, is widely used as an oscillator in microwave distance measuring instruments. In most microwave distance measuring

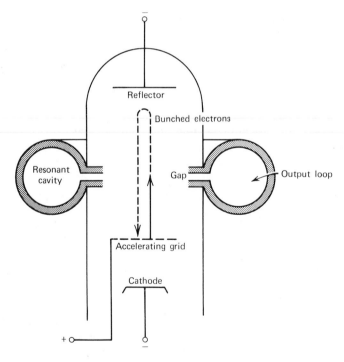

Figure 2.14   Reflex klystron.

equipment, however, a modified and simplified version of the two-cavity klystron called a *reflex klystron* is employed. In the reflex klystron the feedback from the amplified output to the input is obtained by using a single cavity only. Thus the input cavity and the output cavity are the same. The collector plate is replaced by a repeller plate or reflector, to which negative potential is applied. When the electron beam passes through the gap for the first time, it is velocity modulated; when it enters the retarding field caused by the reflector, it reverses direction and passes the second time in bunched form through the gap, delivering energy to the cavity. If the electron bunches arrive at the gap at a proper phase of the resonant oscillation, amplification is achieved and the resonator oscillates strongly at the resonant frequency. The phasing is a question of proper timing, and the operating frequency can be varied over a small range by changing the negative potential of the reflector, since this potential determines the transit time of the electrons between their first and second passages through the gap. Figure 2.14 is a schematic diagram of a reflex klystron. For FM see Section 2.9.

2.9. MODULATION AND DEMODULATION

Oscillators described in the previous section generate continuous sine waves. A continuous wave (CW) does not include any information or intelligence, but it can be effectively used as a carrying agent for a variety of information, such as speech and voices of any kind. The fundamental information transmitted by the carrier wave in electronic surveying is a sinusoidal wave form to be used for measurements. The process of superimposing the desired sine wave or other periodic signal onto the carrier wave is called *modulation*. Three main types of modulation are frequently used today in electronic distance measurements:

1. Amplitude modulation (AM).
2. Frequency modulation (FM).
3. Phase modulation (PM).

In *amplitude modulation* the frequency and phase of the carrier wave do not change in the modulation process, but the strength or amplitude V_c of the carrier wave, $V = V_c \sin \omega_c t$, alternates sinusoidally with an amount of $V'_c = \alpha V_c$. The coefficient α indicates the depth, or the *degree of modulation*; it is defined as V'_c / V_c, where V'_c is the amplitude of the modulation wave (Fig. 2.15). During the modulation, the amplitude of the carrier wave thus alternates between the limits $V_c + V'_c$ and $V_c - V'_c$, or

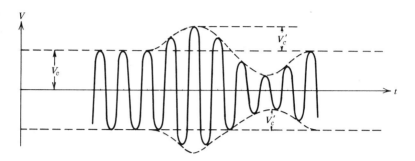

Figure 2.15   Amplitude modulation.

$V_c(1+\alpha)$ and $V_c(1-\alpha)$. If the angular velocity of the modulation wave is designated ω_m, the instantaneous expression for the modulated carrier wave is

$$V = (V_c + V_c' \sin \omega_m t) \sin \omega_c t \qquad (2.21)$$

where $(V_c + V_c' \sin \omega_m t)$ is the variable amplitude of the modulated carrier wave. By trigonometric transformation, Eq. 2.21 can be written in the following form with the provision $V_c' = \alpha V_c$:

$$V = V_c \sin \omega_c t - \frac{\alpha V_c}{2} \cos (\omega_c + \omega_m)t + \frac{\alpha V_c}{2} \cos (\omega_c - \omega_m)t \qquad (2.22)$$

From Eq. 2.22 it can be seen that the modulated carrier wave consists of three components

$$V_c \sin \omega_c t$$

$$\frac{\alpha V_c}{2} \cos (\omega_c + \omega_m)t$$

$$\frac{\alpha V_c}{2} \cos (\omega_c - \omega_m)t$$

with corresponding frequencies of $\omega_c/2\pi$, $(\omega_c + \omega_m)/2\pi$, and $(\omega_c - \omega_m)/2\pi$. The first frequency is the original carrier frequency, the second frequency the *upper sideband*, and the third the *lower sideband*. A simple method of amplitude modulating the carrier is to feed the modulation signal to the grid current of an oscillator.

In *frequency modulation* the amplitude of the carrier wave is kept constant, but the frequency varies according to the amplitude and polarity of the modulation signal. The carrier frequency is increased during one half-cycle of the modulation signal and decreased during the other half-cycle; thus the frequency is lowest when the modulation is least positive and highest when the modulation is most positive (Fig. 2.16). Let ω_c be the angular velocity of the carrier wave at a given instance and $\Delta\omega$ its change. During modulation, the instantaneous angular velocity is

$$\omega = \omega_c + \Delta\omega \sin \omega_m t \qquad (2.23)$$

where ω_m is the angular velocity of the modulation wave. We can write

$$\omega \, dt = \omega_c \, dt + \Delta\omega \sin \omega_m t \, dt$$

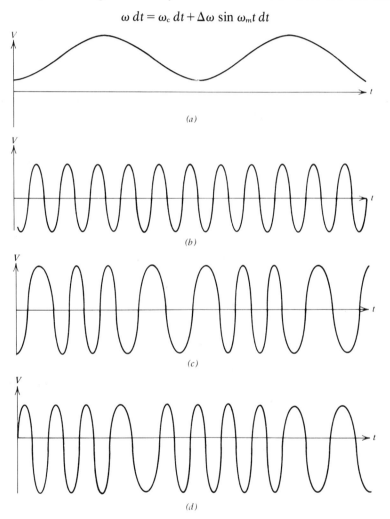

Figure 2.16   Frequency and phase modulation. (*a*) Modulation wave, (*b*) carrier wave, (*c*) frequency-modulated carrier wave, (*d*) phase-modulated carrier wave.

After integration, this yields

$$\omega t = \omega_c t - \frac{\Delta \omega}{\omega_m} \cos \omega_m t \qquad (2.24)$$

The instantaneous expression for the modulated carrier wave is now

$$V = V_c \sin \omega t = V_c \sin\left(\omega_c t - \frac{\Delta \omega}{\omega_m} \cos \omega_m t\right) \qquad (2.25)$$

Since $\Delta \omega = 2\pi \Delta f$, Eq. 2.25 can also be written in the following form:

$$V = V_c \sin\left(\omega_c t - \frac{2\pi \Delta f}{\omega_m} \cos \omega_m t\right) \qquad (2.26)$$

where Δf is the *frequency deviation*, which is proportional to the instantaneous amplitude of the modulation signal.

The usual way to establish FM in instruments utilizing the microwave technique is to feed the modulation signal into the reflector circuit of the reflex klystron.

In *phase modulation* as in frequency modulation, the amplitude of the carrier wave remains constant, and there are several points of resemblence between the two forms of modulation. In phase modulation, however, the phase of the carrier wave is varied according to the phase of the modulation wave. Thus during the top portion of the modulation wave (at 90° phase angle), the phase-modulated (PM) wave looks just like the original carrier wave but its phase is different. During the bottom portion of the modulation wave (at 270° phase angle), the phase-modulated wave again looks like the original carrier wave but in this case its phase is 180° out of phase compared to the previous instance. This kind of phase change causes bunching and spreading of the modulated wave; as Fig. 2.16 indicates, the bunching up of the waves occurs when the modulation increases (reaching the maximum rate of change at 0° phase angle), and the spreading of the waves occurs when the modulation decreases (reaching the maximum rate of change at 180° phase angle). Thus the main difference between the FM and PM is that in the former case the frequency deviation is proportional to the instantaneous amplitude of the modulation signal, while in the latter case the frequency deviation is proportional to both the instantaneous amplitude *and* the frequency of the modulation signal.

The instantaneous expression for the modulated carrier wave can be given as follows:

$$V = V_c \sin (\omega_c t + \Phi \sin \omega_m t) \qquad (2.27)$$

where Φ represents the maximum phase change during the modulation.

Demodulation in the receiver is the separation of the modulation signal from the modulated carrier wave. This can also be called *detection*, which particularly applies to the demodulation of AM signals. In a *detector* an AM signal is rectified to form a pulsating DC; the carrier wave component is filtered out, leaving a DC component that varies in the same way as the modulation of the original signal (i.e., sinusoidally). Finally the varying DC voltage is fed through an amplifier and only the *variation* of voltage is transferred, and the output is the desired AC sinusoidal signal with constant amplitude. The FM and PM demodulations are established in the *frequency discriminator*. When the modulated carrier signal is fed into the discriminator, at any instantaneous moment when the FM signal has no modulation, the modulated carrier equals the original carrier and the discriminator has zero output. When the frequency deviation swings to higher frequency, the output amplitude of the discriminator increases in the positive direction, and when the frequency deviation swings to lower frequency, the output amplitude increases in the negative direction. As a result, the output of the discriminator is a continuous AC at the modulation frequency.

2.10. MEASUREMENT OF PHASE DIFFERENCES

Measurement of phase or time differences is an integral operation in all equipment capable of determining distances or distance differences. The process involved can vary depending on the design of the instruments. The following methods are discussed: (*a*) manual phase delay between the transmitted and received signals, (*b*) manual or automatic phase shift between the signals, and (*c*) the digital method of phase measurement.

Phase Delay. In older model Geodimeters the phase difference between the transmitted and received modulation signals was detected and ultimately compensated by using the so-called *delay line.* The delay line was placed between the oscillating generator, which fed the Kerr cell in the transmission circuit, and the photomultiplier located in the receiving circuit. It was pointed out in Section 2.4 that in the series resonant circuit the effects of capacitance and inductance are 180° out of phase. Thus the delay line can be considered as a variable impedance of such a circuit. By placing a movable pickup connector to the inductor coil or by utilizing inductive coupling, the impedance between the connector and one of the terminals of the circuit can be varied. Variation causes the change in phase value between the alternating current applied to the circuit and the voltage developed across it. Manual adjustment of the position of the pickup connector produces the in-phase condition, which is shown by the *null meter*. Theoretically, the phase difference between the transmitted and received signals is obtained directly from Eq. 2.14: $\tan \Phi = (X_L - X_C)/R$, although in practice the phase difference is calibrated directly in the delay line in metric units.

Phase Shift. Phase shifting between the transmitted and received modulation signals is established by a device called the *resolver* (phase delay, in effect, yields to phase shifting also). Two types of resolvers are discussed here in principle. In Fig. 2.17 two pairs of *stator coils* are located at right angles to each other. If alternating voltages V_1 and V_2 are applied across the coils, electromagnetic fields are formed in the coils, and if the applied voltages are out of phase, an alternating current is induced into the *rotor coil* V_3, which is not directly excited by any applied voltage. The rotor coil can be rotated manually until it is parallel to the resultant field of the applied voltages. At this position of the rotor coil the current is zero in the rotor coil circuit, which is shown again by the zero reading of the null meter. The angle Φ (the amount by which the rotor coil has been turned from the reference 0 direction) is the phase difference between the two applied voltages and can be read in metric units in the

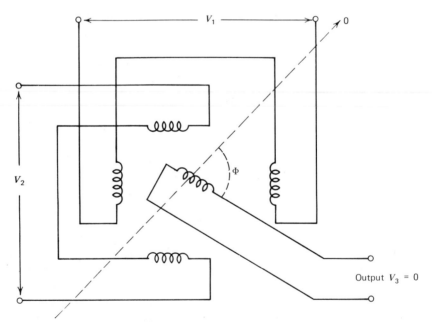

Figure 2.17   Resolver with two input signals.

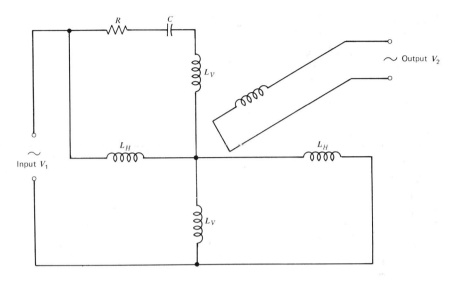

Figure 2.18   Helmholtz resolver.

drum driving the rotor coil. The induced alternating current in the rotor circuit also can be used to drive a servo motor, which, in turn, rotates the rotor coil until the in-phase condition is reached. At this moment no current runs through the coil circuit, the motor stops, and the reading of the turn of the rotor will be displayed directly in metric units (e.g., Distomat DI-10).

Another type of resolver may be used when only one signal is fed through the resolver. The output signal of the resolver is then compared with the other (reference) signal in the phase discriminator for the phase comparison. The basic concept in this type of resolver is the *Helmholtz-coil* phase-shifting circuit (*Radar Electronics Fundamentals*, 1944). This design uses two sets of coils for the primary circuit and obtains the necessary 90° difference in phase by means of a capacitor and a resistor connected as a fixed phase-shifting circuit in series with one set of the coils (Fig. 2.18). Since the horizontal coils L_H are connected directly across the input line, the current through them lags the input voltage by almost 90°. The same condition would exist in the vertical coils L_V if they also were connected directly across the input terminals. But since a capacitor C and an adjustable resistor R are added in series with the vertical coils, this circuit can be adjusted to ensure that the current through the vertical coils is exactly 90° out of phase with the current in the horizontal coils. The alternating current in the stator coils induces a voltage in the rotor coil. The phase relation between the output voltage V_2 and the input voltage V_1 depends on the physical position in which the rotor coil is placed, and the required phase shift in this case can be also read in metric units from the drum driving the rotor. Newer model Geodimeters and most instruments utilizing microwaves as the carrier (e.g., Tellurometers, excluding the models MRA-1 and MRA-2) employ this type of resolver. Physically the devices may differ from the one just described in that both stator and rotor coils consist of two perpendicular elements. Whatever form of resolver is used, it is important that the maximum currents in the coils that make a pair are carefully matched, which means that the wires in the coils must have the same number of turns. Finally, in the phase comparison process the particular position of the resolver control drum representing the in-phase condition must be indicated. This provision is obtained from the *phase discrimination* circuit into which the transmitted or reference signal and the received signal are fed. The phase of the former signal is variable, since it has passed the resolver. The *discriminator* functions in such a way that no current flows through the circuit, and the current meter in that circuit indicates zero when the two signals are in phase. Any deviation from the in-phase condition causes the *null meter* to deviate from its zero position, whereupon it must be rezeroed by changing the phase of the reference signal in the resolver.

Digital Method of Phase Measurements. The digital or pulse-counting method is used only in two instruments employing microwaves as the carrier (Distomat DI-50 and DI-60), but it is very commonly found in instruments designed for short distance measurements and using the

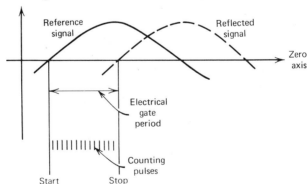

Figure 2.19 Pulse-counting principle in phase measurements (courtesy Keuffel & Esser Co.).

He–Ne laser or Ga–As diode as a light source. It is very suitable for instruments designed for complete automation. As the transmitted modulation or reference (light or microwave) signal crosses the zero axis of the polarity, it opens the gate of a pulse-counter device, and L pulses are formed before the reflected signal crosses the same zero axis and closes the gate (Fig. 2.19). If the total amount of pulses included in one wavelength is designated as k, the phase difference between the transmitted and reflected waves is $\Phi = (L/k)\, 360°$, thus the fractional part of one wavelength is $\Delta\lambda = (L/k)\lambda$. The total measured length then will be

$$D = \frac{1}{2}\left(n + \frac{L}{k}\right)\lambda$$

where n is the total number of modulation wavelengths included in the distance. The total number of pulses k is so chosen with respect to the modulation wavelengths that the unit length of one pulse represents a small integer distance such as 1 mm or 1 cm.

2.11.   UNITS OF AMPLIFICATION

The unit of amplification universally used in electronics, electric, and acoustic work is called the *Bel*, and in practical work the customary unit is the *decibel* (dB), which is one-tenth this amount. The decibel is a logarithmic unit and is equal to $10 \log (P_2/P_1)$, where P_1 and P_2 represent different values of power that are to be compared. The relationship between power, voltage, and current is

$$P = EI \tag{2.28}$$

for DC, and

$$P = E_0 I_0 \cos \omega t$$

for AC. In Eq. 2.28, P is the power stated in watts (W), and E and I are the voltage and the current, stated in volts and amperes, respectively. If we denote input power, current, and voltage by P_1, I_1, and V_1 and the corresponding output quantities by P_2, I_2 and V_2, the amplifications in decibels will be:

Power	Current	Voltage
$10 \log \dfrac{P_2}{P_1}$	$20 \log \dfrac{I_2}{I_1}$	$20 \log \dfrac{V_2}{V_1}$

The numerical values of the amplification ratios stated in decibels are given in Table 2.1.

TABLE 2.1

P_2/P_1	V_2/V_1 or I_2/I_1	Decibels
1	1.0	0
2	1.4	3
10	3.2	10
100	10.0	20
1,000	31.6	30
10,000	100.0	40

A value of amplification less than one, of course, represents attenuation of the signal and is obtained by taking the inverse values of the power, current, and voltage ratios as P_1/P_2, I_1/I_2, and V_1/V_2, respectively yielding to the negative decibel value.

2.12.   RADIATION OF ELECTROMAGNETIC ENERGY

Radiation of electromagnetic energy is actually the propagation of alternating electric and magnetic fields through any medium (atmosphere). For all practical considerations it is assumed to be a sinusoidal wave motion consisting of such finite parameters as velocity, frequency, time (period), and length.

2.12.1.   Classification of Electromagnetic Radiation

The total spectrum of electromagnetic radiation used in electronic and electro-optical distance measurements encompasses wavelengths from the

TABLE 2.2

Multiplication Factor	Prefix	Symbol
10^{12}	tera-	T
10^9	giga-	G
10^6	mega-	M
10^3	kilo-	k
10^{-3}	milli-	m
10^{-6}	micro-	μ
10^{-9}	nano-	n
10^{-12}	pico-	p

TABLE 2.3

Frequency, f	Period, $T = 1/f$	Wavelength, $\lambda = cT$
Hz	1 sec	3×10^8 m
kHz $= 10^3$ Hz	1 msec $= 10^{-3}$ sec	3×10^5 m
MHz $= 10^6$ Hz	1 μsec $= 10^{-6}$ sec	3×10^2 m
GHz $= 10^9$ Hz	1 nsec $= 10^{-9}$ sec	3×10^{-1} m
THz $= 10^{12}$ Hz	1 psec $= 10^{-12}$ sec	3×10^{-4} m

visible light of about 5×10^{-7} to about 3×10^4 m at the radio frequency region. Frequencies are large numbers and always positive powers of the basic unit, the hertz; periods are small numbers and always negative powers of the basic unit, the second. Table 2.2 presents the prefixes and symbols in *SI units** for the ranges of multiples that are frequently used in electronic techniques. Of course the symbols in Table 2.2 accompany the symbols of the basic unit. Table 2.3 lists relationships between the frequency, period, and wavelength when the propagation velocity c is given the approximate value of $c = 300,000$ km/sec. For practical reasons the radio frequency (RF) spectrum is divided into groups or bands. The banding distinguishes different orders of frequencies used for various purposes and provides them with special symbols employed in electronic terminology. Table 2.4 gives a commonly used classification. At present the combination of the last two bands is called the *microwave band*, such that the SHF-band is classified as the *S*-band and the EHF-band is classified as the *X*-band.

TABLE 2.4

Classification	Symbol	Frequency	Wavelength (m)
Very low frequency	VLF	10–30 kHz	30,000–10,000
Low frequency	LF	30–300 kHz	10,000–1000
Medium frequency	MF	300–3000 kHz	1000–100
High frequency	HF	3–30 MHz	100–10
Very high frequency	VHF	30–300 MHz	10–1.0
Ultrahigh frequency	UHF	300–3000 MHz	1.0–0.1
Superhigh frequency	SHF	3–30 GHz	0.1–0.01
Extremely high frequency	EHF	30– GHz	0.01–

* The International System (Système Internationale) was adopted by the Eleventh General Conference on Weights and Measures in 1960.

2.12.2. Applications in Electronic Distance Measurement

Sinusoidal electromagnetic radiation is used in two forms in electronic or electro-optical distance measurements. It can be that of the *continuous wave*, CW, transmission, or the carrier wave can be modulated by a *modulation wave*. Continuous wave transmission is widely used in land-to-sea hydrographic surveying or navigation by instruments utilizing MF, LF, or VLF bands. Carrier frequencies in some cases are in the MF region, but the most commonly used carriers are the 3, 10, and 37.5 GHz frequencies that are shared by microwave distance measuring instruments available on the market to date.

Light is widely used as a carrier, especially in instruments that are built for precision and compactness in short distance measurements. The bands in the light spectrum which are of special interest in electro-optical distance measurements today are the following:

Visible noncoherent light	Mercury vapor lamp	$\lambda = 0.55\ \mu$m
	Tungsten lamp	$\lambda = 0.57 \mu$m
Coherent laser light	He–Ne laser	$\lambda = 0.63\ \mu$m
	Ruby laser	$\lambda = 0.69\ \mu$m
Infrared light	Ga–As diode	$\lambda = 0.875–1.00\ \mu$m

Every one of the classified RF bands (Table 2.4) and also the above mentioned light wave bands are or have been utilized by modern electronic or electro-optical distance measuring devices. Although the reader is not yet familiar with the great variety of instruments available (Chapter 20), samples from each group are given in the randomly made selection of Table 2.5.

Modulation frequency is often called *pattern frequency* or *measuring frequency*, because the actual measurements are made by comparing phases of the modulation signals rather than those of the carrier signals. Two frequencies are commonly used in microwave modulation, namely, 10 and 15 MHz. The first frequency is suitable when *time* is measured, since one period in 10 MHz modulation is equal to 100 nsec. The second frequency is well adaptable in instruments that measure the *distance*, since the modulation wavelength in 15 MHz modulation is equal to 20 m round trip, thus 10 m in direct distance between two instruments making the measurements. Modulation frequencies as high as 500 MHz, corresponding to a *unit length* as short as 30 cm, are not unknown (e.g., the Kern Mekometer, ME 3000).

TABLE 2.5

Bandwidth	Instrument	Applications
VLF	Omega	Worldwide navigation
LF	Decca	Navigation, hydrographic surveying
MF	Raydist	Hydrographic surveying
HF	Gee-H	Land-to-air surveying (obsolete)
VHF	Secor	Satellite positioning system
UHF		Older type radio altimetry
SHF (S-band)	Microwave (e.g., Electrotape)	Land, sea, and air surveying
SHF (X-band)	Tellurometer MRA-4	Land-to-land surveying
Visible, non-coherent light	Geodimeter	Older models, land-to-land surveying
He–Ne laser	Rangemaster	Land-to-land, long distance surveying
Ruby laser		Satellite and lunar ranging
Ga–As diode	Reg Elta	Tachymetric-type land-to-land surveying

2.12.3.   Propagation Properties of Electromagnetic Radiation

The velocity of electromagnetic radiation in vacuum (in space), like the speed of light, is constant, and according to the latest research, its value is $c_0 = 299,792.458 \pm 0.001$ km/sec (see Chapter 6). In the atmosphere the velocity is reduced because of density variations between atmospheric layers. This reduction is indicated by the *refractive index*, which, in turn, can be computed by knowing the dielectric properties of the atmosphere. The velocity can be divided into two categories or classifications, the *group velocity* and the *phase velocity*. Let us illustrate the difference between these two velocities by considering Fig. 2.15, which represents amplitude modulation. The modulation wave or envelope propagates at the group velocity, but the carrier wave within the envelope travels at a slightly higher velocity through the modulation envelope because the carrier wave is composed of variations of field intensity that appear to move faster than the envelope. Thus the carrier is said to propagate at a phase velocity that is higher than the group velocity. A similar effect occurs when VLF transmission (e.g., Omega) travels between two conducting layers, earth and the ionosphere, in which the two layers act to guide the VLF energy. As a result, the propagation velocity of VLF transmission is greater than it would be without the existing surfaces; thus the velocity is the phase velocity. This increased velocity in the atmosphere also affects the apparent increased speed of light, which then will be $c_0 = 300{,}574$ km/sec. In practice, the use of phase velocity has very little significance except for Omega transmissions, since in all systems that use modulation, measurements are made with respect to the modulation wave, thus the group velocity is important. In systems featuring continuous waves, the phase velocity and the group velocity are the same. The velocity of generated radio waves in the atmosphere depends very little on the frequency. In the light wave spectrum, however, the velocity is significantly related to the frequency. The velocity of infrared radiation, for example, is different from the velocity of light emitted by a tungsten lamp. The velocities of different frequency groups in the light spectrum at standard atmospheric conditions are also called group velocities. The group velocities must be determined before the ambient velocities are obtained, and this can be accomplished by observing the atmospheric parameters: temperature, pressure, and humidity.

The *bending* of radio or light "beams" in the atmosphere is also a function of the atmospheric parameters and can be determined easily if the ambient refractive index is known. In the ionosphere the *reflections* of radio frequency transmissions are considerably more complex and unpredictable than the reflections of light waves. In this case the frequency plays a crucial role. In the following conclusion of the propagation properties, two main groups of frequencies are considered. From Table 2.4, VHF, UHF, SHF, and EHF are grouped together and called the *short waves*, and MF and LF are grouped together as the *long waves*. Generally speaking, short waves propagate along a line that is only slightly bent. Therefore, in utilizing this wave group the line-of-sight condition must exist. The long waves propagate along a line whose curvature is equal to that of the earth.

Figure 2.20a presents the propagation and reflections of short waves. If the antennas A and B are oriented toward each other, the direct beam AB is the desired one, but because of the finite *beamwidth*, the so-called *ground reflection* or *ground swing* ACB occurs and must be eliminated. If the antenna A is directed toward D, the beam is split into two components when it reaches the highly conductive layer of free electrons (the ionospheric layer or *ionosphere*). One component penetrates the layer and ultimately reaches space, and the other component reflects back to earth. In

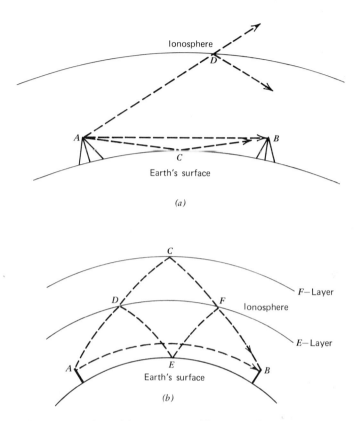

Figure 2.20 Propagation of (*a*) short and (*b*) long radio waves.

Fig. 2.20*b*, which represents the propagation of long waves, the antennas *A* and *B* are usually omnidirectional. This means that after leaving antenna *A* the beam is split into the *ground wave* component *AB* and the *sky wave* component *ACB* at antenna *B*. This interference causes serious errors in surveying systems utilizing phase measuring techniques, but its effect can be eliminated in systems using pulsed techniques. There are several ionized layers in the ionosphere, designated as *E-layer*, F_1-*layer*, and F_2-*layer*, the total thickness of the ionosphere being about 300–400 km. The distance of the lowest *E*-layer from the earth's surface (about 100 km) varies according to the time of day and year. Multireflections may also occur from ionospheric layers such as the path *ADEFB*, shown in Fig. 2.20*b*.

In the satellite tracking process, when the short waves penetrate the ionosphere, their paths are bent while passing through the ionospheric

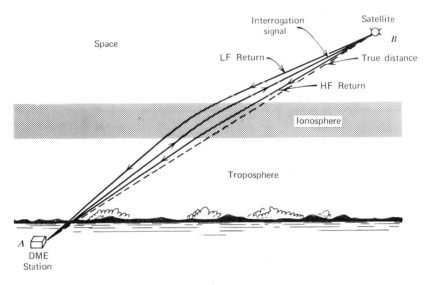

Figure 2.21 Ionospheric refraction (courtesy Cubic Corp.).

layers. This bending, which is highly dependent on the wavelength used, is illustrated in Fig. 2.21, referring to the satellite tracking system Secor described in detail in Chapter 20. There are three different frequencies involved, termed the interrogation signal, the HF return signal, and the LF return signal. All frequencies are within the limit of about 225 and 450 MHz, and LF and HF are only symbols of signals and bear no connection to the classification given in Table 2.4. It is advantageous to use *two* different returning frequencies from the satellite because the magnitude of bending of the two beams is different and from this difference the ionospheric correction to the measured distance can be computed.

THREE

ELECTRONIC SURVEYING SYSTEMS AND BASIC COMPONENTS

3.1. RADIO WAVE TRANSMISSION

In this section we discuss the radio wave spectrum of electromagnetic radiation. As was learned in the previous chapter, modern electronic surveying equipment occupies all radio bands. However, the ways the bands are used and techniques applied vary greatly. Classification here is made along the following lines: *pulse-type* transmissions, *continuous wave* transmissions, and *microwave modulation*. All these systems are in use today.

3.1.1. Pulse-Type Transmission

In electronic surveying systems that utilize pulse-type transmission, the target from which the transmitted pulses are returned can be either passive or active. In the former case the target is either a natural or man-made reflecting surface as in the PPI-radar, linescan radar, or radioaltimeter applications. In the latter case the target is a transceiver-type secondary instrument similar to the primary instrument that originates the pulse transmission. The basic concept in determining the distance is to observe the time t needed for pulses to travel from the primary instrument to the secondary instrument and back. If the average velocity c of the electromagnetic propagation is known, the distance will be

$$D = \frac{ct}{2} \tag{3.1}$$

A skeleton-type layout of a pulse transmitter and receiver station is presented in Fig. 3.1. The bare essentials of the transmitting unit consist of a *pulse modulator*, an *oscillator*, and the *transmitting antenna*. The receiving unit consists basically of *limiter*, *triggering circuit*, *time base generator*, *cathode ray tube* (CRT), and *receiving antenna*.

The transmitter emits a stream of short pulses, whose duration is of the order of one microsecond, having a period of recurrence of about $1000 \, \mu s$. When a pulse leaves the transmitting antenna, a small *synchronizing pulse* passes at the same instant through a short length of cable to the receiving circuit, where it is extended to correspond to the maximum time the transmitted pulse will need to travel to a target and back. The pulse is then applied to the *time base generator*. As the time base is drawn on the face of the CRT by the time base generator, starting with the dispatch of the pulse from the transmitter, the transmitted pulse has traveled out to the target, undergone reflection, and returned to the receiving antenna. The received pulse is first amplified and then applied to the CRT, where its intersection with the time base becomes a visible indication of the distance from the antenna to the target.

3.1.1.1. *Transmitter*

The transmitter produces a continuous train of short pulses, each pulse having an extremely sharp onset (i.e., a steep wave front) and a

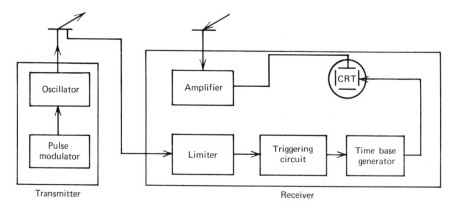

Figure 3.1 Simplified block diagram of pulse transmitter and receiver.

moderately abrupt end. To employ these pulses, a considerable amount of power is required. This power is the *peak power*, and even though pulse transmission calls for a great deal of it, the system is still economical of power when compared with systems involving the continuous radiation of energy. Power is radiated only during the pulse periods, and these occupy an extremely small fraction of the total time of operation.

The oscillator stages of pulse transmitters vary considerably according to the physical size, output power, and working potentials of the equipment. In some applications a simple vacuum tube circuit as described in Section 2.8 is acceptable, but in applications such as PPI-radar, line-scan radar, or radioaltimeters, the wavelength can be of the order of centimeters or millimeters, and vacuum tubes must be replaced by other devices. In addition to the influence of transit time on electrons, capacitance and inductances must be made small; thus the capacitance and inductance of the connecting wires and the electrodes themself cannot be neglected. Special tubes such as acorns, magnetrons, or klystrons are then utilized.

There are different methods available for *pulse modulation*, but whatever approach is used, the oscillator tubes must not draw any current between pulses. Otherwise there will be a steady dissipation of power at the anodes, and the continuous oscillation of the oscillation stage in the transmitter might affect the receiver by direct radiation, rendering impossible the detection of weak signals.

3.1.1.2. *Receiver*

At the same time the transmitted pulse is radiated toward the target, a synchronizing pulse goes to the receiver and is first fed to a simple *rectifier-limiter* circuit from which it appears as a unidirectional negative pulse. The synchronizing pulse is directed to the two circuits that form the time base, namely, the triggering circuit and the time base generator. The "extension" action of the synchronizing pulse is established in the triggering circuit, which for all practical purposes has an infinite resistance until the voltage across its terminal reaches a certain predetermined value. During this period a capacitor C in the time base circuit (Fig. 3.2) charges, and the illuminated spot of an electron beam on the face of the CRT moves and traces out the time base. The triggering circuit T then "triggers" and presents a short-circuit path across the capacitor, which rapidly discharges and allows the spot to fly quickly back to the original position. This periodic movement is repeated with a so-called *timing frequency*, which can be altered according to the group of ranges to be measured.

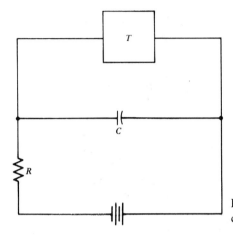

Figure 3.2 Basic time base generator circuit.

CATHODE RAY TUBE AND TIME BASES. *The cathode ray tube* translates the range information obtained from the transmitted and reference pulses into a visual presentation, the time base, which is a visible trace of electrons on the fluorescent screen of the tube. The process of this translation consists of the following steps:

1. Focusing the electron beam.
2. Deflecting the electron beam.
3. Forming the time base.

The *focusing* of electrons into a concentrated, narrow beam can be effected by either electrostatic or electromagnetic means. In the former type the focusing is accomplished by the application of suitable potentials

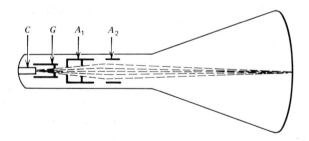

Figure 3.3 Electrostatic focusing (Laurila, 1960).

to the various internal electrodes, while in the latter type the focusing is produced by the action of a direct current flowing through coils mounted externally, through which the electron beam passes. The first type is presented here as an example. In Fig. 3.3 a heated cathode C forms a cloud of electrons around itself, and a cylindrical grid G with slightly negative potential is placed in front of and around the cathode, causing the repulsion of electrons from the grid wall. If an anode A_1 having high positive potential is placed in front of the grid, the electrons emerge through the grid toward the anode but in the form of disorganized bundles. By placing in front of the first anode a second anode A_2, which is also applied by positive potential, an effect similar to that of a lens system is achieved, and the focusing of the electron beam emerging through the second anode can be controlled by altering the ratio of the two positive anode voltages. The assembly consisting of the cathode C, the grid G, and the anodes A_1 and A_2 is called the *electron gun.*

The *deflection* of the electron beam focused as a spot on the screen of the CRT can be accomplished by using either an electrostatic or electromagnetic deflection system from which the electrostatic deflection is presented as a sample. In this case, two small metal plates are mounted in the tube parallel to each other and connected by wires to an external source of potential difference so that one plate is made positive with respect to the other. An electric field then exists between the plates in a direction at right angles to the axis of the tube. When electrons leave the electron gun and enter that field in the form of a converging beam, they are repelled by the negative plate and attracted by the positive plate (Fig. 3.4). On leaving the field, the electrons continue in a straight line that makes an instantaneous angle with the axis of the tube; the angle is directly proportional to the voltage difference applied to the plates and inversely

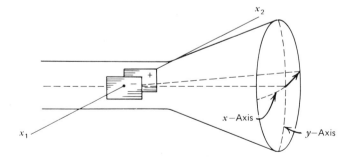

Figure 3.4 Electrostatic deflecting system (Laurila, 1960).

proportional to the spacing between them. There are two such pairs of plates in the tube. One pair, generally known as the X-deflection plates, is mounted vertically and causes horizontal deflection of the beam, the other pair, the Y-deflection plates, is mounted horizontally and causes a vertical deflection.

To make the sweep of the electron beam or *time base* visible, the inside of the large end of the CRT is coated with a thin layer of fluorescent material that emits visible light when its surface is subjected to the impact of the electron stream. Thus the focused beam of electrons striking the end of the tube or screen forms a luminous spot. If the electron stream is suddenly cut off, the spot of light does not immediately disappear but continues for a while. This effect is known as the *afterglow*, and its duration is a function of the type of the fluorescent material used. To produce a *linear time base* across the screen, the voltage V_c applied across the capacitor (Fig. 3.2) must change linearly with time. That condition is obtained when the differential dV_c/dt is equal to a constant K.

The capacitance C is obtained as a function of the voltage V_c and the charge Q in the following way:

$$C = \frac{Q}{V_c} \tag{3.2}$$

where C is in farads, V_c in volts, and Q in coulombs. By application of the constant condition stated previously into Eq. 3.2, and by appreciation that the capacitance for the particular capacitor is constant, the following expression is obtained:

$$\frac{dV_c}{dt} = \frac{dQ}{dt} \frac{1}{C} = K \tag{3.3}$$

Since the charge Q is defined as product of electric current I and time t differentiation of the product with respect to time yields

$$\frac{dQ}{dt} = I$$

which, when substituted into Eq. 3.3, gives the desired condition:

$$I = CK \tag{3.4}$$

This means that to produce a linear time base, the current I must be constant and the resistance in Fig. 3.2 must be replaced by a constant-current device, such as a saturated diode described in Chapter 2. A linear time base is used in the *Loran-A* navigation receiver, for example.

If two alternating voltages are applied simultaneously to the X- and Y-plates, which are situated at right angles to each other, the trace of the

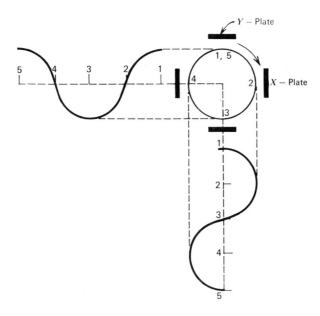

Figure 3.5 Generation of circular time base (Laurila, 1960).

spot on the screen forms a pattern known as a *Lissajous figure*. When the two voltages alternate at exactly the same frequency, the spot will describe a stationary figure, which may be a straight line, an ellipse, or a circle. Which of these figures is generated depends only on the phase relationship of the two voltages applied to the deflecting plates, assuming that their amplitudes are equal. A *circular time base* is obtained by applying equal amplitude voltages 90° or 270° out of phase to the X- and Y- deflection plates. The generation of a circle by these so-called quadrature components of same-frequency sinusoidal voltage is schematized in Fig. 3.5. Instruments that use the circular time bases include the Shoran, Hiran, and Tellurometer models MRA-1 and MRA-2.

3.1.2. Continuous Wave Transmission

Continuous wave (CW) transmission is mostly applied in electronic position fixing involving land-to-sea hydrographic surveying or navigation. In each case the system is mutually active, which means that active instruments (transmitters or receivers or both) are located on ground-based stations and on shipboard. Both circular and hyperbolic modes (Chapter 4) of location are used according to the grouping of the units. As a general rule, when the circular mode is applied both the transmitter and the receiver are located on the survey vessel, whereas with the hyperbolic mode, usually only the receiver is on shipboard and the transmitters (and receivers) are placed on shore.

The basic concept of CW transmission relies on the condition that all measurements at point P ultimately are made by phase comparison between signals transmitted by stations A and B (Fig. 3.6), where D_A is the distance between A and P, and D_B is the distance between B and P. If the instantaneous phase values at A and B are designated ϕ_A and ϕ_B, respectively, then at point P, after the signals have propagated over the distances D_A and D_B, the phase values of these two signals are

$$\phi_{AP} = \phi_A + \frac{2\pi D_A}{\lambda} \quad \text{and} \quad \phi_{BP} = \phi_B + \frac{2\pi D_B}{\lambda} \tag{3.5}$$

where λ is the common wavelength. From Eq. 3.5 the phase difference between the two signals at point P is obtained as follows:

$$\Delta\phi_P = \phi_A - \phi_B + \frac{2\pi}{\lambda}(D_A - D_B) \tag{3.6}$$

The basic equation 3.6 cannot be solved as such for two reasons. First, the equation assumes that both base stations are transmitting with the

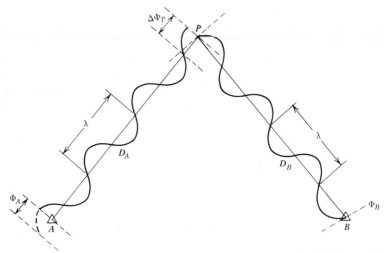

Figure 3.6 Phase comparison in CW transmission. In this simplified case $D_A = D_B$, $\Phi_A = 90°$, $\Phi_B = 0°$; thus $\Delta\Phi_P = 90°$.

same frequency; thus distinctions cannot be made between the receptions from the two base stations without further arrangements of the transmissions. Second, since the initial phase values ϕ_A and ϕ_B are seldom known, they must be "locked" to some standard or their influence on the final phase measurement must be eliminated. There are several systems available that meet the two conditions just stated. They are presented in detail in Section 20.2 (Land-to-Sea Survey) and in Part III (Electronic Navigation), but a review is beneficial at this stage.

The transmission frequencies from two (or three) ground-based stations may be different, but they are harmonically related to a basic frequency. When received in the survey vessel, the frequencies (which in this case are distinguishable from each other) are further multiplied by a set of harmonic numbers to produce a pair or two pairs of so-called *comparison frequencies*. The phase comparison is done and the lattice chart is drawn with respect to the comparison waves, which are significantly longer than the original transmission waves. In this type of arrangement the instantaneous phase values of the ground-based "side" stations are *locked* to the phasing of the "main" station. Therefore, in the commonly used vocabulary, the side stations are termed the *slave stations* and the main station is termed the *master station* (e.g., *Survey Decca*, operating with transmission frequencies of about 100 kHz).

The transmission frequencies from the ground-based master and slave stations may have exactly the same frequency, but they are sent on a time-sharing basis to the survey receiver to achieve distinction. Phase comparison is made between the original transmitted waves, whose instantaneous phase values are phase locked to the master phasing with the aid of an accurate and very stable storage circuit (e.g., *Decca Hi-Fix* series, operating with transmission frequencies of about 2 MHz).

The transmission frequencies from the ground-based master and slave stations may differ by a small amount, say a few hundred hertz in the audio band. These transmissions are received in the survey receiver and are heterodyned down to form a low frequency beat signal or beat note. At a fixed reference station on shore, the same transmissions are also received and heterodyned down to the same beat frequency. The beat note from the reference station is then sent to the survey receiver by way of an FM link. In the survey receiver the phases of the two beat notes are compared, and this phase difference is equal to the corresponding phase difference between the original transmissions. Because of the reference system, the measured phase difference in the survey receiver is independent of the instantaneous phase values at the ground stations (e.g., Raydist N, operating with transmission frequencies of about 1.5 MHz).

We present a simplified example of the use of the *single sideband* (SSB) transmission technique, which in this case refers to the circular mode. A transmitter onboard a survey vessel transmits continuous waves with the same frequency to two ground-based stations. In the ground stations the transmission is heterodyned down to an intermediate frequency (IF) which is about one-half the received frequency. One base station transmitter is tuned a few hundred hertz above and the other about the same amount below the IF frequency, thus producing beat notes. The beat note from the first ground station is returned to the survey receiver in the form of single sideband (i.e., the upper sideband), and the corresponding beat note from the second ground station is returned to the survey receiver in the form of the lower sideband. In the survey receiver heterodyning similar to that done in the ground stations is performed between the received carriers and the locally generated carrier to produce the reference beat note for purposes of phase comparison. The single sidebands obtained by filtering out the opposite sides of a fully modulated AM signal in this type of application serve two purposes: they establish the distinction between the signals, and they make the transmission more effective, allowing the available transmitter power to be used to greater advantage (e.g., Raydist DR-S).

In the worldwide navigation system (Omega, which operates with a frequency band of about 10 kHz), each of eight existing stations transmits with the same frequency on a time-sharing basis for about one second in turn, every 10 sec, providing station identification. In each station the transmission is locked to the system's frequency and phase by a station *cesium standard* or *atomic clock*; thus any shift in phase measured in a vessel is due only to the change in distance from the vessel to any pair of stations.

There is also a system available that combines continuous wave (carrier wave) and pulse-type transmissions. Pulses are used to determine the coarse distance difference from the survey vessel to two ground stations by measuring the difference in time of arrival of synchronized, pulsed signals from the ground stations. Phase differences are measured between synchronized carriers within the pulses to determine the fine distance difference from the vessel to the same two ground stations (e.g., Loran-C, which operates with a carrier frequency band of about 100 kHz).

3.1.3. Microwave Modulation

Since the introduction of the Tellurometer system by T. L. Wadley in the late 1950's instruments utilizing microwave modulation have seen much service in land, sea, and air surveying. The system is composed of two active units called the *master* and the *remote*; in most cases the units are interchangeable in that the *distances* between them can be measured by using either unit as the master where the measurements are made. The carrier frequencies used vary within the S- and X-bands. These bands are suitable for all weather operations but are rather vulnerable to ground reflection. To reduce the effect of ground reflection, quite large parabolic antennas produce beamwidths between 2° and 20°, depending on the carrier frequencies used. Also, in the focal point of the antennas two half-wave dipoles are located at right angles to each other, one for transmission and the other for reception. Both dipoles form 45° angles with the vertical, thus giving better polarizations to reduce the reflection from the ground. Distances to about 150 km can be reached in land surveying with accuracies of a few centimeters. In sea or air applications distances may be much longer but with reduced accuracies of around one meter.

The basic principle in the microwave surveying technique is to *frequency modulate* the carrier wave to form the *modulation* or *pattern* frequencies. The modulation waves are then used in a way similar to that characterizing continuous wave transmissions. Since microwave instruments always measure distances, we can say that the distance is obtained by comparing the phases of the transmitted modulation waves in the master unit to those of the modulation waves returned by the remote unit. The concept of this type of arrangement can be written in the following simple form:

$$D = \tfrac{1}{2}(n\lambda + \Delta\lambda) \tag{3.7}$$

where λ is the modulation wavelength, n is an integer amount of whole wavelengths included in the distance, and $\Delta\lambda$ is the fractional part of the last wavelength, obtained from the phase difference between the transmitted and received modulation waves. Because modulation frequencies and wavelengths serve as "yardsticks" in measurements, they are often called the *measuring frequencies* and *measuring wavelengths*.

From Eq. 3.7 it is seen that to solve the distance D, $\Delta\lambda$ and n must be determined. A detailed analysis of phase propagation is given in Section 20.1.1.4, but a simplified outline can be summarized at this stage. A highly stabilized oscillator in the master unit generates the modulation signal, which is then applied to the reflector of a klystron. The transmitted FM carrier is received at the remote unit, which is tuned to a carrier frequency slightly higher than that of the master transmission, to produce the IF frequency. Also in the remote unit a high precision oscillator generates a modulation signal having a frequency 1.0 or 1.5 kHz (depending on the make of the instrument) lower than the modulation frequency of the master signal, and this signal is applied to the reflector of the remote klystron. The resulting heterodyned 1 kHz beat note now possesses "information" about the phase difference between the two modulation signals *at the remote station.* The phase information is sent to the master station by means of a subcarrier, for example, and is compared there to a similarly heterodyned 1 kHz beat note that possesses information about the phase difference between the two modulation signals *at the master station.* This method of comparing "differences" yields the final phase difference between the transmitted and received *master* modulation waves and is independent of the remote modulation frequency. Depending on the type of equipment used, CRT, resolver, or digital methods can be chosen for phase comparison.

To understand better the determination of n in Eq. 3.7, let us assume that the master modulation frequency is a round number, 10 MHz. The fractional part of the wavelength then can vary between 0 and 15 m presented in distance D. This means that only a distance to 15 m could be measured unambiguously or, if any longer distance is measured, its approximate length should be known within 15 m and the last significant digits would be obtained from the fractional part of the wavelength (phase difference). If another round number—say 10 kHz—had been chosen as the modulation frequency, the fractional part of the wavelength would vary between 0 and 15,000 m. In this case distances up to 15 km could be measured unambiguously, but because of the limited possibilities of measuring phase angles accurately, the last few digits would be in error or doubtful. Therefore, more modulation frequencies are needed between these two extreme samples, spaced at intervals of multiples of tens, so that only the first significant digits from each "yardstick" need be used. This would lead to a complicated design of circuitry, capable of handling such a variety of modulation frequencies. The same result, however, can be obtained by utilizing *differences* of modulation frequencies that are close to each other. For example, the difference between 10.000 and 9.990 MHz modulation frequencies would yield the 10 kHz "simulated" modulation frequency. This example represents the method used by all microwave distance measuring instruments to determine the whole, unambiguous length of a line. The selection of modulation frequencies may vary, but a common conclusion in each case is that n in Eq. 3.7 can be regarded purely as a symbol of theoretical consideration, having no practical meaning.

3.1.4. Cesium Standard; Atomic Clock

In modern electronic surveying and navigation, especially when land-based transmitting stations are separated by large distances, it is of great importance to use methods that allow the transmitting stations to operate independent of each other but in strict synchronization with the "system" (e.g., Omega, Loran-C). In this case the stations are tied to a common standard to regulate the frequency and phase. A popular standard today is the *cesium standard.*

The quantum theory, developed by Planck, Einstein and Bohr in the early twentieth century, proves that an atom can exist in various quantum states and that each of these states has a discrete energy E_i. With the transition of an atom from a higher level to a lower level, electromagnetic energy is radiated and has the frequency of

$$f = \frac{E_1 - E_2}{h} \tag{3.8}$$

where E_1 is the energy at the initial level, E_2 is the energy at the final level, and h is Planck's constant. A cesium atom is an atom of alkali metal that has a single valence electron. This electron may spin in the same direction as the nucleus or in the opposite direction creating energy levels which are dependent on the direction of the spin. The frequency of the transition of a cesium atom between these two energy levels, called the *hyperfine levels*, is 9,192,631,770 Hz, which is the cesium frequency standard.

To understand how the cesium frequency standard is used as a reference frequency, let us consider that each atom behaves like a small magnet in an external magnetic field, where quantum theory requires the atoms to be polarized into two allowed states, the magnetic moments either "up" or "down." When the beam of cesium atoms leaves the oven A (Fig. 3.7), it is first assumed to propagate through a uniform magnetic field (not shown); thus the plane of the spin is maintained perpendicular to the direction of travel of the beam. The beam then crosses through a strong *nonuniform* magnetic field caused by the magnet B_1, and the atoms in the beam are deflected in the plane to opposite directions, depending on the direction of spin of their valence electron. The deflected atoms then pass through an alternating magnetic field generated by a crystal-controlled RF oscillator, for example. If the applied radio frequency is in exact resonance with the cesium frequency, a change of polarization direction of the atoms occurs, and the atoms have undergone the

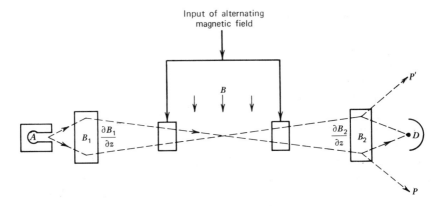

Figure 3.7   Principle of propagation of cesium beam.

transition. Therefore, when the atoms pass through a second nonuniform magnetic field, caused by a similar magnet B_2, they are reflected back toward the axis of the beam and they reach the detector D. If the applied radio frequency had not been in resonance with the cesium frequency, the atoms could not have undergone the transition and would have been reflected further in the same direction as in the first nonuniform magnetic field (i.e., toward points P and P', Fig. 3.7).

In the detector, which is a hot wire of tungsten or platinum-iridium, the atoms are ionized and drawn to a collector plate as an electric current. The current is at its maximum when the number of atoms undergoing the transition is greatest, thus providing means to tune the applied RF in resonance with the cesium frequency by maximizing the current. From this basic frequency other desired frequencies (e.g., transmission frequencies in electronic distance measurements) can be derived by scaling down; the frequencies then can be maintained in synchronization with the cesium standard with an accuracy of about $\pm 5 \times 10^{-12}$.

Because of its high stability, the cesium frequency has been chosen to secure the closest possible agreement with the basic astronomic definition of the unit of time, *the second*. The second of ephemeris time (ET), defined by the International Astronomical Union (IAU) in 1955 as 1/31,566,925.9747 of the tropical year for 1900 January 0 at 12 hr ET, was adopted by the International Committee of Weights and Measures in 1956. The resolution was ratified by the General Conference of Weights and Measures in 1960. Without changing the definition of the second just given, the General Conference defined an *atomic second* in 1964 as follows: "The second is the duration of 9,192,631,770 periods of the radiation corresponding to the transition between the two hyperfine levels of the ground state of the cesium-133 atom." In 1967 this definition became the standard definition of the second in the International System of Units (SI). This definition of the second, which is dependent on the cesium frequency, makes all cesium beam oscillators act as *atomic clocks* that can be transported to synchronize clocks far from each other. The accuracy figure of $\pm 5 \times 10^{-12}$ in maintaining the resonant frequency is equivalent to a discrepancy in time of ± 1 sec in 15,000 years.

3.1.5. Antennas

The strength of an electromagnetic energy radiated by an antenna is termed the *field strength*; it refers to a value of the electric field at a given point and is measured in terms of volts per meter. One volt per meter is equivalent to a potential of one volt induced in an antenna wire one meter long. The variation of the field strength around the antenna system can be shown graphically by so-called *polar diagrams*, which indicate both magnitude and direction of the field strength for a given distance from the antenna.

In low frequency transmissions, tubular antenna masts or towers are used. The height of the antenna assembly, together with an extensively built ground system, is of prime importance in achieving the most effective transmission output power. The *polarization* of such a transmission is vertical; that is, the antenna radiates vertically polarized waves, usually *omnidirectionally* (i.e., radiation in horizontal plane covers 360°). In high frequency or microwave transmission it is important to concentrate the largest possible part of the radiated energy on the receiving antenna of the secondary equipment or on a reflector at the target. A basic form of antenna that serves to radiate electromagnetic waves at the frequency bands stated earlier is the half-wave *dipole*, from which many of the more complex antennas are constructed. Such an antenna operates independently of the ground and therefore may be installed far above the surface of the earth. The directional effects are obtained by phasing and spacing two or more dipoles so that their resultant polar diagram gives maximum value in the desired direction. Such arrays of dipoles may be termed the *plane antennas*. Directional effects are also produced by using *parabolic* reflection surfaces, which focus the radiation much as light is focused in an ordinary automobile headlight.

Although the basic element in short- and microwave antennas is called the half-wave dipole, in practice its physical length is slightly shorter than half a wavelength. This is because the velocity of electromagnetic propagation along the conductor (dipole) is always less than that in the surrounding air (space); also, the dipole is not absolutely isolated from surrounding objects. The physical length of a half-wave dipole can be obtained from the following expression:

$$L = \frac{150}{f} K \tag{3.9}$$

where L is the length in meters, f is the frequency in megahertz, and K is the "velocity coefficient," which is dependent on the wavelength and the thickness of the conductor.

3.1.5.1. Plane Antennas

Two main dipole assembly types are used in connection with VHF transmissions, namely, the *end-fire* system and the *collinear* system. In the end-fire system two identical dipoles are placed parallel to each other, spaced one-quarter wavelength (90°) apart and driven with equal currents having a phase difference of 90°. The radiation pattern is *unidirectional,*

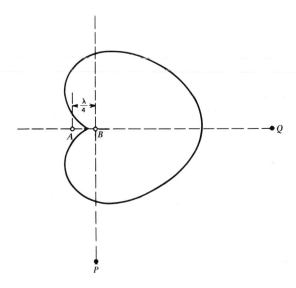

Figure 3.8 Polar diagram of end fire antenna system (Laurila, 1960).

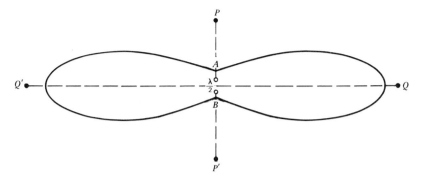

Figure 3.9 Polar diagram of collinear antenna system (Laurila, 1960).

and at point Q (Fig. 3.8), the field strength resulting from the currents in the two dipoles is double the field strength resulting from either of them alone, since by the time the electromagnetic wave leaving dipole A reaches dipole B, the wave leaving dipole B will have the same phase; thus the two waves add to produce maximum radiation to the right. By the time the electromagnetic wave leaving dipole B reaches dipole A, however, the wave leaving dipole A will be 180° out of phase, causing complete cancellation to the left. The polar diagram is a *cardioid*, where the field strength at the point P, at right angles to the line of dipoles is $\sqrt{2}$ times that due to either of the dipoles acting alone. In the end-fire system, maximum directivity is given in the plane formed by the dipoles and along a line normal to their length. An application of the end-fire directivity system is the so-called *Yagi* antenna, where only one dipole is excited. Behind this dipole is placed a *parasitic reflector* dipole, and in front there are several *parasitic director* dipoles.

The basic form of collinear antenna system consists of two horizontal dipoles A and B (Fig. 3.9), separated by distance equal to one-half wavelength and erected to ensure that they are in line, with their free ends directly opposite. Both dipoles are fed in-phase RF currents. The field strength at points Q and Q', which are situated in the line at right angles to the line of dipoles, is double the field strength resulting from either of the dipoles alone, since the currents are in phase, thus additive. At points P and P', which are situated in the plane of dipoles, the field strength is zero, for currents in dipoles A and B are opposite in phase along both extensions of the line AB. An application of the collinear system is the so-called *broadside* antenna, where collinear elements are stacked above and below another set of similar elements on the same

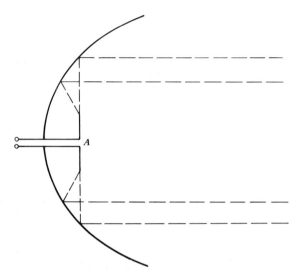

Figure 3.10   Parabolic reflector (Laurila, 1960).

plane, each dipole being spaced one-half wavelength from the adjacent dipole. The directivity of the broadside antenna is perpendicular to the plane formed by the collinear elements.

3.1.5.2.   Parabolic Antennas

Applications of the microwave technique in distance measurements feature reflectors that have the basic shape of a parabola. If a half-wave dipole is placed at the focal point A of a parabolic reflector (Fig. 3.10), the reflector concentrates the radiation into a beam parallel to the central axis of the parabola, just as a searchlight reflector focuses a light beam. The transmitted radio beam produced by the parabolic reflector does, however, diverge much more rapidly than an equivalent light beam. It can be shown that the energy transmitted is directly proportional to the area of the reflector "opening" and inversely proportional to the square of the wavelength. The directivity achieved is directly proportional to the diameter of the paraboloid and inversely proportional to the wavelength used. Thus a large reflector and a short wavelength are necessary to obtain a narrow, concentrated beam. Mathematically the directivity—the *beamwidth* in radians—is obtained by the following expression:

$$\alpha = \frac{\lambda}{D} \tag{3.10}$$

where λ is the wavelength and D is the diameter of the parabola, both given in the same units. By definition, the beamwidth is the apex angle of a cone containing one-half the transmitted power and therefore often called the 3 dB beamwidth. For example, the Tellurometer MRA-1 and MRA-4 both use a parabolic reflector with 30 cm diameter. The carrier wavelength of the former is $\lambda = 10$ cm and that of the latter is $\lambda = 0.8$ cm. The corresponding beamwidths are then approximately 17° and 1.5°, respectively. There are several modifications of parabolic reflectors depending on the need of particular beamwidths in vertical and horizontal planes. Figure 20.36 is a photograph of the rectangular parabolic reflector of the Tellurometer MRB-201, which produces a vertical beamwidth of about 20° and a horizontal beamwidth of about 24°. The figure illustrates the arrangement of the transmitting and receiving dipoles, which are at right angles to each others and at 45° angles to the vertical. The polarization principle is presented in Fig. 3.11, where horizontal polarization is used as an example.

3.1.5.3.   Effect Factors of Antennas

If the transmitting antenna were a point source of energy radiating electromagnetic energy *isotropically* (i.e., uniformly in all directions), the power flow through a unit area at a distance R from the antenna could be found by dividing the total radiated power P by $4\pi R^2$. A *directive* antenna, however, concentrates energy in a certain direction, and the difference observed at some distant point between the power radiated isotropically and the same original power radiated by a directive antenna defines the *effect factor* of the antenna or the *antenna gain*. The effect factor G is obtained from the following expression:

$$G = \frac{4\pi(kA)}{\lambda^2} \qquad (3.11)$$

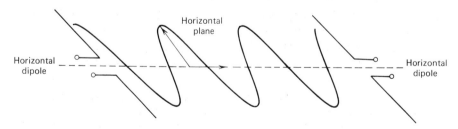

Figure 3.11   Polarization principle.

where A is the geometric opening surface or "aperture" of an antenna, stated in the same units as the wavelength λ, and k is a dimensionless factor that equals 1.0 when the excitation is uniform in both phase and intensity over the whole aperture of the antenna. The magnitude of the k-factor in most parabolic antennas is about $k = 0.65$. The product (kA) is called the *effective surface* of an antenna, A_e. If the gain of an antenna is known, the effective surface can be obtained from Eq. 3.11 as follows:

$$A_e = \frac{G\lambda^2}{4\pi} \tag{3.12}$$

To illustrate the magnitude of the effect factor or gain of different parabolic antennas fed by different radio frequencies, let us examine some typical cases.

1. *NASA/JPL/DNS:*[*] Frequency $f = 8450$ MHz, wavelength $\lambda = 0.0355$ m, antenna diameter $D = 9.15$ m. By utilizing the factor $k = 0.65$ and substituting the data into Eq. 3.11, we find the antenna gain to be 425,150 times the isotropic radiation or $G = 56$ dB.

2. *Tellurometer,* MRA-1: Frequency $f = 3000$ MHz, wavelength $\lambda = 0.10$ m, antenna diameter $D = 0.3$ m. The resulting antenna gain is 58 times the isotropic radiation or $G = 8$ dB.

3. *Tellurometer,* MRA-4: Frequency $f = 35,000$ MHz, wavelength $\lambda = 0.008$ m, antenna diameter $D = 0.3$ m. The resulting antenna gain is 9020 times the isotropic radiation or $G = 39$ dB.

3.1.5.4.   *Passive Reflectors; Corner Reflector*

In many radar applications, such as harbor surveys, navigation through narrow waterways, and artificial satellite and lunar tracking operations, passive reflectors play an important role. They may be natural reflecting surfaces or parts of man-made structures, or they may have been built into special form to collect the maximum amount of radiated energy to reflect it back to the energy source. In each of these cases the reflectors absorb the electromagnetic energy and after absorption act in turn as a radiating energy source. The radiation also has a gain compared to the isotropic radiation, as was the case in the original transmitting antenna,

---

[*] A worldwide lunar radar time synchronization system, Jet Propulsion Laboratory, Pasadena, Calif. Baumgartner, W. S. and M. F. Easterling. "A World-Wide Lunar Radar Time Synchronization System." *AGARD Conference Proc.* No. 28. Advanced Navigational Techniques (Ed. W. T. Blackband), Technivision Services, Slough, England, 1970.

and this gain is often expressed as the *absorbing* or *electric surface* which is related to the geometric surface of the object by its shape and the wavelength used.

For solid objects whose dimensions are large compared with the wavelength, good approximate values of absorbing surfaces have been computed (*Ridenour*, 1947), and are given below.

1. For a flat sheet with geometric surface A_p perpendicular to the incident beam, an absorbing surface is obtained as

$$\delta_p = \frac{4\pi A_p^2}{\lambda^2} \qquad (3.13)$$

2. For a cylinder of radius R and length l, an absorbing surface is obtained as

$$\delta_c = \frac{2\pi R l^2}{\lambda} \qquad (3.14)$$

3. For spherical surfaces very large compared to the wavelength, the following expression can be used to obtain the absorbing surface;

$$\delta_s = \pi R^2 \qquad (3.15)$$

where R is the radius of the sphere.

The gain may be considered as a factor stating how many times the electric surface is larger than the corresponding geometric surface. If the object is either cylindrical or spherical, the geometric surface is that of the cross section, whereas in the case of a plane the geometric surface is the plane surface A_p itself. Thus the gains in the three cases just given will be

$$\Delta_p = \frac{4\pi A_p}{\lambda^2} \qquad (3.16)$$

$$\Delta_c = \frac{\pi l}{\lambda} \qquad (3.17)$$

$$\Delta_s = 1 \qquad (3.18)$$

From Eqs. 3.16 to 3.18 it will be seen first that a sphere reradiates isotropically. Second, if all three geometric surfaces are equal to the same unit area, say one square meter, and the wavelength is given the value $\lambda = 0.1$ m, for example, the gain for the reradiating plane is 1257 compared to 1 for the sphere. The gain for the cylinder depends on the shape of cylinder (length l).

This amazing gain for the plane is more or less theoretical, however since it can be achieved in practice only when the incident beam is exactly perpendicular to the plane. To obtain the high gain of the plane in practice, so that strong reflection is obtained even when the direction of the incident beam deviates greatly from the normal to the plane, a special reflector called a *corner reflector* is used. A corner reflector consists of three mutually perpendicular intersecting metallic plates (Fig. 3.12). If a beam is directed into the corner reflector, triple reflection occurs and sends the beam in the direction from which it came. The maximum geometric surface *BDE* for triple reflection is achieved when the incident beam coincides with the axis of symmetry of the reflector, that is, along the line that makes equal angles with each of the edges. Geometrically speaking, the maximum aperture $BDE = A_r$ is obtained by projecting a regular cube onto a plane perpendicular to its diagonal. If the edge of the corner reflector (i.e., the side of the cube) is symbolized as a, the aperture will be

$$A_r = \frac{a^2\sqrt{3}}{2} \tag{3.19}$$

When substituted into Eq. 3.13, this yields the maximum electric surface of a corner reflector with edge of a:

$$\delta_r = \frac{3\pi a^4}{\lambda^2} \tag{3.20}$$

To illustrate the effectiveness of a corner reflector, let us take the example of a harbor survey or narrow-water navigation. If the targets are

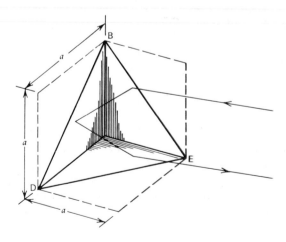

Figure 3.12   Corner reflector.

spar buoys of rods with 0.1 m diameter sticking up vertically 2.2 m from the surface, their geometric surfaces (cross-sections) A will be 0.22 m². By scanning them with a radar that utilizes 3 cm microwaves, we find that the electric surfaces of the buoys, according to Eq. 3.14, are 51 m². If the buoy rods are replaced with corner reflectors having edges $a = 0.5$ m, the maximum geometric surfaces A of the reflectors will be also 0.22 m², according to Eq. 3.19.

Substitution of the a-value, together with the wavelength λ, into Eq. 3.20 yields the electric surface of the corner reflectors $\delta_r = 655$ m², which shows that by using corner reflectors instead of rods, more than 10 times stronger reflection is produced. However, it should be noted that the direction from which the buoy rods are scanned is of no importance. The direction in scanning the corner reflectors is not very critical either. Simple geometry indicates that the reduction of the electric surface is directly proportional to the square of the cosine of the angle between the incident beam and the normal to the plane BCD (Fig. 3.12). Also, to reduce to a minimum the loss of electric surface caused by the erroneous orientation of the reflector, *clusters* of corner reflectors are used, making the target a good reflector in all directions.

3.1.5.5. *Radar Equation*

Now that we need to measure distances to objects far in space, such as artificial satellites and the moon, solution of the free-space distance problem is gaining ever-increasing importance. To analyze the problem, we consider such elements as the quantitative relations between transmitted and received powers, the design of transmitting and receiving antennas, the target characteristics, and the distance between the transmitter and the target. The equation that expresses the solution as function of these factors is called the *radar equation*.

If we consider a transmitter that transmits a peak power of P_t isotropically, the power is radiated equally in all directions, filling a sphere of constantly increasing radius. As was pointed out in the previous section, the power per unit area at a distance R is equal to the radiated power divided by an area $4\pi R^2$ of the sphere of radius R. If the gain of the transmitting antenna is denoted G_t, the power per unit area intercepted by the target will be

$$P'_t = \frac{P_t G_t}{4\pi R^2} \tag{3.21}$$

The radiation received in the target or reflector energizes it, and the target will reradiate an amount of power equal to the arriving power per unit area multiplied by the electric surface δ of the target; thus the reradiated power will be

$$P''_t = \frac{P_t G_t \delta}{4\pi R^2} \tag{3.22}$$

This energy has to travel back to the original transmitting antenna over the same distance R, and if the effective surface of this antenna is A_e, the receiver input power will be

$$P_r = \frac{P_t G_t A_e \delta}{(4\pi)^2 R^4} \tag{3.23}$$

If the transmitting and the receiving antennas are the same, Eq. 3.23 may also be written in another form by substituting G_t from Eq. 3.11 as follows:

$$P_r = \frac{P_t A_e^2 \delta}{(4\pi)\lambda^2 R^4} \tag{3.24}$$

If in lunar applications, where R is large compared to distances on earth, a signal is sent to the moon to be reflected back to another receiving antenna, the effective surface A_e in Eq. 3.23 refers to that of the receiving antenna. Also, when a signal is sent to the moon, the simple expression for δ_s, Eq. 3.15, cannot be used as such in Eq. 3.24. Because of the dielectric characteristics of the moon, δ_s must be multiplied by a reduction factor K, which also depends on the wavelength. For example, in using the wavelength $\lambda = 2.7$ m, the reduction factor is $K = 0.177$ (*Mofenson*, 1946).

3.2. LIGHT WAVE TRANSMISSION

The way light is used as a carrier does not have the variety of applications found in the case in radio wave transmission. All today's instruments use amplitude or intensity modulation for distance measurements. The measurements are made by comparing the phases of transmitted and received modulation signals with techniques similar to those described for the microwave applications. For many years the light source for the carrier was the tungsten lamp, which emits a fairly steady and homogeneous white light; the relatively small amounts of power required can be obtained from regular car batteries. The disadvantage of this light source is the small light density in the visible, incoherent region, allowing daytime measurement of distances of only about one kilometer. The next improvement was the use of mercury vapor lamps, which have a light intensity about 20 times larger than that of the tungsten lamp. Distances up to about 10 km could be measured during the daytime, but the power requirement was large and portable generators were needed as a source of power.

Emphasis in this section is placed on the description of modern light sources, modulation, and detection of light signals.

3.2.1. Lasers

As proposed by Max Planck in 1900, quantum theory stated that radiant energy, such as light waves, consists of definite, elemental quantities of energy. Planck also considered the electron to be an electric oscillator that can emit only whole quanta. This theory was expounded in 1917 by Albert Einstein, who showed that light quanta also are absorbed in entire units, and no fraction of a quantum can be emitted. Einstein gave the radiation quantum the name *photon*, to which he attributed properties of a particle with zero rest mass. At about the same time Niels Bohr proved that only in jumping from one energy level to another does the atom absorb or emit a Planck quantum of radiation. The quantum theory explains the meaning of *luminescence*, which is light emission that cannot be considered to be caused merely by the temperature of the emitting body.

In association with the description of the cesium standard (Section 3.1.4), it was stated that with the transition of an atom from a higher energy level to a lower energy level, electromagnetic energy is radiated. We can now amplify the statement by saying that if electrons in atoms are excited to any higher level, the ultimately occurring return of the electrons to their original energy level is called the *transition*, and it produces the emission of photons with the energy obtained from Eq. 3.8 as follows:

$$E_2 - E_1 = \Delta E = hf \tag{3.25}$$

The "natural," nonstimulated fall of electrons from an excited level E_2 to a lower level E_1 causes an emission of photons called *spontaneous emission*. There can be several energy levels where the photons are absorbed, and the rate at which the electrons fall to a lower level depends on the *lifetime* of the higher level. With reference to a two-level system, this means that the intensity of the emitted light is proportional to the population of electrons on that level and also the *decay* of the population, which is exponential (i.e., more photons are

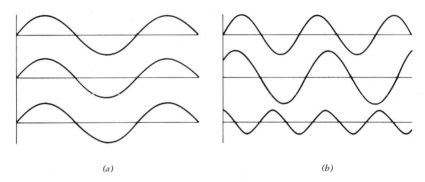

(a)   *(b)*

Figure 3.13 Principle of coherency. (*a*) Coherent light, (*b*) noncoherent light.

emitted at the beginning than later on). This slow process of spontaneous emission can be stimulated by allowing electrons in E_2 level, waiting for the spontaneous drop to level E_1, to be affected by a field of photons having an energy equal to $E_2 - E_1$. The phases and directions of the stimulating photons are the same as those of the photons of spontaneous emission. As a result of these induced transitions, new photons are produced, and by continuation of the process of the induced transitions, it is possible to build up a large radiation field. The summary-type description of stimulated emission may help to make the acronym "laser" (*l*ight *a*mplification by *s*timulated *e*mission of *r*adiation) easier to understand. Since each photon in the process just explained has an energy of exactly $hf_{12} = E_2 - E_1$, thus the same wavelength and frequency, the photons are said to have *temporal coherency*. Also, because of their exact phasing, the photons are said to have *spatial coherency*. Given this combined coherency, the laser emission is able to radiate very narrow, concentrated beams over large distances. Figure 3.13 illustrates the difference between additive coherent waves and nonadditive waves of ordinary light.

The ratio of the number of electrons at levels E_1 and E_2 for a system in *thermal equilibrium* is given by

$$\frac{n_2}{n_1} = e^{-(E_2 - E_1)/kT} \tag{3.26}$$

where n_2 is the number of electrons at level E_2, n_1 is the number of electrons at level E_1, E_2 and E_1 are the corresponding energies, k is the Boltzmann factor, and T is the temperature. The equilibrium is established (*a*) by spontaneous emissions from level E_2 to level E_1 (*b*) by delivering the energy from these transitions to the system and by using the resulting electromagnetic field to cause transitions from level E_1 to level E_2, resulting in absorption at level E_2, and (c) by stimulated emission from level E_2 back to level E_1. Thus in thermal equilibrium the absorption equals the sum of the spontaneous and stimulated emissions, the contribution of the latter being negligible. The foregoing summary and Eq. 3.26 indicate that two conditions must be fulfilled to build up an effective laser system. First, the photon density that affects the stimulation of emissions must be increased by causing the photons to undergo several multireflections between two solid surfaces in the laser. Second, the stimulated emission must be made greater than the absorption. In thermal equilibrium, $n_2 < n_1$ if temperature T is assigned any positive value: it means that there are more electrons at level E_1 than at E_2, thus the absorption is greater than the stimulated emission. To fulfill the second condition, the population in levels E_1 and E_2 must be changed, making $n_2 > n_1$. This is called the *population inversion*, and it is established by pumping either optical or electrical energy to the atoms at level E_1 to excite them into the E_2 energy level. Population inversion is often called "negative temperature state" because according to Eq. 3.26 it is possible only when T is given a negative value. In practice, however, this odd-looking phenomenon has little importance, since Eq. 3.26 simply is no longer valid when the equilibrium is disturbed by an outside energy agitation (pumping).

All practical lasers involve more than two energy levels. In the three-level system presented in Fig. 3.14, E_2 is the upper laser level and E_1 the lower laser level, in this case the ground level. The desired laser radiation operates with the frequency of $f_{21} = (E_2 - E_1)/h$; thus the condition $n_2 > n_1$ must be achieved, and this can be done by proper arrangement of the lifetimes of different levels. First the electrons are excited to the level E_3

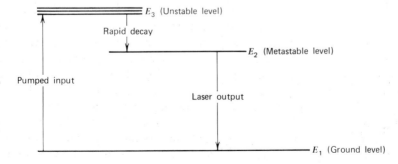

Figure 3.14   Energy levels in three-level laser.

by some pumping process but since this level has a very short lifetime for transitions to level E_2, however, electrons fall very rapidly to that level without radiating light but giving up the energy difference $E_3 - E_2$ of this transition in the form of heat, for example. If the lifetime of level E_2 is long, electrons are reluctant to leave it; therefore this level is called the *metastable level*. The importance of this level becomes quite obvious if the rate of decay from E_2 to E_1 is smaller than the rate of transition from E_1 to E_3, because the electrons tend to accumulate into level E_2, thus producing a large population inversion in that metastable level.

3.2.1.1. *Ruby Laser*

The ruby laser is a *solid state laser*, first demonstrated by *Maiman* (1960). Ruby is crystalline aluminum oxide, in which a small fraction of the aluminum atoms is replaced by chromium atoms. It is a three-level laser, and the stimulated chromium atoms through the transition between the metastable and ground levels emit red light with a wavelength of $\lambda = 6943$ Å. The ruby crystal is used in a form of a rod that is cut so that the ends are parallel to each other and perpendicular to the axis of the rod. The ends are polished and coated with highly reflective material acting as mirrors, but one of the mirrors is made semitransparent, letting a fraction of the light leak out of the system as the laser output. Inside the ruby rod, which functions as a solid cavity, the stimulating photons reflect between the end surfaces and cause the induced transition to the electrons in the metastable level. Spatial coherency is always achieved when an integer number of half-wavelengths are included in the distance separating the mirrors. Because of the shortness of the wavelengths, the mechanical spacing of the mirrors is not critical: there are always some portions of the spacing that satisfy the above-mentioned condition.

Population inversion in a ruby laser is obtained by utilizing *optical pumping*, which in this case means that the atoms at ground level are illuminated by light with a wavelength shorter than that of chromium emission. Light used for the pumping is in the green part of the spectrum. If the levels are numbered 1, 3, and 2 (Fig. 3.14), the energy difference $E_3 - E_1$ refers to the green light, and electrons are raised from ground level (state) E_1 to level E_3, which is very unstable. The lifetime of this level is extremely short, and electrons decay rapidly to the metastable level E_2, giving out the energy difference $E_3 - E_2$ as heat. The mean lifetime of the metastable level is about 5 msec, and great population inversion occurs.

In practice optical pumping is arranged by wrapping a xenon flash tube around the ruby rod; a pulse of very intense light is created by discharging a capacitor through the tube. We have been discussing the desirability of stimulated emission in general, but in the case of the ruby laser an important phenomenon should be observed. During the xenon flash, which has a sharp onset and a more slowly diminishing tail, stimulated emission from the metastable level to the ground level occurs every time the flash intensity is large enough to create population inversion. The level of this occurrence is called the *threshold pumping level,* and the resulting laser outputs consisting of intense short spikes have amplitudes that follow closely the intensity of the flash. The spikes occur in high frequency throughout the flash exposure, exceeding the threshold level. During the formation of these spikes, the population of the metastable level diminishes, thus the creation of controlled, large laser pulses becomes impossible because of lack of large population concentration in the metastable level. Therefore, a means of "spoiling" or preventing the stimulated emission temporarily is required. It is obvious that the discontinuation of the stimulated emission and the continued pumping of more energy into the system creates a giant population inversion, that when released for transition, causes a giant controlled pulse. This process is called the *Q-spoiling* or *Q-switching,* where Q is the quality factor of the resonant structure. The Q-switching effect can be achieved in several ways. Mechanically it is possible if one replaces the back-end mirror of the ruby by a mirror that can be rotated at high speed around an axis perpendicular to the axis of the ruby. By controlling the speed of rotation of the mirror, the ruby can be made to act as a cavity resonator during a desired period only. The same effect is also produced by using a fast electro-optical shutter, such as the *Kerr cell* described in Section 3.2.3.

3.2.1.2.   *He–Ne Laser*

Whereas the ruby laser, a high-powered pulsing-type laser, is used mainly in space applications, many other surveying operations require low-powered continuous wave lasers, whose beams can be modulated like any other steady light source. The solution is the gaseous laser, in which the excitation of atoms is established by electric discharges. This discharge in a *plasma tube* containing mixture of gases can be produced either by a high frequency (RF) voltage or by applying a DC between two electrodes within the plasma tube, the latter method being more popular in electronic distance measuring equipment today. The processes involved give the desired population inversions in the gas to build up the population densities at the different levels necessary for oscillation.

In the *He–Ne laser,* helium and neon gases are mixed, usually in a ratio of about $10:1$. The electric discharge creates in the gas electrons that excite the atoms from the ground state to the metastable levels. Helium atoms in the excited state have a long lifetime, and once they are excited to the metastable level they lose their energy not by spontaneous radiation, but by a process of collisions with neon atoms. With reference to Fig. 3.15, let us consider that through the electrical pumping process the helium atoms are excited to level $2s$, which has nearly the same energy as level $3s$ of neon. Energy is transferred from level $2s$ to level $3s$ by collision of the helium and neon atoms, and because of the long lifetime of level $2s$ and its proximity to level $3s$, large population inversion occurs at level $3s$ especially because $2p$ is not a ground state but a relatively quickly decaying excited state. The inversion condition $n_{3s} > n_{2p}$ is therefore easy to obtain. We omit other levels and transitions in this description, emphasizing only the transition between levels $3s$ and $2p$ of neon atoms giving off stimulated emission at a wavelength of 0.6328 μm or 6328 Å in the visible red spectrum, which is used in modern electronic surveying equipment. Gas lasers are able to operate in the continuous wave mode, the population of the $3s$ level being created at the same rate that the stimulated emission depopulates it. Thus the desired laser emission is produced by transitions in the Ne atoms, whereas the primary purpose of the He atoms in the mixture is to intensify the excitation process. The average energy output of a high-powered gas laser ranks among the best obtained so far.

The He–Ne laser consists of a glass or plasma tube filled with the gas mixture, the electrodes or other means for creating a discharge in the gas

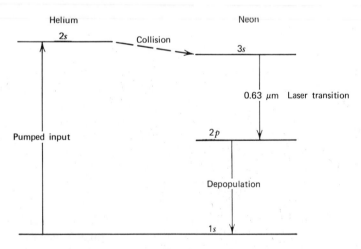

Figure 3.15   Energy levels in He–Ne laser.

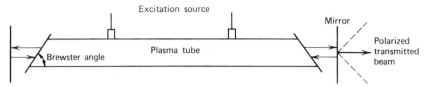

Figure 3.16 He–Ne laser.

mixture, and an electromagnetic resonant system, built by placing mirrors at both ends of the tube. The mirrors may be either plane or spherical and are usually placed outside the tube. The photons inside the tube reflect from mirror to mirror until a certain level in coherency is reached; then about 0.5% of the photons hitting one of the mirrors pass through, providing output beam energy of the laser. This condition is achieved by ensuring that one of the mirrors has a finite transmission. Because the plasma tube is a gas-filled closed tube, the photons must pass through its end windows, and to minimize the reflection of photons from them, the windows are placed at a certain angle called the *Brewster angle* with the axis of the tube. At this angle only plane polarized light is transmitted through the Brewster window and reflected back to gas from the end mirrors (Fig. 3.16). This property can be used in advance in light modulation (Section 3.2.3).

3.2.2. Luminescent Diodes

In the recombination of electrons and holes in a semiconductor or by the deexcitation of carriers from higher impurity levels to the equilibrium states, light is given off by the material. This light emission is called *luminescence*, and it is created independently of heating the material. If the excitation source is current passing through the material, the phenomenon is called *electroluminescence*. When carriers are injected across a forward-biased *p-n* semiconductor junction, the electrons and holes recombine, electrons drop from the conduction band to the valence band, and emission of photons monochromatic within the band width occurs from the junction. The material used to fabricate the semiconductor, thus the energy gap $(E_2 - E_1)$, determines the wavelength of the radiation. In most electronic distance measuring equipment designed for short distances, gallium arsenide alloy (Ga–As) in the luminescent diodes produces the necessary light source for transmission. Since the wavelength of

Ga–As emission is in the infrared region (λ = ca 0.9 μm), the diodes are referred to by the instrument manufacturers as *infrared diodes* or simply *Ga–As diodes*. The emission of such diodes is spontaneous, thus spatially incoherent, and to produce a narrow beam the diode must be placed in the focal point of the optical system of the transmitting instrument. The great advantage in using Ga–As diodes in distance measuring equipment lies in the rapid variation in the intensity of emission in accordance with the applied voltage. Therefore, no separate light modulator is needed because the desired intensity or pattern modulation is obtained directly from the diode by RF-controlled voltage.

The spontaneous recombination radiation of the Ga–As diode also can be made temporally coherent, and diode then acts like a laser—the *semiconductor laser*. To make this possible, population inversion is needed at the junction, and a resonant system must be built within the semiconductor. The first requirement is satisfied by applying a bias current large enough to build up large concentrations of electrons within the conduction band and of holes within the valence band; minority carriers become majority carriers, and population inversion results over an *inversion region*. The second condition can be fulfilled by constructing a (solid) resonant cavity in the junction region by polishing the faces perpendicular to the crystal plane, thus forming a thin (ca. 1 μm) depletion layer, where the photons cause stimulation.

Ga–As diodes emit light in the invisible infrared region of the spectrum suitable for distance measurements even during daytime, but through the proper selection of various impurities, it is possible to construct light-emitting diodes (LED) giving off light in the visible band. These diodes are important as "electronic dials" in pocket and desk calculators and most of all as distance displays in electronic distance measuring equipment. Arrays of LEDs can be used to form letters and numbers that can be switched on and off in nanoseconds and require very little power.

3.2.3. Light Modulation

All electronic distance measuring equipment using light sources other than the infrared emission from the Ga–As diode need a modulator to form the sinusoidal light intensity variation to the light carrier. In instruments utilizing noncoherent, nonpolarized light, the *Kerr cell modulator* is a common component. In the operation of the Kerr cell modulator, sometimes also called an *electronic shutter*, three double refractions occur in sequence. The light beam emerging from the light source L (Fig. 3.17a) passes through a *polarizer P*. Already in the early seventeenth

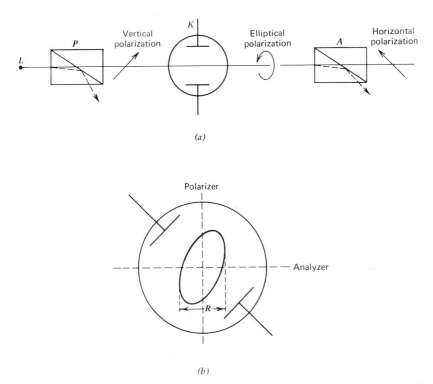

Figure 3.17 Kerr cell as light modulator. (*a*) Principle of Kerr modulation, (*b*) light intensity component leaving modulator.

century, *Bartholin* proved that a ray of light incident on a plate of calcide is split into two rays, the *ordinary ray* and the *extraordinary ray*, which differ mainly in that the former ray obeys Snell's law of sines and the latter does not. This characteristic of double refraction will be observed later in the construction of the polarizer, but the important result of Bartholin's discovery is that the light entering the polarizer vibrates in all planes containing the ray, but after leaving the polarizer the vibration occurs only on two planes that are perpendicular to each other, namely, those defined by the ordinary and the extraordinary rays. The rays are said to be linearly or *plane polarized*. If the polarizer is to serve any practical purpose, one of the polarization planes must be removed, thus creating plane polarization in one direction only, say vertical. *Nicol prisms* are used as polarizers, and they are constructed by cutting two pieces of calcite crystal in a special way and cementing them together with silicone oil. Because the critical angle for total reflection is less for the ordinary ray than for the extraordinary ray, the cutting must ensure that the angles between the unpolarized ray and the incident surface of the prism and between the ordinary ray and the cemented surface of the prism cause a total reflection for the ordinary ray but pass through the extraordinary ray, which is vertically polarized.

Now let us assume that the light beam entering another similar Nicol prism A has undergone double refraction and total reflection, making the prism a horizontally polarizing device called an *analyzer*. Because the polarization planes of prisms P and A are at right angles to each other, no light can emerge from prism A. Since the relationship between the light intensity that enters the prism and leaves it is a function of the angle between the *privileged directions* of prisms P and A, theoretically the intensity could be controlled by rotating the prisms. This, of course, would lead to mechanical impossibilities; besides, a sinusoidal variation is needed for modulation.

In practical arrangement of the Kerr modulator, the Kerr cell K is placed between the two Nicol prisms. The Kerr cell is a closed glass cylinder filled with purified nitrobenzene, which has a high refraction capability. If a strong electric field is applied in a direction perpendicular to the axis of the cylinder (beam) through parallel plates placed at an angle of 45° with the plane of polarization from the first prism, double refraction or birefringence occurs. As a result, the plane-polarized light beam is resolved into two components, one perpendicular and the other parallel to the electric field. In nitrobenzene the index of refraction, thus the velocity of light vibrating in the direction of the electric field, is slightly different from the velocity of light vibrating perpendicular to it. Therefore when the two components have traveled through the Kerr cell, a phase difference is produced, and the following conclusion can be made: if two light components with equal frequency but having a constant phase difference vibrate at planes perpendicular to each other, the combination of the vibrations results in an ellipse in the plane perpendicular to the direction of the light beam. The light is then said to be *elliptically polarized*. To produce sinusoidal modulation, an AC voltage is applied across the electrodes, and the voltage variation changes the characteristics of the ellipse (rotation and shape) analogously to those of the Lissajous figures. Variation of the applied voltage causes variation of the intensity of light passed by the analyzer prism, and the instantaneous intensity is the horizontal component of the elliptically polarized light, which is obtained as the projection R of the ellipse on a horizontal axis (Fig. 3.17b).

The applied voltage $V = V_0 \sin \omega t$ includes the positive and negative half-cycles, but since the intensity of the emitted light cannot have negative values, a bias voltage larger than V_0 can be applied to the electrodes.

Many instruments that utilize the He–Ne laser as the light source have a *Pockels* modulator, consisting of several crystals of potassium dideuterium phosphate (KDP). The velocity of light passing these crystals is different for the two perpendicular oscillation directions of the light when an electric field is applied across the crystal. This means that if linearly polarized light enters the crystal, one of the perpendicular components of light will be slowed down during its passage through the crystal. The polarization will therefore change to elliptical (circular), and if the delay corresponds to one-quarter wavelength, the polarization will be linear but with the direction of polarization rotated 90°. The delay of 90° is produced by placing a thin mica or quartz plate, known as the *quarter-wave plate*, in front of the light beam entering into the modulator. The analyzer has its polarization plane parallel to the linearly polarized beam emerging from the modulator and permits the light to pass through. The intensity of the modulated light follows the applied voltage, thus forming sinusoidal light intensity modulation. The first polarizer used in connection with the Kerr cell is not needed because of linear polarization of the laser beam caused by the Brewster effect. The Pockels modulator today is considered faster and more accurate than the Kerr modulator, which is slightly affected by impurities in the nitrobenzene.

3.2.4. Light Detection

The detection of modulated light signals in electronic distance measuring equipment in which visible light is a carrier is exclusively done by the *photomultiplier*. As the name suggests, the energy impinging on the photomultiplier in the form of photons is converted into energy of electrons and is multiplied enormously by using the energy source of the receiving circuitry. But the energy needed to remove electrons from the metallic surface of the photomultiplier cathode is supplied by the incoming light signal, and the process is called *photoelectric emission*. The energy a photon can give to an electron in a solid when light strikes the surface of the solid is the already familiar value $W = hf$, where h is Planck's constant $(6.63 \times 10^{-34} \text{ J/sec})$ and f is the frequency of light. If electrons are to be removed from metal, a certain minimum energy must be given to the electrons to free them from the solid. This minimum energy is called the *work function*, W_W. Thus if the energy given to electrons by photons is W and the work function of that particular metal is W_w, the electrons escape from the metal with energy of

$$W_e = W - W_w \text{ joules} \qquad (3.27)$$

The multiplication of energy in the photomultiplier is achieved by the so-called *secondary emission*. If energetic electrons strike a surface of a metal, they transmit energy to the metal through collision, and depending on whether the energy given by the *primary electron* to the *secondary electron* in the metal is greater than the work function, one or more secondary electrons may escape the metal. The multiplied number of electrons can then be accelerated to strike another surface, and the process is continued until the desired energy level is reached.

The construction of a photomultiplier consists of the cathode C (Fig. 3.18) coated with a semitransparent layer of photoemissive material. When light impinges on the cathode the emitted electrons are focused on an electrode D_1 called a first *dynode*, which has a positive charge relative to the cathode. The dynode is usually a metal or an alloy having a low work function, such as cesium antimonide, potassium chloride, or silver oxide on silver; the silver alloy yields 8 to 10 secondary electrons for every primary electron (*Pederson et al.*, 1966). The most stable photomultiplication, however, is obtained by producing only 2 or 3 secondary electrons. The multiplied number of electrons is then directed to a second dynode D_2, where again multiplication occurs because the positive charge of the second dynode is larger than that of the first dynode. The total number of dynodes in a photomultiplier can be 9 or 10, and their shape and relative positioning depend on the type of photomultiplier used. The multiplied stream of electrons is finally collected by the collector plate P at the end of the tube. The overall amplification of a photomultiplier is directly proportional to the expression δ^n, where δ is the average number of

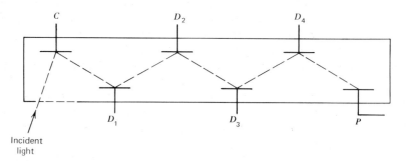

Figure 3.18   Photomultiplier tube.

secondary electrons and n is the number of dynodes, the total amplification ultimately depending on efficiency factors between dynodes. In an average photomultiplier the total gain of an electron stream is of the order of several millions. Through the operation of a photomultiplier, very weak light intensity variations are converted to relatively strong electric signals of the same variation frequency.

In electronic distance measuring equipment in which infrared radiation as a carrier is produced by Ga–As diodes, the modulated light signals are detected by junction diodes called *photodiodes*. In Section 2.7.1 we learned how the electron-hole recombination process in a forward-biased p-n junction diode causes holes act as the majority carrier, the current flowing through the junction from p to n. Also, in Section 3.2.2 it was shown that in a forward-biased luminescent diode the recombination of electrons and holes causes electrons to drop from the conduction band to the valence band, and photons are emitted. If the bias is reversed, the light-emitting diode becomes the photodiode. The principle of *photoconductivity*, which is the capability of a semiconductor to conduct light, is discussed briefly in the following paragraph.

In a p-n junction diode there are very few minority carriers (holes) in the n-type material and very few minority carriers (electrons) in the p-type material. If the diode is made reverse biased (i.e., if the positive terminal of the DC power supply is connected to the n-side of the diode), any holes formed in the junction region itself or in the n-type material close to the junction will be attracted into the p-type material by the electric field across the junction. Similar movement occurs among the electrons in the p-type material, where they are the minority carriers but in the reverse direction. This movement of charge carriers, which is the consequence of *electron-hole pair* (EHP) *production*, results in a flow of current through the diode, which, however, is very small without the presence of an outside energization source because a very limited number of electron-hole pairs are generated. This current is called the "dark current" of a photodiode. If light impinges on the p-n junction, photons having energy larger than that of the energy gap will produce charge carriers. If the energy equation (Eq. 3.25) is satisfied, great concentration of minority carriers occurs, and electrons are raised from the valence band to the conduction band. The current that now flows through the photodiode in the $n \rightarrow p$ direction is the sum of the dark current and the photo current. The increase of this current is proportional to the number of photons that produce EHPs; thus variation in the strength of the output electric current follows the level of illumination or the intensity of the received light.

The main characteristics of photodetectors can be summarized as follows. Photomultipliers offer fast response, good stability, and ease in designing circuitry, because no further amplification is needed. Photodiodes are compact in size and shape, and when equipped with a low noise preamplifier, they detect smaller signals than photomultipliers. They also have good efficiency in the infrared region.

Part

TWO

ELECTRONIC SURVEYING

FOUR

GEOMETRY OF ELECTRONIC SURVEYING

The geodetic sciences aim to measure the shape and size of the earth, to establish a network of points correctly located on the reference ellipsoid, and to fill this framework by producing a correctly scaled map that contains all the details and information needed for various purposes. The execution of these goals calls for teamwork among construction engineers, land surveyors, geodesists, geophysicists, hydrographers, and navigators. Before the development of modern electronic and electro-optical methods and instrumentation, the work was primarily based on measurements of angles by theodolites and measurements of distances by tapes or wires.

In the use of a variety of instruments based on electromagnetic radiation, *distance* can be considered to be the basic unit. Distances serve engineers in testing minute movements of constructions, surveyors in measuring sides in traversing, geodesists in establishing calibration baselines or in measuring the sides in trilateration networks, and geophysicists in observing lateral crustal movements. Distances observed from two known stations to determine the location of a new point can be deemed distance intersection and classified as a *circular* approach. This methodology is commonly used by surveyors, geodesists, and hydrographers to expand the number of known points in a surveyed area: in such applications the procedure is usually called *control point extension*. Finally, *distance differences* can be measured to create two families of hyperbolas for positioning new points. The *hyperbolic* approach is widely used by hydrographers operating off-shore and by navigators.

4.1. GEOMETRY OF HYPERBOLIC LOCATION

The geometry of hyperbolic location, the construction of hyperbolic lattice charts, and the analysis of the errors and variables involved differ essentially from those described in Section 4.2, where we study circular location. In hyperbolic location, three ground-based transmission stations are needed, compared with two in the circular case. Since the hyperbolic principle is primarily utilized in offshore hydrographic surveying and long-distance navigation, the better intersection angles that are obtained, even at long distance from the ground stations, can be counted an advantage. To help the reader to follow more complicated computations and applications later, basic concepts about the characteristics of hyperbolic patterns are presented. A *hyperbola* is the locus of a point that moves so that the difference of its undirected distance from two fixed points is always equal to a constant. The use of this definition in hyperbolic surveying methods is illustrated in Fig. 4.1. Two stationary radio stations are situated at the fixed points A and B, which are called the *foci*, and a mobile station is situated at point P. If the station moves from point P to point P' so that the distance differences $PA - PB \cdots P'A - P'B$ read at the mobile station are kept constant, we know that it has moved along a curve that is a hyperbola. The numbers and intervals of adjacent hyperbolas are dependent on the units preselected to establish the hyperbolic patterns. Usually these units are defined by the electronic characteristics of the systems employed.

In the case illustrated, the AB hyperbolic pattern takes numbers from 10 to 18 and the coordinate value of point P is $P_B = 11.00$. To get a definite location of point P, we need another pattern of hyperbolas. This means that another pair of fixed points must be established. In practice only one extra point C is needed, since both patterns can have mutual point A. Numbering this new set of hyperbolas from 1 to 9, the coordinates of the point P will be

$$P_B = 11.00$$

$$P_C = 3.00$$

It must be noticed, however, that ambiguity still exists in the definite location of point P, since the hyperbolas $B = 11.00$ and $C = 3.00$ also intersect each other at point P'' on the other side of the baselines BAC. In

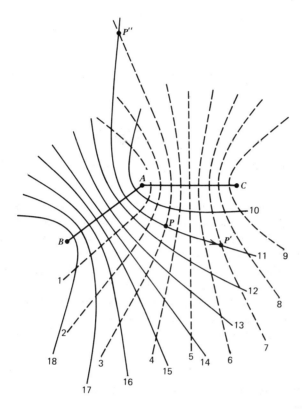

Figure 4.1 Principle of hyperbolic location (Laurila, 1960).

practice this is not a real problem, since the locations of points to be fixed usually are known with an accuracy that is rough, but sufficient to distinguish between these two opposite cases.

Hyperbolic location methods can be divided into two groups according to the electronic principles employed to find the distance differences. Distance differences can be determined either by pulse-type equipment such as Loran, or by comparing phase differences of continuous, unmodulated waves (Decca) or heterodyned signals (Raydist) at the point to be fixed. Figure 4.1 represents an application of control point extension.

Long line determinations by hyperbolic systems based on phase difference measurements are made in the following way. Aircraft or surface vessels carrying phasemeters travel around the base between two points at which the transmitting stations are situated; these points represent

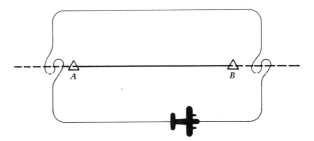

Figure 4.2   Long line measurement by hyperbolic method (Laurila, 1960).

terminals of the line to be measured. Observations are made along both extensions of the line (Fig. 4.2), and according to the general principle of phase difference measurements, a maximum value is obtained along one extension and a minimum value is obtained along the other extension of the baseline. Since the velocity of electromagnetic waves through the medium and the frequency used for transmission are known, the length of the baseline can be calculated. If pulse-type equipment is used, similar observations are made based on time difference measurements along both extensions of the baseline.

4.1.1.   Derivation of Hyperbolic Coordinates

In hyperbolic location the center station in the configuration is called the master station because it controls the functioning of the two other stations, the slave stations. To distinguish the hyperbolic lattices appearing on the map or chart, one set is printed in green and the designated symbol for this slave station is G. The other set is printed in red, and that slave station is designated R. Because of the difference in electronic principles, continuous wave and pulse-type methods are treated separately.

In the continuous wave (and heterodyne) methods, the *hyperbolic coordinate* value N at any point P is the number of hyperbolas of zero phase difference that are numbered starting with a value of zero at the master station. The derivation of the hyperbolic coordinate N_R at point P for the red family of hyperbolas is illustrated in Fig. 4.3, where S_R is the length of the red baseline, r_M is the distance between the master station and point P, and r_R is the distance between the red slave station and point P. When the master transmission reaches points P and R, the phases lag

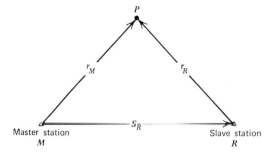

Figure 4.3 Determination of hyperbolic coordinates (Laurila, 1960).

behind their values at the master station by

$$\Phi_{MP} = \frac{2\pi r_M}{\lambda_R} \quad \text{and} \quad \Phi_{MR} = \frac{2\pi S_R}{\lambda_R} \tag{4.1}$$

respectively, where λ_R is the wavelength for the construction of the lattice charts. When the red slave transmission reaches point P, the phase lags behind its value at the slave station by

$$\Phi_{RP} = \frac{2\pi r_R}{\lambda_R} \tag{4.2}$$

The phase of the red slave transmission is phase locked to the master transmission in the red slave station by $-\Phi_{MR}$, and the phase difference (master slave) at P is therefore

$$\Delta\Phi_R = \Phi_{MP} - (\Phi_{RP} - \Phi_{MR}) \tag{4.3}$$

or, when values from Eqs. 4.1 and 4.2 have been substituted:

$$\Delta\Phi_R = \frac{2\pi}{\lambda_R}(S_R + r_M - r_R) \tag{4.4}$$

The phasemeter itself cannot distinguish phase differences that are multiples of 2π, but the phase-indicating pointer, which makes one revolution for a phase change of 2π rad, is geared to a second pointer that gives the total number of revolutions made. The fractional part of the last revolution is now obtained by dividing both sides of Eq. 4.4 by 2π as follows:

$$\frac{\Delta\Phi_R}{2\pi} = N_R = \frac{1}{\lambda_R}(S_R + r_M - r_R) \tag{4.5}$$

or

$$N_R = \frac{f_R}{c}(S_R + r_M - r_R) \tag{4.6}$$

where c is the propagation velocity of electromagnetic waves and f_R is the frequency representing the wavelength λ_R.

In the case of the phase difference principle, the wavelength λ_R used for constructing the hyperbolic lattices served as a "yardstick." If a pulse-type instrument is used, a predetermined *time difference dT* is the "yardstick" that defines the interval between two adjacent hyperbolas. The hyperbolic coordinate in this case can be derived much as before. The time elapsed for the master transmission to reach point P is

$$T_{MP} = \frac{r_M}{c} \tag{4.7}$$

where c is the propagation velocity of electromagnetic radiation. The time elapsed for the master transmission to reach the slave station R plus the fixed coding delay needed before the master signal triggers the slave transmitter into action is

$$T_{MR} = \frac{S_R}{c} + k \tag{4.8}$$

where k is the coding delay in units of time. Finally, the time needed for the slave transmission to reach point P is

$$T_{RP} = \frac{r_R}{c} \tag{4.9}$$

and the time difference at point P will be

$$\Delta T_R = T_{MP} - (T_{RP} - T_{MR}) = \frac{1}{c}(S_R + r_M - r_R) + k \tag{4.10}$$

Since dT defines the interval of adjacent hyperbolas, the following quotient yields the hyperbolic coordinate N_R:

$$\frac{\Delta T_R}{dT_R} = N_R = \frac{1}{dT_R c}(S_R + r_M - r_R) + \frac{k}{dT_R} \tag{4.11}$$

4.1.2. Lanewidth on the Baseline

The interval between two adjacent hyperbolas is called the *lane* and its value on the baseline is designated by a symbol l'. The number of lanes included in the base and the lanewidth on the baseline are obtained from Fig. 4.3 by assuming first that point P is moved at the master station M. Then in the case of continuous wave applications $r_M = 0$ and $r_R = S_R$, and Eq. 4.5 is $N_{R\,min} = 0$. If point P is moved at slave station R, $r_M = S_R$ and $r_R = 0$, and Eq. 4.5 becomes

$$N_{R\,max} = \frac{2S_R}{\lambda_R} \qquad (4.12)$$

which indicates the number of lanes included in the baseline. The lanewidth on the baseline is obtained by dividing the length of the base S_R by the number of lanes $(N_{R\,max} - N_{R\,min})$ as follows:

$$l' = \frac{\lambda_R}{2} \qquad (4.13)$$

which means that the lanewidth on the baseline is equal to one-half the wavelength.

For a pulse-type instrument we use Eq. 4.11 and a procedure similar to the one just described to obtain the total number of lanes by

$$N_{R\,max} = \frac{2S_R}{dT_R c} \qquad (4.14)$$

The lanewidth on the baseline yields

$$l' = \frac{dT_R c}{2} = \frac{dS_R}{2} \qquad (4.15)$$

From Eqs. 4.13 and 4.15 it is seen that the lanewidth on the baseline in all hyperbolic location methods is one-half the original yardstick applied to that particular method. From the derivations of Eqs. 4.12 and 4.14 it also becomes obvious that along the extension of the baseline at the master end, the hyperbolic coordinate N remains as a minimum constant (or zero), and along the extension at the slave end it remains as a maximum constant.

4.1.3. Direction of Hyperbola

The direction of a hyperbola at any point P (Fig. 4.4) is obtained by assuming that a receiver moves from point P to point P' so that both distances r_M and r_R from the point to the master and the slave stations change by an equal amount $dr_M = dr_R$. The lane reading will not change; thus the direction PP' coincides with the direction of a hyperbola.

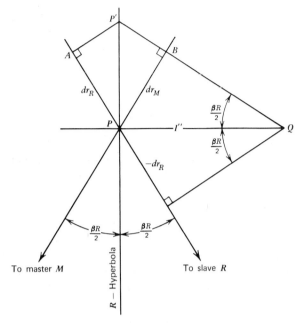

Figure 4.4   Determination of direction and lanewidth of hyperbola at point P (Laurila, 1960).

Since the triangles AAP' and BPP' are symmetrically equal, the line PP' bisects the angles APB and $MPR = \beta_R$, hence the following definition can be set down: *the direction of a hyperbola at any point coincides with the bisector of the angle formed by the lines joining the point to the pair of stations.*

4.1.4.   Lanewidth, General Case

To derive the formula for a lanewidth l (Fig. 4.5) at any point in the hyperbolic system as a function of the lanewidth l' on the baseline, Eq. 4.5 is used. This gives the number of a hyperbolic coordinate when the phase difference system is employed. The result of the derivation, however, is independent of the electronic features, thus is universally applicable to all hyperbolic systems, regardless of whether pulse-type or phase-difference systems are used.

The pattern of hyperbolas formed by the master station and the red slave station is taken as the example used in the analysis. Suppose the

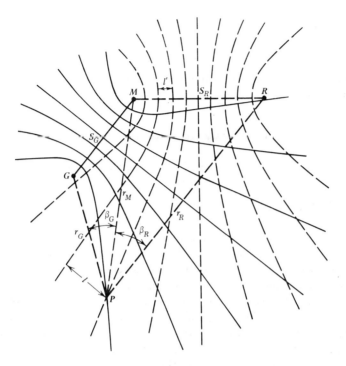

Figure 4.5 Location of point P by two intersecting hyperbolas (Laurila, 1960).

receiver at point P is moved a small distance l'' to a point Q, where PQ forms a right angle with the hyperbola PP' (Fig. 4.4). Thus if lines are drawn from Q perpendicular to MB and RA, the following will be valid:

$$dr_M = l'' \sin \frac{\beta_R}{2} \quad \text{and} \quad dr_R = -l'' \sin \frac{\beta_R}{2} \qquad (4.16)$$

Taking the total differential of Eq. 4.5, we have

$$dN_R = \frac{1}{\lambda_R}(dr_M - dr_R) \qquad (4.17)$$

which, when the values of Eq. 4.16 have been substituted, yields

$$dN_R = \frac{1}{\lambda_R} 2l'' \sin \frac{\beta_R}{2} \qquad (4.18)$$

If the differential dN_R is assigned the value of one, and the corresponding distance l'' is symbolized as l to represent the width of a lane at point P, we can write

$$l = \frac{\lambda_R}{2} \frac{1}{\sin (\beta_R/2)} \tag{4.19}$$

From Eq. 4.13 it is seen that the lanewidth l' on the baseline is equal to one-half the wavelength; thus Eq. 4.19 can be written in general form as follows:

$$l = l' \frac{1}{\sin (\beta_R/2)} \quad \text{or} \quad l = l'G \tag{4.20}$$

where $G = 1/[\sin (\beta_R/2)]$ is called the *lane expansion factor*.

4.1.5. Accuracy Considerations

In evaluating the potential of the hyperbolic principle in surveying and navigation, we must consider economy, accuracy and the general possibility of establishing a feasible fix at all. Especially in navigation, where continuous observations are made to keep a track of the vessel's position, deviations from this assumed position should be promptly available in the same system of coordinates found on the map or chart being used. In hyperbolic surveying and navigation, maps and charts consist of two sets of printed coordinate systems—rectangular or geographic, and hyperbolic. Therefore, two operational steps must be kept in mind when using hyperbolic charts. These are the fixing of points in relation to the hyperbolic system of coordinates and the location of this system or part of it in relation to the map coordinates being employed. Consequently, in checking the accuracy of hyperbolic surveying, the error cannot be expressed as an all over constant in metric units without first analyzing the geometric elements involved. Figure 4.5 indicates that the geometric factors affecting the accuracy of the location of point P are the subtended angles β_G and β_R in which the bases can be seen from P. Accuracy also is affected by the station angle γ, formed by the two hyperbolas intersecting each other at point P.

4.1.5.1. *Accuracy Factor*

If two hyperbolic patterns are generated with a common focus, a hyperbolic lattice is formed as shown in Fig. 4.5. At point P let the baseline S_R subtend an angle β_R, and the baseline S_G an angle β_G. An enlarged view of the lattice around P appears in Fig. 4.6, where the R- and G-hyperbolas intersect each other at point P with an angle of $(\beta_R + \beta_G)/2 = \gamma$. By assuming that the standard errors of location of these hyperbolas at

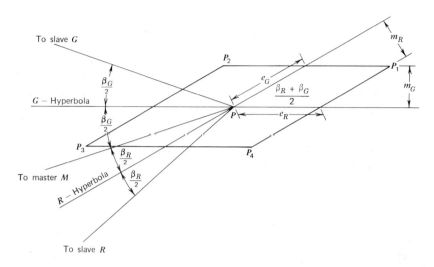

Figure 4.6 Error parallelogram around point P (Laurila, 1960).

this point are $\pm m_R$ and $\pm m_G$ meters, respectively, an error parallelogram $P_1P_2P_3P_4$ is formed.

If the erroneous displacements of point P along the hyperbolic lines R and G are symbolized as e_G and e_R, respectively, their magnitudes are obtained directly from Fig. 4.6 as follows:

$$e_R = \frac{m_R}{\sin\left[(\beta_R + \beta_G)/2\right]} \quad \text{and} \quad e_G = \frac{m_G}{\sin\left[(\beta_R + \beta_G)/2\right]} \quad (4.21)$$

Since the standard errors m_R and m_G can be stated as functions of the corresponding standard errors m'_R and m'_G of the hyperbolas on the baselines expressed in hyperbolic coordinates (Eqs. 4.21) can be written

$$e_R = \frac{m'_R \cdot l'_R}{\sin(\beta_R/2)\sin\left[(\beta_R + \beta_G)/2\right]}$$

$$e_G = \frac{m'_G \cdot l'_G}{\sin(\beta_G/2)\sin\left[(\beta_R + \beta_G)/2\right]} \quad (4.22)$$

by the use of Eq. 4.20. The area of the error parallelogram $P_1P_2P_3P_4$ will be

$$Z = 4 \cdot e_R \cdot e_G \cdot \sin\frac{\beta_R + \beta_G}{2}$$

When the values of Eq. 4.22 have been applied, the foregoing expression yields

$$Z = \frac{16 \cdot m'_G \cdot m'_R \cdot l'_G \cdot l'_R}{4 \cdot \sin(\beta_G/2) \cdot \sin(\beta_R/2) \cdot \sin[(\beta_R + \beta_G)/2]} \qquad (4.23)$$

where the denominator is called the *accuracy factor* and is symbolized by Δ. By a trigonometrical transformation, this factor can be written in a more convenient form:

$$\Delta = \sin\beta_R + \sin\beta_G - \sin(\beta_R + \beta_G) \qquad (4.24)$$

A simple approximation for Δ may be derived if the point to be located is far from the ground stations. Subtended angles β_R and β_G then can be considered to be small and equal, which allows us to write

$$\Delta \simeq \beta^3 \qquad (4.25)$$

The common subtended angle β is approximately inversely proportional to the distance from the master station; thus Δ is inversely proportional to the cube of this distance. In other words, the area of the error parallelogram is directly proportional to the cube of the distance.

4.1.5.2. *Lines of Zero Accuracy*

The accuracy factor is a useful indicator when investigating the regions in the hyperbolic network that fulfill prespecified standards of fixing accuracy. The area of the error parallelogram theoretically becomes infinite, and the system has no accuracy when $\Delta = 0$. It will be seen, therefore, that the system cannot provide a fix under the following conditions:

$$\begin{aligned} (a) & \quad \text{when } \beta_R \text{ or } \beta_G = 0 \\ (b) & \quad \text{when } \beta_R + \beta_G = 0 \end{aligned} \qquad (4.26)$$

In case (a) the point to be fixed is situated along any of the extensions of the baselines. This limitation is more theoretical than practical, since along these lines one pattern of hyperbolas degenerates into straight lines, and fixes on either side of these lines are ambiguous, the instantaneous lanewidth, Eq. 4.20, becoming infinite. In practice, where the ambiguity can be solved from previous observations, the error is not great in the neighborhood of the stations, provided the angle between baselines does not too closely approach 180°.

In case (b) the point is situated along either of the extensions of the line joining the two slave stations. In these circumstances the angles β_R and β_G are equal in magnitude but have opposite signs. The extensions of the line joining the slave stations is the region where the hyperbolic lines run parallel to each other, since the bisectors of angles β_R and β_G have a common direction.

In the initial stage in planning the hyperbolic network, sectors close to the lines of zero accuracy must be carefully inspected to ensure that they do not include any areas that later prove to be important for hydrographic surveying or navigation. Conveniences, such as ideal station locations and good supply roads, must be sacrificed to prevent expensive reerection of stations and redrafting of lattice charts.

4.1.5.3.   *Error Components in the Direction of Rectangular Coordinate Axes*

If the metric errors made in determining the R and G hyperbolic coordinates are of different magnitude, or if the station angle between the hyperbolas deviates from 90°, the errors occurring in position fixing are oriented, thus elliptical. As demonstrated previously, the direction and dimensions of the error parallelogram vary greatly according to the several electronic and geometric parameters involved. For this reason it is difficult to establish a universally valid answer to the question "what is the error resulting from hyperbolic fixing?"

To obviate this ambiguity it is more desirable from the standpoint of surveying operations to state any error in relation to certain fixed directions. It is appropriate to choose the X- and Y-axes of the rectangular coordinate system superimposed on the hyperbolic chart as the stationary reference directions.

In Fig. 4.6 e_G and e_R represent standard errors of the located point P along the hyperbolic coordinate axes R and G, respectively. If an ellipse is inscribed within the parallelogram $P_1P_2P_3P_4$, it becomes the standard error ellipse for the fixed point P with e_R and e_G as conjugate semiaxes (Fig. 4.7). If the error components e_R and e_G along the hyperbolic lines are mutually independent, the corresponding components along the X- and Y-axes are obtained by projecting e_R and e_G onto the said axes, and we have

$$e'_{R_y} = e_R \cdot \sin \alpha_G \qquad e'_{R_x} = e_R \cdot \cos \alpha_G$$
$$e'_{G_y} = e_G \cdot \sin \alpha_R \qquad e'_{G_x} = e_G \cdot \cos \alpha_R \tag{4.27}$$

By adding the error components quadratically as follows, the metric standard errors are obtained:

$$m_X = \pm\sqrt{(e_R \cos \alpha_G)^2 + (e_G \cos \alpha_R)^2}$$
$$m_Y = \pm\sqrt{(e_R \sin \alpha_G)^2 + (e_G \sin \alpha_R)^2} \tag{4.28}$$

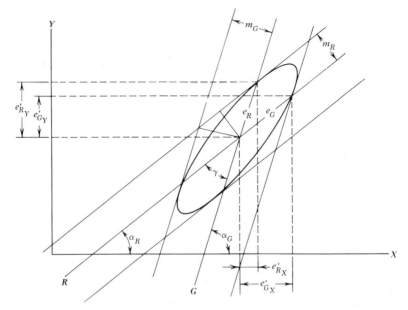

Figure 4.7   Standard error ellipse in hyperbolic location (Laurila, 1960).

where α_R is the angle between the direction of the R hyperbola and the X-axis and α_G is the angle between the direction of the G hyperbola and the X-axis (*Laurila*, 1953).

By substituting values of Eqs. 4.20 and 4.22 into Eqs. 4.28, the metric standard errors along the X- and Y-axes can be written as follows:

$$m_X = \pm\operatorname{cosec}\gamma\sqrt{(m'_R l_R \cos\alpha_G)^2 + (m'_G l_G \cos\alpha_R)^2}$$

$$m_Y = \pm\operatorname{cosec}\gamma\sqrt{(m'_R l_R \sin\alpha_G)^2 + (m'_G l_G \sin\alpha_R)^2}$$

$$(4.29)$$

where $\gamma = (\beta_R + \beta_G)/2$ is the station angle, l_R and l_G are the lanewidths of the R and G lattices at point P, and m'_R and m'_G are the standard errors of the corresponding hyperbolas in the units of hyperbolic coordinates.

4.1.6.   Construction of Hyperbolic Lattice Charts on a Plane

In long-distance navigation (Loran-C and Omega), hyperbolic lattice charts are constructed with reference to an ellipsoidal surface as a function of geographic latitudes and longitudes. Since this requires long and complex computations, lattice charts drawn on certain projection planes with X- and Y-coordinates are preferred in offshore hydrographic surveys and medium distance navigation. In this chapter the *universal transverse Mercator* (UTM) projection is the reference plane for the red family of hyperbolas as example, based on the phase difference principle. Even if the final lattice chart is presented on the projection plane, the ellipsoidal parameters must be taken into consideration since, after all, the propagation of long waves follows closely the geodetic line. To construct hyperbolas in areas close to the ground stations where their curvature is highest, a graphical method is fast and sufficiently accurate. At longer distances the physical dimensions of the drafting device sets limitations, and since the curvature of the hyperbolas there is smoother, a numerical interpolation method is preferred.

4.1.6.1. *Graphical Method*

If the coordinates of the master station M and the slave station R are given as X_M, Y_M and X_R, Y_R, the length of the baseline on plane is obtained from the following equation:

$$S_R = \sqrt{(X_M - X_R)^2 + (Y_M - Y_R)^2} \tag{4.30}$$

To find out the integer amount of wavelengths, and especially the last fractional value included in the baseline, the length of Eq. 4.30 must be reduced to the ellipsoid by

$$s_R = S_R\left(\frac{1}{0.9996} - \frac{X_M^2 + X_M X_R + X_R^2}{6MN}\right) \tag{4.31}$$

where s_R is the ellipsoidal distance, X_M the distance of the master station M from the central meridian, X_R the distance of the slave station R from the central meridian, M the meridian curvature radius at the midpoint of the baseline, and N the curvature radius in prime vertical at the midpoint of the baseline.

Master station M as the center concentric circles are drawn at intervals of λ, 2λ, 3λ, etc. If the fractional value of λ, radiated from the master station at the slave station, is $n\lambda$ (where $n < 1$), concentric circles are drawn at intervals of $n\lambda$, $(1 + n)\lambda$, $(2 + n)\lambda$, etc., indicating slave station R as the center. In the case illustrated in Fig. 4.8, $n = \frac{7}{8}$ and the corresponding intervals will be $\frac{7}{8}\lambda$, $1\frac{7}{8}\lambda$, $2\frac{7}{8}\lambda$, etc. All circles drawn from both stations now represent loci of phase angles $\Phi_{(M,R)} = 0°$. At the intersections of all these circles, the phase angles of both M- and R-waves are zero; consequently their *phase differences* are also zero.

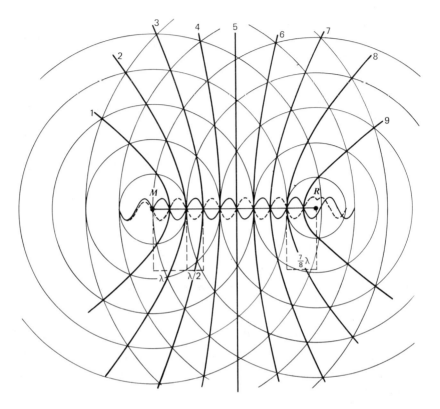

Figure 4.8   Construction of hyperbolas by graphical interpolation.

If smooth curves are drawn through the intersections of the circles, it is obvious that these curves of constant phase difference are hyperbolas, since the difference in distances from any point along the curves to the M and R stations is constant. This graphical presentation (Fig. 4.8) reveals that the lanewidth on the baseline equals one-half the wavelength, as shown in Eq. 4.13.

Since all ground stations and hyperbolas are located on the UTM projection, radii of the circles must be corrected to fit that projection. Equation 4.42 may be used without causing any significant error, although the distances of points along the circles to the central meridian vary.

The drawing device can be about 4 m long, made of duraluminum or invar, pivoted at one end to the stations concerned, and having a sliding vernier with a stylus at the other end. The reading and drawing accuracy should be about 0.1 mm. If the map to be plotted is small compared to the length of the drawing bar (e.g., the ground station lies outside the map), the distances and directions from the ground station to at least two points on the map must be known, to orient the map with respect to the ground station.

4.1.6.2. *Numerical Interpolation Method*

In working with numerical interpolation, it is best to divide the survey area into, say, 50×50 km squares, each provided with the UTM rectangular grid. Using the known plane coordinates of the ground stations M and R, distances are computed from the ground stations to each point where the X- and Y-axes intersect each other at 5 km intervals within the 50×50 km worksheet.

These distances, together with the baseline S_R, are reduced to the ellipsoid by Eq. 4.31, and the hyperbolic coordinates N_R for the points of intersection are computed by Eq. 4.6. The computed hyperbolic coordinates obviously are not integer; therefore an interpolation is necessary to find out the coordinates X_n or Y_n, where a hyperbola with integer value intersects either the X- or the Y-axis. The selection of the reference axis depends on the incident angle of the hyperbola with that axis. In Fig. 4.9 the directions of the hyperbolas are close to right angle with the Y-axis, and N_{-2}, N_{-1}, N_0, N_{+1}, and N_{+2} are computed hyperbolic coordinates in the red pattern of hyperbolas with reference to the corresponding coordinates Y_{-2}, Y_{-1}, Y_0, Y_{+1}, and Y_{+2}. By interpolation, Y_n is found where the first integer-numbered hyperbola beyond N_0 intersects the Y-axis.

The following interpolation formula by *Bessel* may be used:

$$N_n - N_0 = n\Delta^{\mathrm{I}}_{1/2} + B^{\mathrm{II}}(\Delta^{\mathrm{II}}_0 + \Delta^{\mathrm{II}}_1) + B^{\mathrm{III}}\Delta^{\mathrm{III}}_{1/2} + B^{\mathrm{IV}}(\Delta^{\mathrm{IV}}_0 + \Delta^{\mathrm{IV}}_1) + \cdots \quad (4.32)$$

where Δ^{I}, Δ^{II}, Δ^{III}, and Δ^{IV} are the first, second, third, and fourth differences of the computed N_0-values, and B^{II}, B^{III}, and B^{IV} are interpolation coefficients that can be found from tables such as *Nautical Almanac* (1974) as a function of n. In Eq. 4.32

$$n = \frac{Y_n - Y_0}{Y_1 - Y_0} \quad (4.33)$$

which means that the so-called inverse interpolation must be utilized to finally obtain Y_n. Since the coefficients B are functions of n, an iteration procedure must be performed in Eq. 4.32. After n has been determined, Y_n is computed by

$$Y_n = Y_0 + 5000n \text{ meters} \quad (4.34)$$

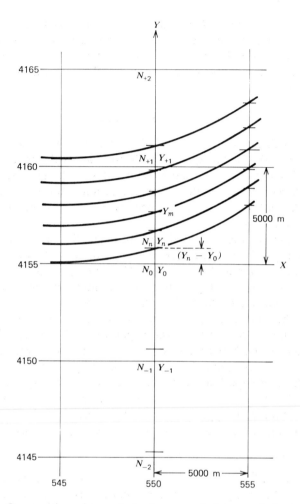

Figure 4.9 Construction of hyperbolas by numerical interpolation.

because in our example, the intervals between points along the Y-axis are 5 km. The locations of the four intermediate hyperbolas along the Y-axis (short lines in Fig. 4.9) are obtained by direct interpolation

$$Y_m = Y_n + m \Delta^I_{1/2} + B^{II}(\Delta^{II}_0 + \Delta^{II}_1) + B^{III}\Delta_{1/2} + B^{IV}(\Delta^{IV}_0 + \Delta^{IV}_1) + \cdots \quad (4.35)$$

where the differences Δ are now differences of the computed Y values and the coefficients B are functions of m and in this case have values of

0.2, 0.4, 0.6, and 0.8. The hyperbola itself can be constructed by drawing a smooth line through all points where the hyperbola is computed to intersect the adjacent Y-axes, spaced 5 km apart.

The practical procedure for computing the integer N-value by using numerical interpolation is as follows. For any hyperbolic application in hydrographic surveying or navigation, a drafting accuracy of about ± 0.005 lane is sufficient. Therefore, the fourth term in Eq. 4.32 can be neglected and the third term need be used only for checking the iteration level. The differences of Δ can be found directly from the computed N-values by the following scheme:

Computed N-Value	Differences		
	First	Second	Third
N_{-1}			
	$\Delta^I_{-1/2}$		
N_0		Δ^{II}_0	
	$\Delta^I_{+1/2}$		$\Delta^{III}_{+1/2}$
N_{+1}		Δ^{II}_{+1}	
	$\Delta^I_{+1\,1/2}$		
N_{+2}			

$$(4.36)$$

which yields

$$\Delta^I_{+1/2} = N_{+1} - N_0$$
$$\Delta^I_{-1/2} = N_0 - N_{-1}$$
$$\Delta^{II}_0 = \Delta^I_{+1/2} - \Delta^I_{-1/2} = N_{+1} - 2N_0 + N_{-1} \qquad (4.37)$$
$$\Delta^{II}_{+1} = \Delta^I_{+1\,1/2} - \Delta^I_{+1/2} = N_{+2} - 2N_{+1} + N_0$$
$$\Delta^{III}_{+1/2} = \Delta^{II}_{+1} - \Delta^{II}_0 = N_{+2} - 3N_{+1} + 3N_0 - N_{-1}$$

The first iteration of n is now obtained from Eq. 4.32

$$n_1 = \frac{N_n - N_0}{\Delta^I_{+1/2}} \qquad (4.38)$$

and the coefficient B^{II} is found from Table 4.1 as a function of n_1. If the product $B\Delta^{III}_{+1/2}$ is larger than 0.005, it can be included in the second iteration. Otherwise, the second iterated value of n can be considered to be the final one; it is computed by

$$n_2 = \frac{(N_n - N_0) - B^{II}(\Delta^{II}_0 + \Delta^{II}_{+1})}{\Delta^I_{+1/2}} \qquad (4.39)$$

TABLE 4.1[a]

n	B	n	B	n	B	n	B	n	B	n	B
0.0000	0.000	0.1101	0.025	0.2719	0.050	0.7280	0.049	0.8898	0.024	0.0	0.000
0.0020	0.001	0.1152	0.026	0.2809	0.051	0.7366	0.048	0.8949	0.023	0.1	+0.006
0.0060	0.002	0.1205	0.207	0.2902	0.052	0.7449	0.047	0.9000	0.022	0.2	0.008
0.0101	0.003	0.1258	0.028	0.3000	0.053	0.7529	0.046	0.9049	0.021	0.3	0.007
0.0142	0.004	0.1312	0.029	0.3102	0.054	0.7607	0.045	0.9098	0.020	0.4	+0.004
0.0183	0.005	0.1366	0.030	0.3211	0.055	0.7683	0.044	0.9147	0.019	0.5	0.000
0.0225	0.006	0.1422	0.031	0.3326	0.056	0.7756	0.043	0.9195	0.018		
0.0267	0.007	0.1478	0.032	0.3450	0.057	0.7828	0.042	0.9242	0.017	0.6	−0.004
0.0309	0.008	0.1535	0.033	0.3585	0.058	0.7898	0.041	0.9289	0.016	0.7	0.007
0.0352	0.009	0.1594	0.034	0.3735	0.059	0.7966	0.040	0.9335	0.015	0.8	0.008
0.0395	0.010	0.1653	0.035	0.3904	0.060	0.8033	0.039	0.9381	0.014	0.9	−0.006
0.0439	0.011	0.1713	0.036	0.4105	0.061	0.8098	0.038	0.9427	0.013	1.0	0.000
0.0483	0.012	0.1775	0.037	0.4367	0.062	0.8162	0.037	0.9472	0.012		

	Diff.		Diff.		Diff.		Diff.		Diff.
0.0527		0.1837		0.5632		0.8224		0.9516	
	0.013		0.038		0.061		0.036		0.011
0.0572		0.1901		0.5894		0.8286		0.9560	
	0.014		0.039		0.060		0.035		0.010
0.0618		0.1966		0.6095		0.8346		0.9604	
	0.015		0.040		0.059		0.034		0.009
0.0664		0.2033		0.6264		0.8405		0.9647	
	0.016		0.041		0.058		0.033		0.008
0.0710		0.2101		0.6414		0.8464		0.9690	
	0.017		0.042		0.057		0.032		0.007
0.0757		0.2171		0.6549		0.8521		0.9732	
	0.018		0.043		0.056		0.031		0.006
0.0804		0.2243		0.6673		0.8577		0.9774	
	0.019		0.044		0.055		0.030		0.005
0.0852		0.2316		0.6788		0.8633		0.9816	
	0.020		0.045		0.054		0.029		0.004
0.0901		0.2392		0.6897		0.8687		0.9857	
	0.021		0.046		0.053		0.028		0.003
0.0950		0.2470		0.7000		0.8741		0.9898	
	0.022		0.047		0.052		0.027		0.002
0.1000		0.2550		0.7097		0.8794		0.9939	
	0.023		0.048		0.051		0.026		0.001
0.1050		0.2633		0.7190		0.8847		0.9979	
	0.024		0.049		0.050		0.025		0.000
0.1101		0.2719		0.7280		0.8898		1.0000	

n	B
0.0	0.000
0.1	+0.004
0.2	0.007
0.3	0.010
0.4	0.011
0.5	+0.012
0.6	−0.011
0.7	0.010
0.8	0.007
0.9	+0.004
1.0	0.000

ᵃ From The American Ephemeris and Nautical Almanac for the Year 1976, Government Printing Office, Washington, D.C., 1974, Table XIII, page 524.

The following numerical example illustrates the procedure just described. On a rectangular plane coordinate system, the coordinates of the master and red slave stations are given in kilometers as follows:

Axis	Master Station	Slave Station
X	280.000	180.000
Y	140.000	290.000

Along the Y-axis, with $X = 50.000$, four points P were selected at intervals of 5000 m.

The length of the red baseline computed from the given coordinates of the master and the red slave stations is $S_R = 180.277$ km. In this example we use a Survey Decca instrument having a comparison wavelength of $\lambda_R = 846.45$ meters. By the use of Eq. 4.5, the N-values for the preselected points P were computed and are given in Table 4.2 together with other pertinent data $(Q = S_R + r_M - r_R)$.

The problem is to find the coordinate Y_n, where the integer red hyperbola $N_n = 396.000$ intersects the Y-axis, by interpolating upward from the computed value $N_0 = 395.480$. The differences are found from Eq. 4.37 and Table 4.2 as follows:

$$\Delta^I_{+1/2} = 2.536 \qquad \Delta^{II}_0 = -0.154 \qquad \Delta^{II}_{+1} = -0.150 \qquad \Delta^{III}_{1/2} = +0.004$$

The first approximation for n will be obtained from Eq. 4.38:

$$n_1 = \frac{0.520}{2.536} = 0.2050$$

The coefficient B^{II}, which is always negative, is now found from Table 4.1 as $B^{II} = -0.041$, and the second iteration will be

$$n_2 = \frac{0.520 + 0.041(-0.154 - 0.150)}{2.536} = 0.2001$$

Since the product $B^{III}\Delta^{III}_{+1/2}$ is only $+0.000032$, it is neglected, and Y_n is

TABLE 4.2

P	Y (km)	r_R (km)	r_M (km)	Q (km)	N
−1	305.000	130.863	283.064	332.478	392.790
0	310.000	131.529	286.007	334.755	395.480
+1	315.000	132.382	289.007	336.902	398.016
+2	320.000	133.417	292.062	338.922	400.402

obtained from Eq. 4.34:

$$Y_n = 310.000 + 0.2001 \times 5.000 = 311.001 \text{ km}$$

To check the validity of the interpolation, the red hyperbolic coordinate N_n of the new point P_n ($X_{P_n} = 50.000$, $Y_{P_n} = 311.001$) can be computed by Eq. 4.5. As a result, the values

$$r_R = 131.685 \text{ km}$$

$$r_M = 286.603 \text{ km}$$

$$Q = 335.195 \text{ km}$$

$$N_n = 395.9994$$

were obtained, and the computed N_n-value proves that the choice of Eq. 4.39 is fully justified.

Although the original hyperbolic charts are usually produced by programming electronic plotters, the method described through Eqs. 4.32 to 4.34 is useful in field operations where a survey area not included in the hyperbolic chart must be supplied with hyperbolic coordinates. Similarly if a hyperbolic grid denser than that provided by the available chart is needed, Eq. 4.35 can be employed.

The numerical interpolation method also can be applied in the reverse problem in hyperbolic location, that is, when rectangular coordinates X and Y of point P are to be computed from hyperbolic coordinates N_R and N_G. In this case, two Y-axes with the known values of X_1 and X_2, respectively, are selected on both sides of point P. The Y-coordinates are found from Eqs. 4.32 and 4.33, where the red and green hyperbolas intersect the preselected Y-axes. If these axes are close together, the hyperbolas between them may be considered as straight lines, defined by two pairs of coordinates, $R(X_{1_R}, Y_{1_R}$ and $X_{2_R}, Y_{2_R})$ and $G(X_{1_G}, Y_{1_G}$, and $X_{2_G}, Y_{2_G})$. The rectangular coordinates of point P are now obtained as the intersection of the lines R and G in the following way.

The equations for the two lines are

$$Y = Y_{1_R} + \frac{Y_{2_R} - Y_{1_R}}{X_{2_R} - X_{1_R}} X$$

$$Y = Y_{1_G} + \frac{Y_{2_G} - Y_{1_G}}{X_{2_G} - X_{1_G}} X$$

(4.40)

When solved for X, these equations yield

$$X = \frac{Y_{1_G} - Y_{1_R}}{(Y_{2_R} - Y_{1_R})(X_{2_G} - X_{1_G}) - (Y_{2_G} - Y_{1_G})(X_{2_R} - X_{1_R})}$$

(4.41)

Finally, Y can be solved from either of the expressions in Eq. 4.40.

A hypothetical sample computation was made using FORTRAN programming in the case of a test point located at an average distance of 250 km from the ground stations. The grid size used for the Bessel interpolation was one kilometer. After the X- and Y-coordinates of the point were computed, the reverse check computation showed no discrepancy in the fourth decimal of the red lane and a discrepancy of 2 in the fourth decimal of the green lane.

4.2. GEOMETRY OF CIRCULAR LOCATION

The designation *circular* can be accepted by anticipating that locations of points or measurements of lengths of lines will be determined by the intersection of distance circles in space, in air, on an ellipsoid, or in a plane. These circles are obtained by simultaneous distance measurements made between the vehicle carrying out the survey and two fixed ground stations. The electronic or electro-optical equipment used operates on wavelengths from the light spectrum to the LF radio band. Also, a great variety of electronic principles come into play. Generally speaking, any instrument capable of measuring distances is a component of circular location.

4.2.1. Control Point Extension

The location of new points to expand the point density in a survey area or control point extensions by use of a circular approach can be utilized on a projection plane in land-to-land surveys by surveyors and geodesists and in land-to-sea surveys by hydrographers. In navigation, where long distances are usually involved, the reference ellipsoid must serve as a basis for computations and an inverse solution such as *Sodano's* (1958) or the *chord method* (see Chapter 17) can be applied. One very common application of control point extension with the circular approach involves photogrammetric methods in connection with aerial photography.

The position of an aircraft (aerial camera) is determined at the times of exposure by simultaneous distance measurements from two fixed ground stations. The vertical projection of the camera position on a projection plane or on the ellipsoid yields the coordinates of the ground detail seen on the photograph where the plumb line from the lens intersects the photo plane. This point is called the *nadir point,* and in truly vertical photography it coincides with the *principal point,* which is found with the

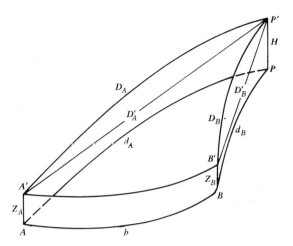

Figure 4.10   Control point extension by circular location (Laurila, 1960).

aid of calibration marks called *fiducial marks* engraved on the photo frame and shown on the photograph. Various methods, such as gyro-stabilized camera mounts, can be used to make the photography truly vertical. To find the location of the nadir point on a tilted photograph with respect to the principal point, one might select an instrument such as a gyrostabilized *verticality finder*.

A skeleton-type layout of the geometry involved in a general case is given in Fig. 4.10. Points A and B are two fixed ground instruments and A' and B' are their respective antennas, having elevations Z_A and Z_B from the mean sea level. At point P' an aircraft is flying its mission during the fix at an altitude of H above mean sea level, and point P is its vertical projection on the ellipsoid. Because of refraction in the atmosphere, the observed distances D_A and D_B are curved lines, which can be assumed to be arcs of circles. The average curvature radii of these circles are first determined (Chapter 5), and the spatial arc distances D_A and D_B are reduced to spatial chord distances D'_A and D'_B and ultimately to ellipsoidal arc distances d_A and d_B.

4.2.1.1.   *Construction of a Circular Grid on a Plane*

In ground-to-ground topographic (reconnaissance) surveys or ground-to-sea hydrographic surveys, where many points must be located in rapid succession, the best solution is to construct a circular coordinate system superimposed on an existing topographic map or sea chart. Although this method is less accurate than the separate computation of the coordinates of points, it serves its purpose within the limits of the drafting accuracy of any map.

If the scale of the map is small enough to allow the use of a drawing device (described previously), the procedure is very simple. With both ground stations as the centers, concentric circles are drawn at intervals representing the desired grid density. The actual propagation of the electromagnetic energy used (microwave, medium, or long waves) either traverses approximately along the ellipsoidal surface or is reduced onto it. Therefore, since the ground stations and the drawn distance circles are located on a projection plane or map, radii of the circles must be corrected to fit that projection.

If again the UTM is the reference projection, the following expression can be used:

$$S = s\left(0.9996 + \frac{X_1^2 + X_1 X_2 + X_2^2}{6MN}\right) \tag{4.42}$$

where S is the length of the radius on the projection plane and s is its length on the ellipsoid, X_1 and X_2 are the distances of the terminals of the radius from the central meridian, M is the meridian curvature radius, and N is the curvature radius in prime vertical at the midpoint of the radius.

If the distance from the survey area to the ground stations is too long to permit the use of any mechanical drawing devices, another procedure can be followed to construct the concentric circles: an arbitrarily chosen point P' (Fig. 4.11) having known coordinates X' and Y' is selected as the point of reference, and the line connecting P to the ground station is constructed by using

$$Y = \frac{Y' - Y_1}{X' - X_1}(X - X') + Y' \tag{4.43}$$

where X_1 and Y_1 are the coordinates of the ground station G.

The distance between point P' and the ground station G is determined as follows:

$$D'_G = \sqrt{(\Delta X)^2 + (\Delta Y)^2} \tag{4.44}$$

This distance is not an integer number, and the line GP' is extended by an amount of $\Delta D'_G$ to obtain the desired integer value D_G at point P. At point P a normal to the line GP is drawn either graphically or by resorting to the following expression:

$$Y = -\frac{X_2 - X_1}{Y_2 - Y_1}(X - X_2) + Y_2 \tag{4.45}$$

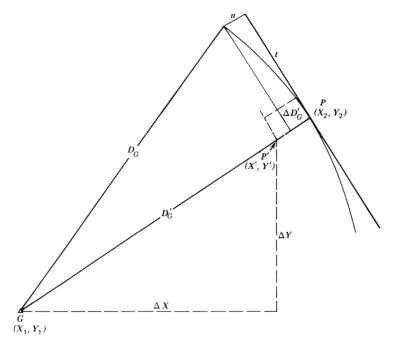

Figure 4.11   Construction of circles in circular grid.

This normal is tangent to the reference circle at point P and by the use of the well-known formula

$$n = \frac{t^2}{2D_G} + \frac{t^4}{8D_G^3}$$ (4.46)

the reference circle can be drawn, taking the parameter n as a function of the parameter t. All additional circles can then be constructed by applying the desired distance values to D_G.

4.2.1.2.   *Error Components in the Direction of Rectangular Coordinates*

The procedure for computing standard error components in the direction of rectangular coordinates in the case of the circular approach is similar to the one described in Section 4.1.5.3, featuring hyperbolic coordinates. The main difference is in the direction of the original input errors. In the hyperbolic case the original errors were considered to be lateral errors around the hyperbolas, as is the case, for example, in observing directions

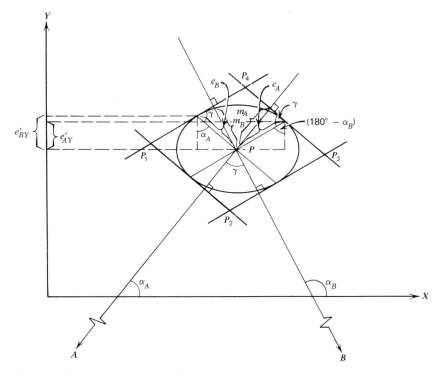

Figure 4.12   Standard error ellipse in circular location.

by theodolites whereas in the circular case the original errors may be deemed longitudinal errors along the radii of the circles.

In Fig. 4.12, α_A and α_B are angles formed by the radii from P to stations A and B with the X-axis, and γ is the station angle at point P. The original error components along the axes are $\pm m_A$ and $\pm m_B$ in meters. Obviously, these error components are not the conjugate semiaxes needed for the solution, but rather e_A and e_B, which are obtained by drawing lines parallel to P_2P_3 and P_1P_2 through point P. The lengths of these conjugate axes are

$$e_A = \frac{m_A}{\sin \gamma} \quad \text{and} \quad e_B = \frac{m_B}{\sin \gamma} \qquad (4.47)$$

The projections of e_A and e_B on the Y-axis are

$$e'_{AY} = -e_A \cos \alpha_B \quad \text{and} \quad e'_{BY} = e_B \cos \alpha_A \qquad (4.48)$$

Similarly on the X-axis, we have

$$e'_{AX} = e_A \sin \alpha_B \quad \text{and} \quad e'_{BX} = e_B \sin \alpha_A \qquad (4.49)$$

By the substitution of Eqs. 4.47 into Eqs. 4.48 and 4.49 we finally have

$$m_Y = \pm \text{cosec } \gamma \sqrt{(m_A \cos \alpha_B)^2 + (m_B \cos \alpha_A)^2}$$
$$m_X = \pm \text{cosec } \gamma \sqrt{(m_A \sin \alpha_B)^2 + (m_B \sin \alpha_A)^2}$$

$$(4.50)$$

4.2.2.   Long Line Measurement by Line-Crossing Procedure

Long line measurement is a special case of the control point extension in that the station angle γ between the radii from two ground stations is 180°. The outline of this method, which is widely used in establishing long sides for trilateration networks, is illustrated in Fig. 4.13. This approach is

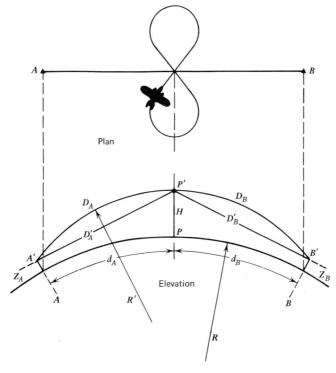

Figure 4.13   Long line measurement by line crossing (Laurila, 1960).

called the line-crossing method because an aircraft crosses the line several times at a constant altitude with a path shaped like a figure eight (8). This shape can eliminate any backlash error from the readings if certain older instruments are used, in which servo or mechanical devices are connected to distance dials. If the figure-eight shaped path is selected when approaching the line, reverse readings will not occur at the crossing point because the partial distance to one ground station continues to increase and the partial distance to the other station continues to decrease. For more modern instruments, which have electronically controlled reading systems, the approach to the line is made at right angles.

While nearing the line, the aircraft starts measuring continuously at short intervals and in strict synchronization the partial distances D_A and D_B to the two ground stations A and B. After crossing the line, the aircraft continues the measurements for a little while, and a set of summations $(D_A + D_B)$ is obtained. These summations obviously reach the minimum value when the aircraft is exactly above the line being measured. Since the flying height H is small compared with the distance D, the minimum sum $(D_A + D_B)_{min}$ can be assumed to be the minimum point of a parabola.

If ΔY (Fig. 4.14) designates the difference between a measured and an approximate sum distance $\bar{Y}$, selected so that the values of ΔY are all

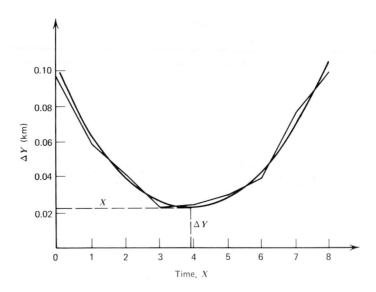

Figure 4.14   Determination of shortest sum distance.

small but positive, and X designates the time at which measurements are recorded, the following expression can be written:

$$\Delta Y = AX^2 - BX + C \qquad (4.51)$$

where A, B, and C are constants that can be determined by a least squares curve-fitting adjustment.

The value of X that causes the function ΔY, Eq. 4.51, to be a minimum is found by forming the first derivative of this function and setting it equal to zero

$$\frac{d\Delta Y}{dX} = 2AX - B = 0 \qquad (4.52)$$

When solved for X, Eq. 4.52 yields

$$X = \frac{B}{2A} \qquad (4.53)$$

The minimum value of ΔY is now found by substituting the value X of Eq. (4.53) into Eq. 4.51:

$$\Delta Y_{min} = -\frac{B^2}{4A} + C \qquad (4.54)$$

To find the values of A, B, and C the equation of observations (Eq. 4.51) can be used, and the error equations will have the form

$$v = AX^2 - BX + C - \Delta Y \qquad (4.55)$$

From these error equations we can form the following normal equations:

$$
\begin{array}{lc}
(A) & [X^4] - [X^3] + [X^2] - [X^2\,\Delta Y] = 0 \\
(B) & [X^2] - [X] + [X\,\Delta Y] = 0 \\
(C) & n \;-\; [\Delta Y] = 0 \\
(\Delta Y) & [\Delta Y^2]
\end{array}
\qquad (4.56)
$$

where n is the number of observations.

Designating the approximate sum distance as $\bar{Y}$, the final minimum sum distance is

$$Y_{min} = \bar{Y} + \Delta Y_{min} \qquad (4.57)$$

Before we can calculate the final ellipsoidal distance d (Fig. 4.13), several reductions for the partial arc distances D_A and D_B must be applied separately. The following is a summary of the steps to be taken to make these reductions possible.

1.   Tabulate the recorded distances D_A and D_B from the aircraft to the two ground stations.

2. Compute the sums $(D_A + D_B)$ from these tabulated distances.
3. Compute the minimum sum, Y_{min}, by Eq. 4.57.
4. Select from step 2 the sum $(D_{A_i} + D_{B_i})$ whose value is closest to that of Y_{min}.
5. Use the partial distance D_{A_i} from step 4 and compute the other partial distance D'_B, without changing the adjusted Y_{min} value, from the following expression:

$$D'_B = Y_{min} - D_{A_i}$$

6. By the use of D_{A_i} and D'_B, make geometric corrections to find the partial ellipsoidal distances d_A and d_B, from which the total ellipsoidal distance between the ground points A and B can be obtained:

$$d = d_A + d_B \qquad (4.58)$$

A nongeometric velocity correction due to the refraction of atmosphere is common to both applications and is discussed at length in Chapters 6 and 7.

4.2.3.   Reduction of Arc Distance to Chord Distance

The reduction of an arc distance to a chord distance is significant in handling the data obtained from electronic surveying instruments. Especially in applications of the circular approach, featuring primarily the light spectrum, microwaves, or short waves, the selection of the proper formula (see next chapter) is essential to reach the most accurate results possible.

In the following, purely geometric aspects are considered in the derivation of the basic formula. A circular arc D_A between points P_1 and P_2 is drawn with O as the center and R as the radius (Fig. 4.15). From O a line is drawn perpendicular to the chord D_C at point P_3. We call the central angle Φ and express it as a function of the arc D_A and the radius R as follows:

$$\Phi = \frac{D_A}{R} \qquad (4.59)$$

From the triangle OP_1P_3

$$D_C = 2R \sin \frac{\Phi}{2} \qquad (4.60)$$

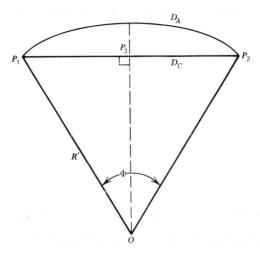

Figure 4.15 Reduction of arc distance to chord distance.

and by substituting Eq. 4.59 into Eq. 4.60:

$$D_C = 2R \sin \frac{D_A}{2R} \qquad (4.61)$$

Now, if $\sin (D_A/2R)$ is expressed as a series expansion

$$\sin \frac{D_A}{2R} = \frac{D_A}{2R} - \frac{D_A^3}{48R^3} + \cdots \qquad (4.62)$$

and only the first two terms are used from this series expansion, Eq. 4.61 can be rewritten

$$D_C = 2R \left(\frac{D_A}{2R} - \frac{D_A^3}{48R^3} \right) \qquad (4.63)$$

The chord distance can then be expressed as

$$D_C = D_A - \frac{D_A^3}{24R^2} \qquad (4.64)$$

and the arc-chord correction ΔD is thus

$$\Delta D = D_A - D_C = \frac{D_A^3}{24R^2} \qquad (4.65)$$

which is always negative.

FIVE

CURVATURE OF THE RAY PATH OF ELECTROMAGNETIC RADIATION

The distance measured between two points by means of electromagnetic radiation is not the shortest distance or chord distance between those points; rather, it is a function of the shape of the ray path concerned. When the measurements are made, the ray path is dependent on atmospheric conditions in that when the ray passes from one atmospheric layer to another, it changes its direction. The curvature of the path can be determined by computing the *refractive index n*, using meteorological observations (Chapter 6), or by direct measurement of the zenith angles at both ends of the line used to compute the *coefficient of refraction k*.

The radius R in the arc-chord correction (Eq. 4.65), applies to any general case, but in electromagnetic radiation applications, the curvature radius R' of the ray path must be first found and substituted into that equation. For medium or long waves, the curvature radius of the ray path equals the curvature radius of the earth; thus

$$R' = R \qquad (5.1)$$

and the low frequency ground waves follow the earth's surface. In the light-, micro-, and short-wave regions the curvature radius of the ray path must be determined separately for each line measured. Several methods are developed for this determination depending on the length of the line, the required accuracy, and the technique applied. In the following, grouping is made according to the length of the line ($D_A < 100$ km and $D_A > 100$ km).

5.1. CURVATURE ON SHORT DISTANCES

Although the curvature of the ray path continuously changes its magnitude when radiation propagates between two points, the shape of the path in all practical considerations is assumed to be a circular arc having a radius equivalent to the average curvature radius of the ray path.

According to the basic concept of the coefficient of refraction k, we can write

$$k = \frac{R}{R'} \tag{5.2}$$

where R is the curvature radius of earth and R' is the average curvature radius of the ray path.

To derive a correction for reducing the data from the measured arc length along the ray path to the ellipsoidal (spheroid) arc length, the correction for the ray path curvature is first obtained by substituting R' from Eq. 5.2 into Eq. 4.65

$$\Delta D_1 = -\frac{k^2 D_A^3}{24R^2} \tag{5.3}$$

where D_A is the measured arc length. If R in Eq. 4.65 represents the curvature radius of earth, the total correction will be the sum of Eqs. 4.65 and 5.3:

$$\Delta D_2 = \frac{1-k^2}{24R^2} D_A^3 \tag{5.4}$$

Saastamoinen (1962) has derived an additional correction ΔD_3 arising from the observation that since the curvature of earth is normally much greater than that of the ray path, the latter dips into the atmosphere, which has a greater average refractive index than the mean value obtained at the terminals of the line:

$$\Delta D_3 = -\frac{k(1-k)}{12R^2} D_A^3 \tag{5.5}$$

The sum of Eqs. 5.4 and 5.5 now yields the final correction:

$$\Delta D_4 = \frac{(1-k)^2}{24R^2} D_A^3 \tag{5.6}$$

When a sloped line measured with a high precision instrument is serving as a comparator to calibrate other instruments of lower order accuracy, the only correction required is due to the curvature of the ray path and its dipping into the lower atmosphere. It is obtained by combining the corrections of Eqs. 5.3 and 5.5, which yields:

$$\Delta D_5 = -\frac{k(2-k)}{24R^2} D_A^3 \tag{5.7}$$

If the refractive indexes adopted for a line are determined only at the terminal points of the line, it is common practice to assign to the coefficient of refraction the value $k = 0.25$, which, according to Eq. 5.2, means that

$$R' = 4R \tag{5.8}$$

and Eq. 5.6 becomes

$$D_4 = \frac{(0.75)^2}{24R^2} D_A^3 = \frac{D_A^3}{43R^2} \tag{5.6a}$$

For high precision surveys, this value is not accurate enough to meet the requirements set forth by the best available He–Ne lasers. To improve the accuracy and to make k represent more realistically the ambient atmospheric conditions, zenith angles of the line may be observed simultaneously with the distance measurements at both terminals. The value for k thus obtained will be (*Meade*, 1969)

$$k = 1 - (h - D_A \cos \psi) \frac{2R}{D_A^2} \tag{5.9}$$

where h = difference in elevation between the terminal points
$\quad\quad D_A$ = slope (arc) distance between the terminal points
$\quad\quad \psi$ = zenith angle
$\quad\quad R$ = curvature radius of the earth

The mean value of the zenith angles observed at both ends of the line can be used.

5.2. CURVATURE ON LONG DISTANCES

The most accurate method for computing the curvature of the ray path at short distances, and the best possible one to be employed in long distances, is to determine the refractive index at many points along the ray path. This is done by observing the atmospheric parameters of temperature, pressure, and humidity, to obtain meteorological constants A, B, and C, which, in turn, define the refractive index model (Eq. 6.29)

$$n = 1 + A + Bh + Ch^2$$

where h is an elevation along the ray path and A, B, and C are constants (see Chapter 6).

To obtain an integrated average curvature radius along the entire ray path, *Jacobsen* (1951) developed the following method. He based his derivation on the fact that the instantaneous curvature Q of the ray path is related to the refractive index n, the elevation angle α of the ray path above the horizontal, and the refractive index lapse rate dn/dh as follows:

$$Q = -\frac{\cos \alpha}{n} \frac{dn}{dh} \qquad (5.10)$$

For survey applications when the ray path elevation angle is small, it can be assumed that $\cos \alpha = 1$. Since the value of n also is close to 1, Eq. 5.10 can be written as

$$Q = -\frac{dn}{dh} \qquad (5.11)$$

The lapse rate dn/dh is obtained by differentiating Eq. 6.29 with respect to n and h as follows:

$$\frac{dn}{dh} = B + 2Ch \qquad (5.12)$$

which, when substituted into Eq. 5.11, gives an *instantaneous* curvature at the altitude h:

$$Q = -(B + 2Ch) \qquad (5.13)$$

Since we require the *average* curvature $\bar{Q}$ along the whole ray path however, h in Eq. 5.13 must be expressed as a function of the known quantities K, H, and D_A, where K is the elevation of the initial terminal, H is the elevation of the end terminal, and D_A is the total distance to be measured.

The first step in this process is to determine the altitude (elevation) h of any point along the ray path as a function of the aforementioned elements. From Fig. 5.1, making slight simplifications, we can obtain

$$h = K + D \tan \alpha + \Delta R - \Delta R' \qquad (5.14)$$

where ΔR and $\Delta R'$ are distances between tangents and corresponding circles at a given distance D from the tangent points (in this case the circles are both the surface of the earth and the ray path). It is easy to show that the values of ΔR and $\Delta R'$ are $D^2/2R$ and $D^2/2R'$, respectively. By substituting these values into Eq. 5.14 and symbolizing the *approximate* average curvature of the ray path as $\bar{\bar{Q}}$, Eq. 5.14 can be written as

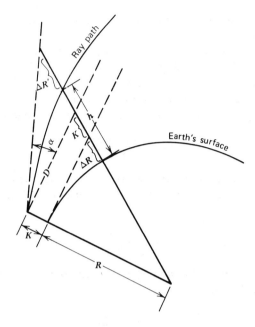

Figure 5.1 Determination of elevation h along ray path (Laurila, 1960).

follows:

$$h = K + D \tan \alpha + \frac{D^2}{2}\left(\frac{1}{R} - \bar{\bar{Q}}\right) \tag{5.15}$$

where R is the curvature radius of earth.

The term $\tan \alpha$ may be determined as a function of K, H, D_A, R, and $\bar{\bar{Q}}$ from Fig. 5.2 in the following way.

$$\tan \alpha = \left(H - K - \frac{D_A^2}{2R} + \frac{D_A^2}{2R'}\right)\Big/ D_A \qquad \text{or}$$

$$\tan \alpha = \left[\frac{H-K}{D_A} - \frac{D_A}{2}\left(\frac{1}{R} - \bar{\bar{Q}}\right)\right] \tag{5.16}$$

Substituting Eq. 5.16 into Eq. 5.15 gives

$$h = K + D\left[\frac{H-K}{D_A} - \frac{D_A}{2}\left(\frac{1}{R} - \bar{\bar{Q}}\right)\right] + \frac{D^2}{2}\left(\frac{1}{R} - \bar{\bar{Q}}\right) \tag{5.17}$$

The instantaneous curvature at altitude h along the ray path can be found in Eq. 5.13, which, when Eq. 5.17 has been substituted into it, gives

$$-Q = B + 2CK + \frac{2CDH}{D_A} - \frac{2CDK}{D_A} - \frac{CD_A D}{R} + CD_A D\bar{\bar{Q}} + \frac{CD^2}{R} - CD^2\bar{\bar{Q}} \tag{5.18}$$

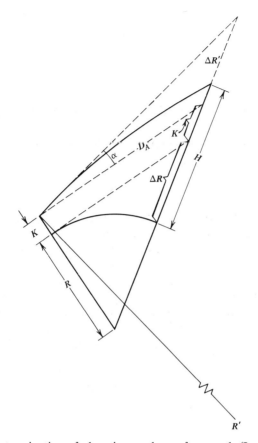

Figure 5.2 Determination of elevation angle α of ray path (Laurila, 1960).

The approximate average curvature $\bar{\bar{Q}}$ in the right-hand side of Eq. 5.18 may be replaced without any significant error by the value of Eq. 5.13, where h is given the mean value of K and H. Thus we have

$$\bar{\bar{Q}} = -B - C(H + K) \qquad (5.19)$$

and

$$-Q = B + 2CK + \frac{2CDH}{D_A} - \frac{2CDK}{D_A} - \frac{CD_A D}{R} - CD_A D[B + C(H + K)]$$

$$+ \frac{CD^2}{R} + CD^2[B + C(H + K)] \qquad (5.20)$$

Finally, the average curvature of the ray path over the whole distance D_A is obtained by taking the average of the function of Eq. 5.20 by applying the integration

$$-\bar{Q} = \frac{1}{D_A} \int_0^{D_A} Q \, dD \qquad (5.21)$$

and collecting terms as follows:

$$-\bar{Q} = B + C(H+K) - \frac{CD_A^2}{6}\left[\frac{1}{R} + B + C(H+K)\right] \qquad (5.22)$$

Consequently, the average curvature radius of the ray path found as a result of this method will be

$$R' = \frac{1}{\bar{Q}} \qquad (5.23)$$

from which the value of the coefficient of refraction k can be obtained by Eq. 5.2. To get an idea of the magnitude of the ray path arc-chord correction ΔD_1 and also the expected error $d\Delta D_1$ when the constant value $k = 0.25$ is used instead of the value based on actual atmospheric conditions, the following example is given. A reference line 65 km long, whose difference in elevation between the terminal points is $h = 2799$ m was chosen (see Section 7.1.3). With the aid of *U.S. Standard Atmosphere 1962* values of temperature, pressure, and humidity, the coefficients B and C were computed and applied to Eq. 5.22, which yielded

$$\bar{Q} = 0.0000247$$

The coefficient k was then obtained from Eq. 5.2

$$k = 0.157$$

and when substituted into Eq. 5.3, k gave the arc-chord correction

$$\Delta D_1 = -0.7 \text{ cm}$$

If the commonly accepted value $k = 0.25$ had been used, the arc-chord correction would have been

$$\Delta D_1 = -1.8 \text{ cm}$$

with the difference (error) $d\,\Delta D_1 = 1.1$ cm, or 2×10^{-7}. In high precision geodetic or geophysical surveying, this magnitude of error is big enough to disturb the ultimate accuracy.

In the final reduction of the observed distance onto the ellipsoid, in addition to the correction ΔD_4, geometric corrections to reduce the line to horizontal and to ellipsoidal level are needed. These are discussed in detail in Chapter 15.

SIX

PROPAGATION VELOCITY OF ELECTROMAGNETIC RADIATION

When utilizing electromagnetic energy in surveying or navigation, a thorough knowledge of the propagation velocity of emitted radiation is most important. This velocity is constant *in vacuo* (at the speed of light) but when affected by the atmosphere it is retarded in direct proportion to the density of air. When radiation propagates through the infinite "layers" of the atmosphere, it changes its direction and speed because of refractivity, which is expressed by the so-called *refractive index*, usually symbolized as n, and is related to the *dielectric constant* μ of the air in the following way:

$$n = \sqrt{\mu} \qquad (6.1)$$

The instantaneous velocity c of the radiation at any point (altitude) within the atmosphere is a function of the speed of light c_0 and the refractive index n; it can be given as

$$c = \frac{c_0}{n} \qquad (6.2)$$

where c_0 is constant. Since measurements are made through variable atmospheric layers, a mean refractive index must be found to obtain the mean velocity over the entire line to be measured (Chapter 7).

In vacuo, n has the constant value of 1.000000, but inside the atmosphere it increases as the altitude decreases. The refractive index is also affected by the frequency of radiation in that within the radio band it is almost constant but within the light spectrum it varies. With reference

125

to the *U.S. Standard Atmosphere 1962*, the sea level values for n when utilizing radio waves and ruby laser radiation, respectively, are as follows:

$$n_1 = 1.000321$$
$$n_2 = 1.000282$$

(6.3)

To avoid large numbers in computations, the sixth decimal of n is designated N. Thus $N = (n-1) \times 10^6$ and from Eq. 6.3 we have $N_1 = 1.0$ and $N_2 = 2.0$.

During the past decade sophistication of instrument design and manufacture has forced scientists and practical users of the multitude of available instruments to develop methods of computing the velocity of electromagnetic radiation to an accuracy that will match that already achieved by instrumentation. Until recent years the absolute velocity was not known accurately enough to measure true distances to the moon or artificial satellites, or to establish calibration baselines exceeding the accuracy obtainable by "classical" methods. On the other hand, users such as construction engineers do not need to know the absolute distance in measuring minute changes caused by strain on dams or bridges. Also, in most geophysical applications such as fault measurements or long-term observations of lateral crustal movements, knowledge of absolute distance is not essential. In these cases the precision or long-term stability of the instrumentation is of prime importance. However in geodetic or space science applications, the "true velocity" is material to keep the scale of the survey in accordance with the existing definition of the standard of length. Therefore it is equally important to know the absolute and the ambient propagation velocities of electromagnetic radiation.

6.1.   PROPAGATION VELOCITY *IN VACUO*

To discuss the speed of light, which is stated in kilometers per second, we must define the international standard of length, the meter. In 1960 the Eleventh General Conference of Weights and Measures declared the meter to be equal to 1,650,763.73 vacuum wavelengths of the orange radiation emitted by the krypton atom of mass 86. Krypton is a gaseous element belonging to the helium family of the periodic system and ^{86}Kr is a pure isotope having highly monochromatic emission. The small fractions of a meter frequently used in the terminology of electronic surveying are: micrometer (μm) = 10^{-6} m; nanometer (nm) = 10^{-9} m, and angstrom unit (Å) = 10^{-10} m.

It is a long way from the first attempt to determine the speed of light during the late seventeenth century to the present sophisticated methods, showing an accuracy figure on the order of 10^{-9}. Understandably, the first interest in this problem was shown by astronomers (*Kondraschkow*, 1961). In 1676 a Danish astronomer *Römer* obtained the value $c_0 = 215,000$ km/sec by observing the lunar eclipse of one of the Jupiter moons. In 1728 an English astronomer *Bradley* used the observations to determine the annual parallax of stars. With the help of aberration constants, he obtained the value $c_0 = 303,000$ km/sec.

The first attempt to determine the speed of light along the earth's surface was made in 1849 by the French physicist *Fizeau*, using a revolving toothed wheel. A light beam was sent through the cogs of the wheel to a distant mirror and, after reflection, it was observed through the cogs again. At a certain angular speed of the wheel, the reflected light beam was eclipsed and the observer could not see it. Thus the velocity of light could be computed by knowing the distance to the mirror, the angular speed of the wheel, and the dimensions of the cogs. This experiment yielded a result of $c_0 = 315,000$ km/sec.

The French physicist *Foucault* in 1862 developed a method to determine the speed of light by way of rotating mirrors, which yielded $c_0 = 298,000 \pm 500$ km/sec. Then in 1879 *Michelson* obtained $299,910 \pm 50$ km/sec and in 1882 *Newcomb* reported $299,860 \pm 30$ km/sec. In 1926 *Michelson* with his light interferometer obtained $c_0 = 299,796 \pm 4$ km/sec, which is surprisingly close to the presently accepted value $c_0 = 299,792.5$ km/sec.

During the 1920s electronic or electro-optical principles were developed to measure the speed of light. Since that time the Kerr cell and the Kerr cell–photo cell combination (e.g., the Geodimeter and the SWW-1) have been designed. Cavity resonators and microwave interferometers are often used to measure wavelength λ. By knowing frequently f, the speed of light can be obtained as $c_0 = f\lambda$. Microwave distancers such as Tellurometers and VHF systems such as Hiran (Shoran) are also used for direct measurement of c_0. A sample of the results of different methods used to determine the speed of light is given in Table 6.1 in chronological order rather than in order of values compared to the most probable value. The listing is by no means complete.

One of the most accurate determinations of the speed of light made in field conditions was carried out by the Finnish Geodetic Institute in 1971 (*Parm*, 1973). The instrument chosen was the Helium-Neon Laser Geodimeter model 8 as modified by the manufacturer AGA, so that only one frequency, 29,970,000 Hz can be used. This causes a higher constancy for the crystal. Measurements were carried out over a Niinisalo

TABLE 6.1

Year	Investigator	Method	Speed of Light (km/sec)	Standard Deviation (km/sec)
1929	Mittelstaed	Kerr cell	299,786.0	±20.0
1941	Anderson	Kerr cell–photo cell	299,786.0	14.0
1946	Bergstrand	Geodimeter	299,793.0	2.0
1947	Essen, Gordon-Smith	Cavity resonator	299,792.0	3.0
1949	Aslakson	Hiran	299,792.0	2.4
1950	Essen	Cavity resonator	299,792.5	1.0
1950	Hansen, Bol	Cavity resonator	299,789.3	0.4
1950	Bergstrand	Geodimeter	299,793.0	0.3
1951	Aslakson	Hiran	299,794.2	1.4
1952	Bergstrand	Geodimeter	299,793.1	0.3
1952	Froome	Microwave interferometer	299,792.6	0.7
1952	Froome	Microwave interferometer	299,793.0	0.3
1953	Mackenzie	Kerr cell–photo cell	299,792.3	0.5
1954	Welitschko, Wassiljew	SWW-1	299,793.9	1.0
1955	Florman	Microwave interferometer	299,796.1	3.1
1955	Schöldström	Geodimeter	299,792.4	0.4
1956	Wadley	Tellurometer	299,792.5	0.3
1957	Bergstrand	Geodimeter	299,792.85	0.16
1958	Froome	Microwave interferometer	299,792.50	0.1

calibration baseline 22.2 km long. In addition to the terminal towers, two extra towers were built for meteorological observations at 10.7 and 16.8 km from the initial point. The observation height of each tower was determined to follow the actual ray path by the formula for trigonometric leveling using the observed vertical angles and the known heights. The modulation frequency of the geodimeter was measured with a counter Beckman Eput and Timer Model 6146, which was compared before and after the field period to a Hewlett-Packard quartz clock.

The mean of 138 determinations yielded

$$c_0 = 299,792.479 \pm 0.010 \text{ km/sec} \tag{6.4}$$

The standard error of the mean, ± 0.010 km/sec, is an indication of precision rather than accuracy. Therefore, a detailed analysis was made, and by adding such error sources as the error caused by the refractive index formula and the error of the reference baseline itself, a conservative value of m_{v_0} ± 0.050 km/sec was published.

The latest development in computing the speed of light is based on measurement of both wavelength and frequency with extremely high accuracy. The capability of modern technique to count frequencies up to 10^{-10} (*Evenson et al.*, 1972) in the light spectrum has opened up new vistas in the finding of the "absolute" velocity of light. *Bay, Luther,* and *White* (1972) reported that the frequency of the 633 nm red laser line allowed them to determine the corresponding wavelength $\lambda = 632.99147 \pm 10^{-5}$ nm. Using the actual value of $f = 473,612,166 \pm 29$ MHz (*Evenson et al.*, 1973), the speed of light was found to be

$$c_0 = 299,792.462 \pm 0.018 \text{ km/sec} \tag{6.5}$$

In addition *Barger* and *Hall* (1972) reported the determination of the wavelength of the 3.39-μm line of methane, which was measured with respect to the ^{86}Kr 6057 Å standard, using a frequency-controlled interferometer. The frequency of the 3.39 μm line from a He–Ne laser oscillating at the methane absorption frequency has been found to be $f = 88.376,181,627 \pm 5 \times 10^{-8}$ THz (*Evenson et al.*, 1972). The reported measured wavelength was $\lambda = 3.392,231,404 \pm 1.2 \times 10^{-8}$ μm; which gave

$$c_0 = 299,792.4587 \pm 0.0011 \text{ km/sec} \tag{6.6}$$

for the speed of light. This is equivalent to a standard error of $\pm 3.7 \times 10^{-9}$.

All these experiments have been carried on in the Quantum Electronics Division and Laboratory Astrophysics Division, National Bureau of Standards, in cooperation with the Joint Institute for Laboratory Astrophysics, Boulder, Colorado.

After the definition of "meter" was adopted in 1960, an intrinsic asymmetry of the krypton emission line was discovered. This anomaly causes confusion about whether the defined krypton wavelength should be applied to the maximum intensity point or to the center of gravity point. In (6.6) reference is made to the maximum intensity point. To avoid this uncertainty the Comité Consultatif pour la Définition du Métre (*Terrien*, 1974) backed by the *International Astronomical Union* (1974) recommended that the following value be adopted as a constant for the propagation velocity of light *in vacuo*:

$$c_0 = 299,792,458 \text{ m/sec} \tag{6.7}$$

with an uncertainty of $\pm 4 \times 10^{-9}$, and that the meter be redefined accordingly. With such accuracy and the present definition of the meter, absolute distance to the moon could be measured to within ± 1.5 m.

6.2. REFRACTIVE INDEX FROM ATMOSPHERIC DATA

After the refractive index n has been determined, the ambient velocity of electromagnetic radiation can be obtained from Eq. 6.2. The refractive index is a function of temperature, pressure, and humidity, which usually are observed on the ground along the ray path or are obtained from meteorological soundings. Depending on the system used and the formula adopted to compute n, the unit of these three parameters can be selected in one of several ways. Temperature T can be given in centigrade (C), Fahrenheit (F), or in absolute or Kelvin (K) units. The relationship between these units is as follows:

$$°C = \tfrac{5}{9}(°F - 32)$$
$$°F = \tfrac{9}{5}°C + 32$$
$$°K = °C + 273.15 \tag{6.8}$$

Pressure P is stated in millibars (mb) or in millimeters of mercury (mm Hg). The relationship of these units with reference to standard atmosphere is as follows:

$$\text{standard atmosphere} = 1013.25 \text{ mb} = 760 \text{ mm Hg} = 760 \text{ torr} \tag{6.9}$$

Torr refers to the unit stated in millimeters of mercury, and often a smaller unit, millitorr (mtorr $= 0.760$ mm Hg) is used.

In all formulas humidity is given as the partial pressure of the water vapor of air. In field observations it is almost always obtained by the simultaneous observations of wet and dry bulb readings of a psychrometer. These readings and the observed pressure permit the use of Eqs. 6.10 and 6.11. (*Smithsonian Meteorological Tables*, 1966):

Temperature (Centigrade)

$$e = e' - 0.00066(1 + 0.00115T')P(T - T') \tag{6.10}$$

Temperature (Fahrenheit)

$$e = e' - 0.000367\left(1 + \frac{T' - 32}{1571}\right)P(T - T') \tag{6.11}$$

where T' is the wet bulb temperature and e' is the saturation vapor pressure. The pressures P, e, and e' are in the same units. The saturation vapor pressure e' can be computed by a lengthy logarithmic formula not suitable for field use. In practice it is obtained from Smithsonian Meteorological Tables (1966) or similar tables as a function of wet bulb temperature. If humidity is given in terms of relative humidity, as is the case in meteorological soundings, e can be obtained from

$$e = \frac{Ue''}{100} \tag{6.12}$$

where U is the relative humidity and e'' is the saturation vapor pressure at the dry bulb temperature.

Since the refractive index has different characteristics depending on whether light or radio frequencies are used, its determination in each case is analyzed separately in subsequent chapters.

6.2.1. Propagation of Light Emission

In electro-optical instruments the light forms groups of waves of slightly different lengths, even if it is monochromatically filtered. Since the velocity of such a group waves differs a little from the velocity of the equivalent or effective wave, a special group index of refraction must be determined. *International Association of Geodesy General Assembly* (1963) recommended that the *Barrell* and *Sears* (1939) formula be adopted with the following constants:

$$(n_g - 1)10^{-7} = N_g \times 10 = 2876.04 + \frac{3 \times 16.288}{\lambda^2} + \frac{5 \times 0.136}{\lambda^4} \tag{6.13}$$

where n_g is the group index of refraction and λ is the equivalent or effective wavelength of the radiation in micrometers. These constants have been computed for dry air at 0°C, 760 mm Hg, and 0.03% CO_2. For a ruby laser with $\lambda = 6943$ Å, the group refractive index will be $N_g = 298.0$, and correspondingly for the He–Ne laser with $\lambda = 6328$ Å, $N_g = 300.2$.

To compute the ambient refractive index for light waves, the *Barrell* and *Sears* (1939) formula is usually applied as follows:

$$N = \left[\frac{n_g - 1}{1 + \alpha t} \frac{P}{760} - \frac{5.5 \times 10^{-8}}{1 + \alpha t} e \right] 10^6 \tag{6.14}$$

where P is the total pressure and e is the partial pressure of water vapor in millimeters of mercury, α is the heat expansion coefficient of air 0.00367, and t is the temperature in degrees centigrade. Since in meteorological soundings the total pressure is given in millibars, *Laurila* (1968, 1969) has shown that to make Eq. 6.14 more suitable for calculator use or computer programming, it can be modified as follows.

By expressing P and e in millibars, utilizing the conversion formulas 6.8 and 6.9, and writing $\alpha = 1/T_0$, where $T_0 = 273.15°K$, Eq. 6.14 can be rewritten as

$$N = \frac{N_g}{1 + \alpha(T - T_0)}\frac{P}{1013.25} - \frac{5.5 \times 10^{-2} \times 760}{1 + \alpha(T - T_0)}\frac{e}{1013.25} \qquad (6.15)$$

and

$$N = \frac{N_g P}{\alpha T 1013.25} - \frac{5.5 \times 10^{-2} \times 760 e}{\alpha T 1013.25} \qquad (6.16)$$

and finally

$$N = \frac{N_g P - 41.8 e}{3.709 T} \qquad (6.17)$$

where T is the temperature in Kelvin units. In the derivation of Eq. 6.17 T_0 was assigned the standard value of $273.15°K$, which yields the heat expansion coefficient $\alpha = 0.003661$.

6.2.1.1. Sensitivity of the Atmospheric Parameters

To investigate the sensitivity of the atmospheric parameters T, P, and e, or their partial effect on N when the light wave propagation is concerned, Eq. 6.17 is differentiated as follows:

$$dN_T = \frac{1}{T^2}\left(-\frac{N_g}{3.709}P + \frac{41.8}{3.709}e\right)dT$$

$$dN_P = \frac{N_g}{3.709 T}dP$$

$$dN_e = -\frac{41.8}{3.709 T}de \qquad (6.18)$$

For example, if the He–Ne laser having a group index of refraction $N_g = 300.2$ is used at sea level, where the standard values of the parameters are $T = 288.15°K$, $P = 1013.25$ mb, and $e = 10.87$ mb, Eq. 6.18 gives

$$dN_T = -0.998\, dT$$

$$dN_P = 0.280\, dP$$

$$dN_e = -0.039\, de \qquad (6.19)$$

If one part per million (ppm) is set as the reference criterion satisfying most of the surveying applications, the allowable standard errors in observing T, P, and e are

$$m_T = \pm 1.0°C$$

$$m_P = \pm 3.6 \text{ mb} \tag{6.20}$$

$$m_e = \pm 25.6 \text{ mb}$$

From Eq. 6.20 it is seen that uncertainty in observing humidity has a very small effect in the determination of N. Since the average humidity, except in tropical regions, is usually less than its allowable standard error m_e, it can be neglected in most applications of Eq. 6.17, which then becomes

$$N = \frac{N_g}{3.709} \frac{P}{T} \tag{6.21}$$

or for the He–Ne laser, for example:

$$N = 80.9 \left(\frac{P}{T} \right) \tag{6.22}$$

6.2.2. Propagation of Emission in the Radio Spectrum

In the radio spectrum the refractive index is independent on the wavelength except near 60 GHz (*Liebe* and *Welch*, 1973) and around 22 GHz (*Liebe*, 1969), where there is dispersion similar to that of the light emission. Also, by the use of LF and VLF bands ($\lambda = 1000$–30,000 m), computation of variation of the refractive index becomes almost irrelevant because of the great influence of soil conductivity on velocity (see Chapter 10). In all other bands the procedure for computing the refractive index is similar to that for light waves.

The three formulas used to determine the refractive index as a function of T, P, and e are

Essen formula

$$N = \frac{77.62}{T} P - \left(\frac{12.92}{T} - \frac{37.19 \times 10^4}{T^2} \right) e \tag{6.23}$$

Smith-Weintraub formula

$$N = \frac{77.6}{T} \left(P + 4.81 \times 10^3 \frac{e}{T} \right) \tag{6.24}$$

Essen-Froome formula

$$N = \frac{103.49}{T} (P - e) + \frac{86.26}{T} \left(1 + \frac{5748}{T} \right) e \tag{6.25}$$

TABLE 6.2

Altitude (km)	Essen N-value	Smith-Weintraub N-value	Difference, ΔN
0	321.0	321.6	−0.6
1	285.6	286.1	−0.5
2	253.0	253.3	−0.3
3	223.0	223.2	−0.2
4	195.7	195.9	−0.2
5	171.1	171.2	−0.1

In Essen and Smith-Weintraub formulas, T is in Kelvin units, and P and e are in millibars. The Essen-Froome formula is actually the Essen formula obtained by rearranging terms and stating pressures in millimeters of mercury instead of millibars. At the meeting of the International Union of Geodesy and Geophysics (1960), it was recommended that the Essen-Froome formula be used to achieve uniformity in calculations of electronic distance measurements.

The Essen and Smith-Weintraub formulas were compared by applying to each formula standard T, P, and e values (*Laurila*, 1968) as in Table 6.2. The largest difference at sea level was 6×10^{-7}. Consequently, the use of either formula is justified in all measurements involving radio frequencies.

6.2.2.1. Sensitivity of the Atmospheric Parameters

To investigate the sensitivity of the atmospheric parameters where radio wave propagation is concerned, the Essen formula (Eq. 6.23) is differentiated as follows:

$$dN_T = \left[-\frac{KP}{T^2} + \left(\frac{L}{T^2} - \frac{2M10^4}{T^3} \right) e \right] dT$$

$$dN_P = \frac{K}{T} dP \qquad (6.26)$$

$$dN_e = \left(-\frac{L}{T} + \frac{M10^4}{T^2} \right) de$$

where $K = 77.62$, $L = 12.92$, and $M = 37.19$.

By the application of the same standard T, P, and e values used in the case of light wave propagation, the following relationships are obtained:

$$dN_T = -1.282\ dT$$

$$dN_P = 0.269\ dP \qquad (6.27)$$

$$dN_e = 4.434\ de$$

Again, if ppm is set as reference criterion, the following allowable standard errors of each parameter are valid:

$$m_T = \pm 0.8°C$$

$$m_P = \pm 3.7\ mb \qquad (6.28)$$

$$m_e = \pm 0.23\ mb$$

From Eqs. 6.20 and 6.28 it becomes fairly obvious that the allowable uncertainty in observing temperature and pressure is of the same magnitude in each case. Humidity is the most critical parameter to be observed in connection with measurements made by using radio waves. The required accuracy by which humidity must be known is about 100 times greater in radio wave applications than it is when light emission is utilized. In high precision geodetic or geophysical applications, therefore, light is the carrying agent of preference.

The large dispersion in the refractive index caused by water vapor has recently been used to determine the integrated vapor refractivity directly. For the purpose a refractometer has been designed (*Thompson* and *Wood*, 1969) that transmits a radio signal simultaneously with an optical signal over a related propagation path. At the far end of the path these signals are received and the differences in their arrival times are measured. The time difference is used to determine the average or integrated contribution of water vapor to the refractive index of a selected radio propagation path.

In addition to this radio-optical combination, a two-wavelength technique has been developed within the light spectrum. Because of the differences in group refractive indexes under standard conditions, the actual refractive indexes in nonstandard conditions also vary. When the path is measured using two different wavelengths of light, different optical distances will be obtained because of the dispersion. From these distances, the average refractive index can be computed (*Thompson* and *Wood*, 1965).

6.3. MODEL ATMOSPHERES

In Sections 6.2.1 and 6.2.2 formulas were given to compute the refractive index n at any point (altitude) at which T, P, and e were observed. In practice, however, the average refractive index along the whole ray path is needed to obtain the distance D. Based on the well-known practice in air physics of expressing n as a parabolic approximation, we can write the following equation for the refractive index as a function of elevation:

$$N = (n-1)10^6 = A + Bh + Ch^2 \qquad (6.29)$$

where h is the altitude of the ray path at a given point, and A, B, and C are meteorological constants.

The mean refractive index then becomes

$$\bar{N} = \frac{1}{h} \int_0^h n\, dn \qquad (6.30)$$

or

$$\bar{N} = A + \frac{h}{2} B + \frac{h^2}{3} C \qquad (6.31)$$

The coefficients A, B, and C for any standard or empirical atmospheric conditions are computed by curve-fitting least squares adjustment. The observed or given (standard) values of K, P, and e at various altitudes are placed in Eqs. 6.17 or 6.23, for example, and the observed values of $N_0 = l$ are used in the error equations as follows:

$$v_i = A + h_i B + h_i^2 C - l_i \qquad (6.32)$$

which yields the simple form of normal equations

$$
\begin{array}{lll}
(A) & n + [h] + [h^2] - \quad [l] = 0 & \\
(B) & [h^2] + [h^3] - \quad [hl] = 0 & \\
(C) & [h^4] - [h^2 l] = 0 & \\
(l) & [l^2] &
\end{array}
\qquad (6.33)
$$

In comparing the strength of $\bar{N}$ obtained from different model atmospheres, the most realistic criterion is the computation of the standard error of the function (Eq. 6.31). According to the general error propagation, this will be

$$M_{\bar{N}} = \pm \mu \sqrt{[FF]} \qquad (6.34)$$

where μ is the standard error of the weight unit and

$$[FF] = [\alpha\alpha] + h[\alpha\beta] + \tfrac{2}{3}h^2[\alpha\gamma] + \frac{h^2}{4}[\beta\beta] + \frac{h^3}{3}[\beta\gamma] + \frac{h^4}{9}[\gamma\gamma] \quad (6.35)$$

The bracketed terms in Eq. 6.35 are the weight and correlation numbers of the unknowns A, B, and C.

Several model atmospheres can be computed or created which are characterized by different A, B, and C, and, ultimately, $\bar{N}$-values.

6.3.1. Standard Atmosphere

The modern standard atmosphere was first developed in the 1920s in the United States and in Europe to satisfy a need for standardization of aircraft instruments and air performance. At the present the standard atmosphere, which is based on theoretical aspects of the physics of air, is very suitable for reference when testing empirical model atmospheres based on actual observations and soundings. This is especially important in land-to-air or land-to-space applications.

The United States standard atmosphere was generated by the National Advisory Committee for Aeronautics (NACA); and the European standard atmosphere was produced by the International Commission for Aerial Navigation (ICAN). There were slight differences between the independently developed ICAN and NACA standards. These differences were reconciled and international uniformity was achieved through adoption of a new standard atmosphere by the International Civil Aviation Organization (ICAO) on November 7, 1952. The computed tables extended from 5 km below to 20 km above mean sea level. The U.S. Committee on Extension to the Standard Atmosphere (COESA) was organized in 1953, and in Janaury 1961 a Working Group was convened to develop a new standard atmosphere up to an altitude of 700 km. The *U.S. Standard Atmosphere 1962* agrees for all practical purposes with the ICAO standard atmosphere and is used in this text.

The standard atmosphere is fundamentally defined in terms of an ideal air obeying the perfect gas law and by assumption that the atmosphere is static with respect to the earth. It is based on acceptance standard values of air density, temperature, and pressure at sea level as follows:

$$\text{density } \rho_0 = 1.2250 \text{ kg/m}^3$$
$$\text{temperature } T_0 = 288.15°\text{K} \qquad (6.36)$$
$$\text{pressure } P_0 = 1013.25 \text{ mb}$$

The density for tabulated altitudes is obtained from

$$\rho = \frac{MP}{RT} \tag{6.37}$$

where $M = 28.9644$ is the molecular weight of air and is considered to be a constant up to an altitude of 90 km, P is the pressure, R is the universal gas constant, and T is the absolute temperature.

The temperature variation is defined as a series of connected segments, linear in altitude. The general formula for each linear segment is

$$T = T_n + L(H - H_n) \tag{6.38}$$

where L is the gradient of the temperature with altitude, dT/dH, and T_n and H_n are values at the initial point of each segment. Within the altitude segments 0 to 11 km, $L = -6.5°K/km$; for 11 to 20 km, $L = 0.0°K/km$; for 20 to 30 km, $L = +1.0°K/km$; for 32 to 47 km, $L = +2.8°K/km$.

Within an atmospheric layer through which T is a linear function of H, the pressure is computed by the following formula when $L \neq 0$;

$$\log_e P = \log_e P_n - \frac{g_0 M}{LR} \log_e \frac{L(H - H_n) + T}{T} \tag{6.39}$$

where $g_0 = 9.80665 \text{ m/sec}^2$ and P_n is the pressure at the initial point of each altitude range. If $L = 0$, the following expression for P is used:

$$\log_e P = \log_e P_n - \frac{g_0 M}{RT}(H - H_n) \tag{6.40}$$

To determine partial pressure of water vapor, which is not readily obtainable from the U.S. Standard Atmosphere, constants A, B, and C must be taken from an available empirical atmosphere, placed in Eq. 6.29, and solved simultaneously for e with Eq. 6.17 or 6.23 (*Laurila,* 1968).

If T, P, and e are taken from the U.S. Standard Atmosphere tables through the troposphere from the mean sea level to an altitude of 9 km at 0.5-km intervals, 19 "observations" are available to compute the three unknowns A, B, and C. Let us assume that a ruby laser with a wavelength $\lambda = 6943$ Å is used for radiation. The point-to-point N-values are then computed with Eq. 6.17, where $N_g = 298.0$. By the curve-fitting least squares adjustment, the coefficients representing this particular model atmosphere will be

$$A = 281.8$$
$$B = -26.4 \tag{6.41}$$
$$C = 0.79$$

These coefficients, when substituted into Eq. 6.31, yield the following mean refractive index over the 0–9 km altitude range:

$$\bar{N} = 184.3 \qquad (6.42)$$

Thus any model atmosphere can be tabulated to include temperature, pressure, and humidity, which in electronic surveying applications may be identified as the basis for computing the coefficients A, B, and C and finally yields the mean refractive index $\bar{N}$. Such standard values are very useful when analyzing the feasibility of new methods or approaches. These coefficients can be used as such in surveying, where the required accuracy can be achieved without knowledge of the local atmospheric conditions.

6.3.2. Supplementary Atmospheres

The standard atmosphere is based solely on theoretical air physics, and actual conditions may differ from those described by *U.S. Standard Atmosphere 1962*. Day-by-day changes and latitudinal differences can be depicted by a family of supplementary atmospheres. From such supplementary atmospheres and the *U.S. Standard Atmosphere 1962*, a set of modified atmospheres has been computed by AFCRL, the Air Force Cambridge Research Laboratory (*Cole* and *Kantor*, 1965; *Cole, Court, and Kantor*, 1965).

These atmospheres represent conditions in the tropics, in the subtropics, at midlatitudes, and in the Arctic for summer and winter climates. Table 6.3 summarizes the summer and winter models at the latitudes 30°N and 45°N through the troposphere from sea level to 9 km (*Schnelzer*, 1972).

It is interesting to note that in spite of the relatively widely scattered values of the regional and seasonal refractive indexes, their arithmetic

TABLE 6.3
AFCRL MODEL ATMOSPHERES

Latitude	Season	A	B	C	$\bar{N}$
30°N	July	270.0	−24.5	0.73	179.4
30°N	January	283.0	−27.6	0.92	183.7
45°N	July	275.0	−25.9	0.82	180.5
45°N	January	299.0	−31.4	1.13	188.2

TABLE 6.4

Altitude (km)	Lihue July	Lihue, February	Fairbanks, July	Fairbanks, February	U.S. Standard 62
0.5	21	17	23	−8	12
5.0	0	−5	−15	−30	−17
10.0	−38	−35	−43	−54	−50
15.0	−69	−71	−45	−46	−57
17.0	−69	−75	−45	−47	−57

mean $\bar{N} = 183.0$ differs only by $d\bar{N} = 1.3$ from the value (Eq. 6.41) obtained from the *U.S. Standard Atmosphere 1962*. This is equivalent to 1.3×10^{-6}.

6.3.3. Empirical Atmosphere

Since the troposphere is not stable, best results are attainable, especially in land-to-air and land-to-space applications, when actual meterological soundings in the vicinity or proximity of the survey site are used. To investigate this phenomenon, four widely separated climatological samples were selected. Sounding data were obtained from the National Weather Service stations located at Fairbanks, Alaska (65°N), and Lihue, Hawaii (22°N), on two dates, February 3 and July 2, 1966, at 00.00 Greenwich time (*Laurila*, 1968).

These samples represented hot and humid, cool and dry, and cold and dry conditions, with a rather anomalous thermal distribution given in degrees centigrade and extended to 17 km as in Table 6.4.

Table 6.5 gives the parameters A, B, C, and $\bar{N}$ for these empirical sample atmospheres from sea level to 9 km.

TABLE 6.5

Location	Season	A	B	C	$\bar{N}$
Lihue	July	274.4	−26.3	0.89	180.1
Lihue	February	276.7	−26.7	0.88	180.6
Fairbanks	July	268.9	−22.0	0.41	181.0
Fairbanks	February	307.1	−33.8	1.34	191.1

All values except the last one computed from the Fairbanks data in February agree fairly well with the U.S. Standard values. The anomalous change in $\bar{N}$ from Fairbanks data is due to strong thermal inversions in the lower troposphere as seen from the following actual soundings:

Altitude (km)	T (°K)	
0.146	254.1	
0.539	266.6	
0.960	267.9	(6.43)
1.406	265.6	
1.877	263.3	
2.376	259.9	

In Section 6.3, the values of A, B, C, and $\bar{N}$ are based on the particular wavelength emitted by a ruby laser. Other wavelengths in the light spectrum yield slightly different values, but if radio waves are used a significant change occurs.

SEVEN

VELOCITY APPLICATIONS
TO SURVEYING

The basic theory presented in the last two chapters yields a variety of modifications, combinations, and practical arrangements depending on the task at hand. Instrumentations operating with wavelengths ranging from the light spectrum to 30 km are manufactured, to serve surveying best under land-to-land, land-to-sea, land-to-air, and land-to-space conditions. In each case, of course, methods and approaches are bound to be different.

7.1. LAND-TO-LAND SURVEY

When measurements are made over solid land, the propagation characteristics of electromagnetic radiation must be viewed in the light of the purpose of the survey, the identity of the user of the instruments, and the kind of instrumentation.

7.1.1. Construction Survey

Engineers involved in high precision building of bridges, open cast mines, dams, and other massive projects are also in charge of detecting any minute changes in such structures caused by strain or structural settlement. This work calls for high precision instruments with long-term stability. To achieve the ultimate in accuracy, instruments whose radiation source is a gallium-arsenide diode or a helium-neon laser are selected. The error independent of distance (instrument error) can be of the order of ± 1 mm, which suggests the need for extreme care in determining the refractive index. However, the distances involved are usually relatively short, about 500 to 1000 m. The instrument error at the lower limit equals 2×10^{-6}, which is easily matched by observing the temperature and pressure at both ends of the line before and after the survey: an inexpensive aneroid barometer and a mercury-filled tubular thermometer serve this purpose. Because light emission is used, Eq. 6.21 can be applied, neglecting the influence of the humidity (see results of the analysis, Eq. 6.20).

7.1.2. Land Survey

Land surveys including land division, cadastral survey, highway layouts, and small trilateration configurations can be performed with equal efficiency with microwave or light wave techniques. The instruments have an accuracy of the order of a few centimetres, but the range varies from a few hundred meters up to 30 km. Since the accuracy requirement in this kind of survey rarely exceeds 5×10^{-6}, a procedure similar to the construction survey can be applied. If microwave technique is used, humidity also must be observed. One of the simplest ways to obtain dry and wet bulb readings calls for two tubular thermometers, one provided with a wet wick around the receptacle. When the wet bulb thermometer is whirled in a sling, aspiration occurs and both dry and wet bulb temperatures can be read simultaneously.

More expensive but better suited for field use are aspirated psychrometers, such as the Bendix Psychron. Two tubular thermometers, one having the wet wick, are housed parallel and horizontally in the same chassis. To maintain constant aspiration, a battery-operated fan draws air through the removable air intake at the rate of about 5 m/sec. The Psychron has a built-in illumination system and a special thermal shield to avoid the radiation effects of sunlight.

In all practical field surveying work, the pressure and dry and wet bulb temperatures are read at both ends of the line before and after the measurements, and then the mean is used to compute the refractive index.

7.1.3. Geodetic and Geophysical Survey

A geodetic survey by electronic means consists of first-order trilaterations, checking the scale of existing first-order triangulations, and establishing highly accurate reference calibration baselines. A geophysical survey supplies measurements over faults and long-term observations of lateral crustal movements. High precision stability of the measurements is essential in both these applications. To achieve the goals stated, the precision of the measurements must be of the order of 10^{-7}. The best He–Ne lasers presently available are able to satisfy this requirement, which means that great care must be taken to determine the mean refractive index.

Since the refractive number N is not linear but rather a parabolic function of height, only in rare cases is ultimate accuracy obtained by observing temperature, pressure, and humidity at several points along the line and taking the arithmetic mean of the computed N-values. The requirements for obtaining highly accurate results (*Parm*, 1973) are as follows: the line must be level, observations must be made at altitudes equivalent to the height of the ray path, and proper weighting must be applied to the computed values. A parabolic approach should be chosen especially for sloped lines.

Equation 6.31 refers to the vertical propagation of electromagnetic radiation (i.e., when the zenith angle $\psi = 0$). It does not take into account the elevations of the terminal points of the line, the length of the line, and the average curvature of the ray path. For the general case, *Jacobsen* (1951) derived the following formula:

$$\bar{N} = A + \frac{B(H+K)}{2} + \frac{C}{3}[(H+K)^2 - HK] - \frac{D_A^2}{12}\left[\frac{1}{R} - \bar{Q}\right]$$

$$\times \left[B + C(H+K) - \frac{CD_A^2}{10}\left(\frac{1}{R} - \bar{Q}\right)\right] \tag{7.1}$$

where K is the elevation of the initial point of the line, H is the elevation of the terminal point of the line, D_A is the measured arc length, R is the curvature radius of the earth, $\bar{Q}$ is the average curvature of the ray path, and A, B, and C are the meteorological constants discussed in Chapter 6. This formula was derived primarily for long lines (ca. 300–400 km) to be measured by land-to-air instrumentation such as Hiran or Shiran. For the land-to-land survey, where distances seldom exceed 70 km, the last term in Eq. 7.1 can be neglected because at that distance it never has a larger effect on $\bar{N}$ than 10^{-8}. The equation then yields the following form:

$$\bar{N} = A + \frac{B(H+K)}{2} + \frac{C}{3}[(H+K)^2 - HK] - \frac{D_A^2}{12}\left[\frac{1}{R} - \bar{Q}\right]$$
$$\times [B + C(H+K)] \tag{7.2}$$

The average curvature $\bar{Q}$ of the ray path can be taken from any of the three Eqs.—5.8, 5.19, or 5.22—depending on the length of the line and the accuracy required.

To summarize the procedures used in obtaining the mean refractive index over a high precision laser reference line, the following steps must be taken.

1. Temperature, pressure, and humidity are observed at several points with known elevation, simultaneously with the making of distance observations. The points must be located in proximity to the line, to avoid the influence of lateral temperature inversions and tilting of the isobaric surface.
2. By the use of Eq. 6.17, the refractive indexes at these points are computed.
3. Based on the least squares curve-fitting principle, the coefficients A, B, and C are computed.
4. The curvature $\bar{Q}$ of the ray path is computed by Eq. 5.22.
5. The mean refractive index $\bar{N}$ is computed by Eq. 7.2.

This method is particularly suitable for lines having high elevation differences between their terminals. Good results have been obtained also by flying a small airplane along the line and observing the atmospheric parameters (*Meade*, 1969).

A simplified but still accurate method is appropriate for lines whose terminal elevations are approximately of the same order. Only three points are needed for meteorological observations—namely, the terminal points and one point with a known elevation close to the middle of the line. The altitude h of any point along the ray path is obtained from Eq. 5.17 as follows:

$$h = K + D\left[\frac{H-K}{D_A} - \frac{D_A}{2}\left(\frac{1}{R} - \bar{\bar{Q}}\right)\right] + \frac{D^2}{2}\left(\frac{1}{R} - \bar{\bar{Q}}\right)$$

where $\bar{\bar{Q}}$ is the approximate curvature of the ray path at an altitude h, and D is the distance from the initial terminal. Since in this application the midpoint altitude is needed, $D = D_A/2$ and Eq. 5.17 can be rewritten

$$h_m = K + \frac{H-K}{2} - \frac{D_A^2}{8}\left(\frac{1}{R} - \bar{\bar{Q}}\right) \tag{7.3}$$

Because the coefficients A, B, and C are not known in this application, $\bar{\bar{Q}}$ must be assigned the standard value $\bar{\bar{Q}} = 0.0000392$ from Eq. 5.8, and Eq. 7.3 will be

$$h_m = K + \frac{H-K}{2} - \frac{D_A^2}{8} 0.000118 \tag{7.4}$$

where R has been assigned the value of 6371 km.

The temperature, pressure, and humidity values observed on the ground proximal to the midpoint of the line are now elevated exactly to the midpoint of the ray path. This is done with the aid of h_m, Eq. 7.4 and the elevation h_p of the ground point by using the proper gradient for temperature and proper scales for pressure and humidity, which can be obtained from the observed differences between the three points.

Given the proper equation, the refractive indexes N_1 and N_2 at the terminal points and N_m at the elevated midpoint can be computed and placed into *Simpson's* formula as follows:

$$\bar{N} = \tfrac{1}{6}(N_1 + 4N_m + N_2) \tag{7.5}$$

To illustrate the validity of the two methods specified by Eqs. 7.2 and 7.5, the following example of test calculations is given. The test analyzes how big an error the uncertainty in determining the atmospheric refraction brings to the measurement of a calibration line A, 65 km long (Fig. 7.1). Lines A to F are planned to be used to investigate the lateral crustal movements within the principal islands of Hawaii.

Temperature, pressure, and humidity are assumed to be observed simultaneously at a benchmark on top of Haleakala, Maui, and at a

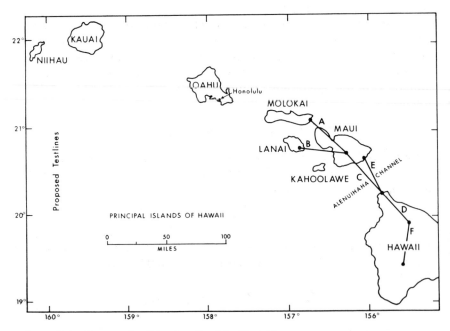

Figure 7.1   Proposed calibration lines in Hawaii.

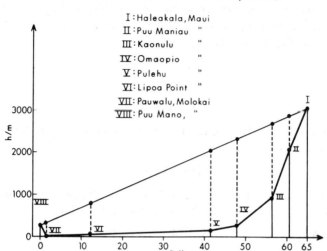

Figure 7.2 Maui-Molokai test line.

triangulation point, Puu Mano, Molokai. In addition, similar observations should be made at six other points II to VII on the ground along the line (Fig. 7.2). Since the primary intention of this experimental calculation is to check the strength of the mathematics involved in these methods, meteorological data are taken from the *U.S. Standard Atmosphere 1962* as given in Table 7.1.

TABLE 7.1

Point	Elevation (km)	T (°K)	P (mb)	e (mb)	N	Residuals, ΔN
I	3.085	268.12	693.67	4.0	215.09	+0.02
II	2.088	274.59	786.38	5.9	238.04	−0.03
III	0.914	282.21	908.17	8.3	267.42	±0.00
IV	0.272	286.38	981.01	10.1	284.63	±0.00
V	0.103	287.48	1000.93	10.6	289.28	+0.01
VI	0.041	287.88	1008.34	10.8	291.01	±0.00
VII	0.000	288.15	1013.25	10.9	292.15	±0.00
VIII	0.286	286.29	979.36	10.1	284.23	+0.01

The N-values for each observation point were computed by the use of Eq. 6.17, where N_g was assigned the value of 308.6; they are tabulated in Table 7.1. The least squares adjustment to obtain A, B, C was performed according to Eq. 6.33, yielding

$$A = 292.153$$
$$B = -27.938 \tag{7.6}$$
$$C = 0.961$$

The average curvature of the ray path was then computed by Eq. 5.22 and proved to be

$$\bar{Q} = 0.0000247 \tag{7.7}$$

The final mean refractive index $\bar{N}$ and its standard error were computed by Eqs. 7.2 and 6.34, giving the following values:

$$\bar{N} = 249.57 \pm 0.02 \tag{7.8}$$

After adjustment, the standard error of the mean (which represents a proportional error as small as 2×10^{-8}) together with the residuals (Table 7.1) suggest that the parabolic assumption is extremely strong and can be used in any high precision survey.

The validity of the use of Simpson's rule can be tested by comparing its result to the value of $\bar{N}$ (Eq. 7.8). The substitution of the elevations of the terminal points, $K = 3.085$ km and $H = 0.286$ km from Table 7.1, together with $D_A = 65$ km into Eq. 7.4, yields the elevation at the midpoint of the line:

$$h = 1.623 \text{ km} \tag{7.9}$$

The sea level values of T, P, and e, after being elevated to the height h become

$$T = 277.61°K$$
$$P = 833.03 \text{ mb}$$
$$e = 6.3 \text{ mb}$$

which, when applied to Eq. 6.17, give the refractive index at that point:

$$N_m = 249.41 \tag{7.10}$$

The substitution of N_I and N_{VIII} from Table 7.1, together with N_m into Eq. 7.5, finally yields the mean refractive index:

$$\bar{N}_s = 249.47 \tag{7.11}$$

It is interesting to note that the difference between the two N-values of Eqs. 7.8 and 7.11 is only 0.10 or 10^{-7}. If the constant value of $\bar{\bar{Q}} = 0.0000392$ had been replaced by the actual local value of $\bar{Q} = 0.0000247$ in Eq. 7.3, the elevation of the midpoint of the line would have been $h = 1.616$ km. The influence of the elevation change on the refractive index is obtained by differentiating Eq. 6.29 as follows:

$$dN = (B + 2Ch)\, dh \qquad (7.12)$$

In our case $dh = -0.007$ km; if we place B and C from Eq. 7.6 into Eq. 7.12, the N value at the midpoint of the line becomes $N_m = 249.58$. After the application of the Simpson's rule, the final mean refractive index yields the value

$$\bar{N}_S = 249.61 \qquad (7.13)$$

which indicates that the most critical error source in sloped lines when using this approximate method is the determination of the elevation of the midpoint. From Eq. 7.12 it is seen that an error of ± 4 m in determining the elevation of the midpoint of the line yields a proportional error in N as great as $\pm 10^{-7}$. An improvement in accuracy may be expected if $\bar{\bar{Q}}$ is assigned a value based on any known model atmosphere.

The results of our test calculations reflect the validity of the mathematics only. The ultimate accuracy depends on the quality of the instruments and the care maintained in observing the meteorological data. In geodetic and geophysical surveys, precision instruments such as the Assman Aspirated Psychrometer and a digital aneroid barometer should be used. Based on the analysis of Eq. 6.20, it seems possible that an overall accuracy of about 10^{-7} is attainable.

In a land-to-land survey, the light wave spectrum or the microwave radio band are used almost exclusively as the carrying agent. In some applications for which lower order accuracy is acceptable (e.g., reconnaissance surveys), medium or long radio waves can be chosen.

7.2. LAND-TO-SEA SURVEY

The modern land-to-sea survey offers a large variety of instrumentation and frequencies for various applications. The wavelengths vary from 3 cm to 30 km, and the electronic principles involved include almost all known designs in electronic surveying. Circular and hyperbolic geometries are used equally. Factors other than atmospheric refraction, such as soil conductivity and ionospheric reflection, affect the velocity of the electromagnetic radiation. Also, when high frequencies are used, the distances are restricted by line-of-sight limitations. Therefore, grouping according to the purpose of operations is preferred.

7.2.1.   Close-to-Shore Hydrographic Survey

Microwave modulation techniques employing frequencies of $f = 3$ GHz or $f = 10$ GHz ($\lambda = 10$ cm or $\lambda = 3$ cm) have become very popular in recent years for close-to-shore hydrographic or harbor survey. Instrumentation (e.g., Cubic Corporation and Tellurometer Pty. Ltd.) has an effective range of about 100 km, although longer distances can be reached depending on the antenna installations. The accuracy of the survey is of the order of ± 1 m which is equivalent to 10^{-5} at that distance. The instruments have been designed so that the readings are referenced to a "built-in" refractive index, usually $n = 1.000321$, which is the standard sea level value. If local conditions vary greatly from standard conditions, temperature, pressure, and humidity are observed simultaneously on land and aboard ship, and the mean values thus obtained are used to find the correction from simple templates or graphs.

The use of medium-long waves with frequencies of around 2 MHz ($\lambda = 150$ m) is a classic procedure in close-to-shore hydrographic survey. It sometimes has the form of a continuous wave technique (e.g., Decca IIi-Fix), or it may be based on a heterodyne principle (e.g., Raydist). The line-of-sight limitation does not exist in this wave band and at an average distance of 200 km from a ground station, a standard error of several meters can be anticipated. Therefore it is customary to use a constant value for the propagation velocity of the radio waves in constructing hyperbolic charts or computing distances. This value may be computed with the aid of the proper equation, corresponding to the average atmospheric conditions in the area concerned. In this case, seasonal changes are ignored.

7.2.2.   Offshore Hydrographic Survey

Distances of up to 400 km can be reached in offshore hydrographic surveys. Heterodyne principles are used as before, but greater output power is supplied. The Decca Survey System has long been acknowledged to be one of the leading systems featuring the continuous wave technique at this distance range. Its low frequency band of around 100 kHz (λ = 3 km) is very vulnerable to such external factors as sky wave reflection and soil conductivity, which can cause the actual velocity to vary as much as 466 km/sec when measurements are made over water and land. *Pressey et al.* (1952) obtained an average velocity over typical geological structures in England of $c = 299,230$ km/sec, and *Laurila* (1956) reported $c = 299,696$ km/sec as a result of extensive tests made over the seawater in the Gulf of Bothnia. Because of the magnitude of the external factors affecting the velocity of electromagnetic waves at this band, it is useless to try to observe any change in the velocity. Best results are obtained by determining a constant velocity for a particular area from direct observations using the least squares principle (Section 7.2.4).

7.2.3. Long-Distance Navigation

All long-distance navigation equipment are based on hyperbolic geometry. Decca, as described previously but with greater output power, can operate at distances up to 1000 km. *Loran-C*, which is a pulse-type instrument utilizing the same low frequency band as Decca, can operate at distances of up to 2000 km using ground waves and up to about 4000 to 5000 km using sky waves. In both methods a worldwide constant of the velocity of electromagnetic waves is applied.

Omega is a VLF radio navigation network that when fully implemented (ca. 1976) will offer continuous, all-weather, worldwide position information. Transmitters from eight locations will broadcast time-sequenced 10.2, 13.6, and 11.33 kHz electromagnetic radiation with an average wavelength of $\lambda = 30$ km. The Omega signals are propagated in the wave guide formed between the earth and the ionospheric D-region. Transmissions are modeled as transverse magnetic (TM) modes. The TM_1 mode propagates with a basic phase velocity

$$c_p = \frac{c_0}{0.9974} \tag{7.14}$$

which yields

$$c_p = 300,574 \text{ km/sec} \tag{7.15}$$

or more than the speed of light.

Inhomogeneities exist in the modeled wave guide; thus a propagation correction table is published to correct the charted lines of position for variations in phase velocity from this nominal value. To compute the corrections, such geophysical path characteristics as ground conductivity, magnetic dip angle, ionospheric conditions, and the azimuth of the propagation path must be considered. For more detailed information on Omega, see Chapter 26.

In offshore survey and long-distance naviagation applications, a fixed station called the monitor "forces" the location results to be correct at that station and also prevents any variations at the same location.

7.2.4. Computation of Velocity from Direct Observations

The best condition for computing velocity directly from observed hyperbolic coordinates is obtained when a test survey is made over a gulf or a narrow sea. Transmitting stations are placed on one shoreline and locations are made close to the opposite shore, where each hyperbolic fix can also be determined simultaneously from known points on shore by way of angular resections. In this case, the distances r_M and r_S from the observed points to the master and slave stations are known; the sum expression $(S + r_M - r_S)$, where S is the base between the master and slave stations, is also known and can be symbolized as Q (Chapter 4). If we assume that the hyperbolic location is based on a phase-measuring technique, a general equation for the hyperbolic coordinate can be written:

$$N - \frac{f}{c} Q \qquad (7.16)$$

Since the frequency f used in the computation of the hyperbolic coordinates is assumed to be a known constant and the hyperbolic coordinate value N is an observed value, the velocity based on observations can be expressed as

$$c_O = \frac{fQ}{N} \qquad (7.17)$$

which is referred to as the observed velocity.

The true velocity to be determined is the observed velocity plus a small velocity correction dc_0 caused by a constant instrument error dN (error in phase locking of Decca system, e.g.). The value of dc_0 is obtained by differentiating Eq. 7.17 with respect to c_0 and N as follows:

$$dc_0 = -\frac{fQ}{N^2} dN = -\frac{c_0}{N} dN \qquad (7.18)$$

The equation of observations thus can be written as

$$c_T = c_0 - \frac{c_0}{N} dN \qquad (7.19)$$

To avoid large numerical values in the adjustment process, an approximate value $\bar{c}$ is added to both sides of Eq. 7.19. The error equations then can be written as follows:

$$v = \frac{c_0}{N} dN + \Delta c - l \qquad (7.20)$$

where $\Delta c = c_T - \bar{c}$ and $l = c_0 - \bar{c}$.

A hyperbolic location is an intersection of two hyperbolas and if these hyperbolas are defined as red and green, the following double-error equations can be written:

$$v_R = \frac{c_{0R}}{N_R} dN_R + \Delta c - p$$

$$\qquad (7.21)$$

$$v_G = \frac{c_0 G}{N_G} dN_G + \Delta c - q$$

where $p = c_{0R} - \bar{c}$ and $q = c_{0G} - \bar{c}$.

The normal equations in this case will then be

$(dN_R) \qquad \left[\frac{c_{0R}^2}{N_R^2}\right] + 0 + \left[\frac{c_{0R}}{N_R}\right] - \left[\frac{c_{0R}}{N_R} p\right] = 0$

$(dN_G) \qquad \left[\frac{c_{0G}^2}{N_G^2}\right] + \left[\frac{c_{0G}}{N_G}\right] - \left[\frac{c_{0G}}{N_G} q\right] = 0$

$(\Delta c) \qquad\qquad 2n - [p + q] = 0 \qquad\qquad (7.22)$

$(l) \qquad\qquad\qquad [p^2 + q^2]$

where n is the number of points observed.

Let us consider a sample computation of the velocity of electromagnetic radiation over seawater with 0.5% salinity, using the Decca system operating on the frequency band around 100 kHz (*Laurila*, 1953). The survey was carried out over open stretches of water in the Gulf of Bothnia between Sweden and Finland. The transmission stations were located along the eastern shoreline in Sweden; the survey ship, with the receivers on board, fixed its position at 12 points off the western shore in Finland. The average distance to the ground stations was 220 km. The pertinent data used for the least squares adjustment appear in Table 7.2.

TABLE 7.2

No.	Red				Green			
	Q	N	c_0	p	Q	N	c_0	q
1	63.507	75.049	299,612	−54	133.617	118.398	299,682	+16
2	67.025	79.214	299,583	−83	132.037	117.002	299,671	+5
3	72.378	85.540	299,585	−81	129.167	114.447	299,702	+36
4	79.794	94.298	299,606	−60	124.490	110.309	299,686	+20
5	86.844	102.605	299,677	+11	119.835	106.174	299,715	+49
6	93.190	110.111	299,655	−11	115.901	102.700	299,681	+15
7	99.535	117.607	299,658	−8	111.434	98.737	299,696	+30
8	104.861	123.908	299,639	−27	106.926	94.750	299,673	+7
9	110.006	129.983	299,649	−17	103.745	91.933	299,667	+1
10	115.842	136.878	299,651	−15	99.059	87.766	299,716	+50
11	120.417	142.282	299,655	−11	95.541	84.645	299,731	+65
12	124.781	147.440	299,651	−15	92.053	81.554	299,733	+67

The approximate value $\bar{c} = 299{,}666$ km/sec used in Eq. 7.21 was the arithmetic mean of all observed c_0 values. As the result, the values of the unknowns dN_R, dN_G, and Δc were

$$dN_R = -0.022 \text{ lane}$$

$$dN_G = +0.001 \text{ lane} \tag{7.23}$$

$$\Delta c = +30 \text{ km/sec}$$

with standard errors of

$$m_{NR} = \pm 0.009 \text{ lane}$$

$$m_{NG} = \pm 0.009 \text{ lane} \tag{7.24}$$

$$mc = \pm 26 \text{ km/sec}$$

The propagation velocity thus determined was

$$c = 299{,}696 \pm 26 \text{ km/sec} \tag{7.25}$$

This value agrees very closely with the propagation velocity for electromagnetic radiation of the 100 kHz frequency band over seawater, which is generally accepted to be $c = 299{,}700$ km/sec by the International Hydrographic Survey Office, Monaco.

7.3. LAND-TO-AIR SURVEY

Land-to-air surveys almost exclusively use circular geometry. The task can be either ascertaining the control point extension, which means establishing controlled photography, or measuring long lines for trilateration purposes (Chapter 4). Grouping the instrumentation according to the operational distance ranges, two categories can be anticipated.

For distances usually not exceeding a total of 100 km, having two legs each approximately 50 km long, the microwave modulation technique is used at the carrier frequency of about 3 GHz and $\lambda = 10$ cm (e.g., *Tellurometer*). In Tellurometer's model MRB-301, the primary pattern frequency of $f = 1.498470$ MHz is chosen to give a direct range readout in meters at an atmospheric refractive index of $n = 1.000330$. In practice, meteorological observations are made both in the aircraft and on the ground at the beginning and end of each survey. To achieve better accuracy, observations can also be made at specific intervals during the survey. The mean of the refractive index obtained from these observations is used to correct the range readout.

In long line measurements made with line crossing techniques, *Hiran* (high accurate Shoran) was used until 1966. Hiran is a pulse-type instrument operating in the VHF frequency band, having a wavelength around $\lambda = 1.0$ m. It was replaced by *Shiran* (*S*-band Hiran: *Cubic Corporation*), which is a microwave modulation instrument whose wavelength is around $\lambda = 10$ cm. Depending on the altitude of the aircraft, the maximum distance obtainable is about 1400 km, with two 700 km legs. Because of the long distances involved, the electromagnetic radiation must travel through the troposphere, where almost all the humidity of the atmosphere is concentrated. Therefore, special care must be taken in computing the mean refractive index.

The procedure is as follows. The airplane carrying out the line crossing mission acts as a sounding device. Before the mission starts, the aircraft levels off at an altitude of about 200 m above the one ground station. Temperature, pressure, and humidity are then observed. By ascending to the mission altitude at the middle of the line, the plane levels off at certain intervals at least 3 minutes apart before a new set of observations is made. The same procedure is repeated during the descent to the other station after the mission is completed. Also, while the actual line crossings are taking place, observations are made at the constant operational altitude. Employing least squares adjustment (Eq. 6.33), the meteorological constants A, B, and C are computed for that particular line crossing.

The average curvature of the ray path is then computed from Eq. 5.22, and the mean refractive index $\bar{N}$ is finally obtained from Eq. 7.1. The elevations h of the observation points are obtained from a barometric altimeter that is calibrated before and after the mission against a radio altimeter over known terrain, usually seawater level. The weakness of this method is that the elevation h is derived from barometric pressure which, in turn, is one of the factors involved in computing the coefficients A, B, and C. This is true for all meteorological soundings in general.

7.4. LAND-TO-SPACE SURVEY

Land-to-space survey is the newest application in the field of electronic surveying. The tracking of artificial earth satellites relies on either coherent light or VHF radio transmission. Lunar ranging through retroreflective prism systems placed on the surface of the moon by the Apollo crews is accomplished exclusively by the use of coherent light. In both applications the electromagnetic radiation propagates through the entire atmosphere. The methods of computing the mean refractive index differ essentially from those described in the previous sections. There is only one terminal of the line inside the atmosphere, the tracking station on the earth. The other terminal is the "edge" of the atmosphere, which is a function of the already familiar parameters T, P, and e. In addition, the deviation of the sighting to the moon or to the satellite from the vertical must be considered; this is known as the zenith angle or zenith distance and is designated ψ.

The retardation of the velocity of electromagnetic propagation diminishes rapidly at an altitude of about 36 km. The total range correction due to the atmosphere from sea level to that altitude is about 240 cm. From 36 to 72 km the range correction (*Schnelzer*, 1972) is only 1.2 cm, and beyond that altitude it is a fraction of a millimeter. Because of the different propagation characteristics, light waves and radio waves are treated separately in the following sections.

7.4.1. Utilization of Light Waves

The atmosphere up to 36 km consists of two parts, the troposphere from sea level to about 9 km and the stratosphere from 9 to 36 km. As mentioned earlier, most of the humidity of the total atmosphere is concentrated in the troposphere. In the stratosphere, the temperature gradient changes its magnitude and direction at intervals of approximately 9 km. Therefore, for the computation of the mean refractive index through the entire atmosphere, it is advisable to divide the atmosphere into zones rather than assuming it to be a continuous medium. *Bean* (1960) expresses the refractive index through the atmosphere as a three-part model. He assumes a linear decrease in N for the first kilometer above the station:

$$N_1 = N_s + \Delta N (h - h_s), \qquad h_s \leqq h \leqq (h_s + 1) \qquad (7.26)$$

where N_s is the observed surface value, ΔN is an empirically derived refractivity gradient, and h_s is the elevation of the station in kilometers. From $(h_s + 1)$ to 9 km, Bean uses the exponential function

$$N = N_1 e^{-c(h - h_s - 1)}, \qquad (h_s + 1) \leqq h \leqq 9 \qquad (7.27)$$

where

$$c = [1/(8 - h_s)] \ln (N_1 / 105)$$

Above 9 km, the following formula reflecting the exponential decay of atmospheric density is used:

$$N = 105 e^{-0.1424(h - 9)}, \qquad h \geqq 9 \qquad (7.28)$$

The N-values in Eqs. 7.26 to 7.28 are point-to-point values with reference to given altitudes h.

To give full benefit to the discontinuity of the stratosphere due to the changes of the temperature gradients, *Laurila* (1968) recommended the use of a four-part model through the atmosphere up to 36 km. The second-order parabolic development (Eq. 6.29) of the basic exponential function is applied as one altitude range through the troposphere and as three separate altitude ranges through the stratosphere. From those four layers of 9 km each, the mean refractive indexes can be computed.

The strength of this approach was again tested by the use of data from the *U.S. Standard Atmosphere 1962* at the zenith distance of $\psi = 0°$. The results are given in Table 7.3, where $\bar{N}$ is the mean refractive index through one altitude range and $m_{\bar{N}}$ is its standard error, $\Delta S = 237.69$ cm is the total negative range correction from sea level to the altitude of 36 km, and $m_{\Delta S} = \pm 0.12$ cm is its standard error. The small standard error over the total altitude range suggests that the method of adding separate parabolic adjustments on top of each other is valid.

To investigate the feasibility of this method under actual meteorological conditions, the sounding data (Section 6.3.3) from approximately sea level to about 33 km were studied. The final results are listed in Table 7.4.

TABLE 7.3

Altitude Range (km)	$\bar{N}$	$m_{\bar{N}}$	ΔS (cm)	$m_{\Delta S}$ (cm)
0–9	184.5	±0.04	166.05	±0.04
9–18	61.0	0.11	54.90	0.10
18–27	15.0	0.05	13.50	0.05
27–36	3.6	0.02	3.24	0.02
0–36	—	—	237.69	0.12

Since the results of these sample computations agreed so well with each other and with the U.S. Standard (U.S. 62) value, the research was continued by *Schnelzer* (1972) to find an empirical "standard" value for ΔS applicable to the continental United States. The eight observation sites that provided radiosonde data were chosen for their proximity to the United States triangulation points of the proposed North American geodetic satellite observation program (Fig. 7.3). To obtain representative coverage, radiosonde data were collected from each site twice each quarter—January, April, July, and October, 1969. One set of data was taken at the beginning and the other during the middle of the month, totaling 64 complete soundings through all four altitude ranges.

Based on the standard errors in computing $\bar{N}$ at each site and through each altitude range, the weights were determined to obtain the weighted mean values N_w and their standard errors $m_{\bar{N}_w}$. The results, together with the corresponding range corrections and their standard errors, are given in Table 7.5.

The value $\Delta S = 238.5 \pm 0.5$ cm thus can be considered to be a year-round standard for the continental United States in measurements made at sea level and at the zenith distance $\psi = 0$.

TABLE 7.4

Altitude Range (km)	Location	Season	ΔS (cm)	$m_{\Delta S}$ (cm)
0–36	Lihue	February	238.1	±0.3
	Lihue	July	239.2	0.5
	Fairbanks	February	239.0	1.5
	Fairbanks	July	235.3	0.4

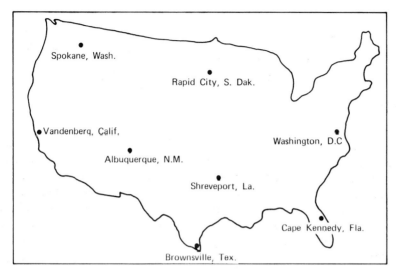

Figure 7.3 Radiosonde observation sites.

Similar computations were made also from the Air Force Cambridge Research Laboratory Model Atmospheres and the summarized results of all computations can be found in Table 7.6, with the difference from the empirical standard value. As Table 7.6 indicates, the empirical standard value represents the mean of all the given samples fairly well if the July Fairbanks value is disregarded. If we add the 1.2 cm that accumulated between the altitude ranges 36 and 72 km to the computed empirical standard value $\Delta S = 238.5$, the range correction through the entire atmosphere seems to be a constant within the continental United States at sea level:

$$\Delta S_0 = 239.7 \pm 0.5 \text{ cm} \qquad (7.29)$$

TABLE 7.5

Altitude Range (km)	$\bar{N}_w$	$m_{\bar{N}_w}$	ΔS (cm)	$m_{\Delta S}$ (cm)
0–9	181.98	±0.43	163.76	±0.39
9–18	63.73	0.28	57.36	0.25
18–27	15.58	0.05	14.02	0.05
27–36	3.37	0.01	3.38	0.01
0–36	—	—	238.52	0.46

TABLE 7.6

	U.S. 62	AFCRL July (30°N)	AFCRL January (30°N)	AFCRL July (45°N)	AFCRL January (45°N)	Lihue July	Lihue February	Fairbanks July	Fairbanks February
ΔS	237.7	238.2	240.0	237.8	238.9	239.2	238.1	235.3	239.0
$d\,\Delta S$	−0.8	−0.3	+1.5	−0.7	+0.4	+0.7	−0.4	−3.2	+0.5

Since tracking of the moon or artificial earth satellites very seldom is made at altitudes higher than 3000 m, an additional set of least squares adjustments can be made at certain intervals from sea level to 3 km to obtain the corresponding total corrections. Results can be tabulated and the needed ΔS values are obtained by interpolation from the table. In practice, measurements are made at various zenith angles; and Eq. 7.29 can be expressed in the following general form:

$$\Delta S = (\Delta S_h) \sec \psi \qquad (7.30)$$

where ΔS_h is the vertical range correction obtained from the table with reference to the station height h and ψ is the zenith angle. In Eq. 7.30, earth and its atmosphere are assumed to be parallel planes, and the slight bending of light rays due to reflection is neglected. These approximations yield an error of less than 1.0 cm up to the zenith angle of $\psi = 50°$.

In all radiosonde data, altitudes are given as geopotential heights, whereas in Eq. 7.29, geometric heights are used. Following is the conversion formula from geopotential height to geometric height:

$$H = \frac{MZ}{g_\psi M/9.8 - Z} \qquad (7.31)$$

where Z is the geopotential height, g_ψ is the actual acceleration of gravity at latitude ψ, and M is the meridian curvature radius of the ellipsoid at

TABLE 7.7

Station Height Above Sea Level (km)	B (mb)	Station Height Above Sea Level (km)	B (mb)
0	1.156	2.0	0.874
0.5	1.079	2.5	0.813
1.0	1.006	3.0	0.757
1.5	0.938	4.0	0.654

TABLE 7.8   VALUES OF δ_L(m)

Apparent Zenith Distance	Station Height Above Sea Level (km)						
	0	0.5	1.0	1.5	2.0	3.0	4.0
60°	+0.003	+0.003	+0.002	+0.002	+0.002	+0.002	+0.001
66°	+0.006	+0.006	+0.005	+0.005	+0.004	+0.003	+0.003
70°	+0.012	+0.011	+0.010	+0.009	+0.008	+0.006	+0.005

that latitude. The conversion is also tabulated in the *Smithsonian Meteorological Tables* (1966).

Based on an extensive study, *Saastamoinen* (1973) has derived a standard formula for the atmospheric range correction through the troposphere and stratosphere as a function of pressure and humidity observed at the station height:

$$\Delta S = 0.002357 \sec \psi (P + 0.06e - B \tan^2 \psi) + \delta_L \qquad (7.32)$$

where ψ is the apparent zenith distance of the satellite or the moon, P is the total barometric pressure in millibars, e is the partial pressure of water vapor in millibars, T is the absolute temperature in degrees Kelvin, B is a correction factor due to the curvature of earth and its atmosphere, and δ_L is a correction term due to the curvature of the ray path.

Tables 7.7 and 7.8 present standard values of the correction factor B and the correction term δ_L, respectively.

To compare the validity of the method based on the four-layer parabolic adjustment to the one given in Eq. 7.32, standard values of $P = 1013.25$ mb, $e = 10.86$ mb, and $h = 0$ km were placed into Eq. 7.32 and Tables 7.7 and 7.8 for zenith distances $\psi = 0°$, $\psi = 50°$, and $\psi = 60°$. When compared with values obtained from Eq. 7.30 and Table 7.3, the results were as follows:

Eq. 7.32	Eq. 7.30	Differences (cm)
$\Delta S_0 = 238.98$ cm	$\Delta S_0 = 238.89$ cm	+0.09
$\Delta S_{50} = 371.18$ cm	$\Delta S_{50} = 371.65$ cm	−0.47
$\Delta S_{60} = 476.63$ cm	$\Delta S_{60} = 477.78$ cm	−1.15

If the station height had been above sea level, the differences would have been even smaller. The close agreement at the zenith distance $\psi = 0°$

was anticipated, since Saastamoinen's formula (Eq. 7.32) is equivalent to the numerical integration of the refractive index from given data.

The Smithsonian Astrophysical Observatory (SAO) uses the following formula, based on the observed temperature and pressure at the tracking site altitude but neglecting the influence of humidity (*Lehr et al.*, 1969):

$$\Delta S = \frac{2.238 + 0.0414 PT^{-1} - 0.23 h_s}{\sin a + 10^{-3} \cot a} \tag{7.33}$$

where P is the total barometric pressure in millibars, T is the temperature in degrees Kelvin, h_s is the station height from sea level in kilometers, and a is the elevation angle of the sighting ($a = 90° - \psi$). If the standard data ($T = 288.15\,°K$, $P = 1013.25$ mb, and $h_s = 0$ km) are placed into Eq. 7.33, the SAO results for zenith angles $\psi = 0°$, $\psi = 50°$, and $\psi = 60°$ are:

$$\Delta S_0 = 238.4 \text{ cm}$$

$$\Delta S_{50} = 370.2 \text{ cm}$$

$$\Delta S_{60} = 475.2 \text{ cm}$$

In this section all coefficients and constants dependent on wavelength are based on the utilization of a ruby laser with $\lambda = 6943$ Å.

7.4.2.   Utilization of Radio Waves

In radio wave instrumentation, Secor (sequential collation of range), manufactured by Cubic Corporation, is used in satellite triangulations. It operates with two carrier frequencies, $f_1 = 499$ MHz and $f_2 = 225$ MHz, which are phase modulated at four transmitting stations on ground. The principle for determining the average refractive index is similar to the one for light wave propagation but the vulnerability of humidity in lower atmosphere renders the overall accuracy poorer and gives the range correction wider spread. On the other hand, the instrument accuracy is of the order of meters rather than centimeters, as in the case of laser applications.

The sample radiosonde data of the previous section were used to compute the range corrections and their standard errors and differences from the U.S. 62 standard value. For the curve-fitting adjustment Essen formula, Eq. 6.23 was utilized to compute the N-values. The results from sea level to an altitude of 72 km are given in Table 7.9 (*Laurila*, 1968).

TABLE 7.9

Location	Season	ΔS (cm)	$m_{\Delta S}$ (cm)	$d\,\Delta S$ (cm)
U.S. 62	—	244.2	±0.2	0
Lihue	February	241.1	1.9	−3.1
Lihue	July	248.3	1.9	+4.1
Fairbanks	February	232.5	0.9	−11.7
Fairbanks	July	234.6	0.4	−9.6

Saastamoinen (1973) derived the following formula for the range correction when radio waves are employed:

$$\Delta S = 0.002277 \sec \psi \left[P + \left(\frac{1225}{T} + 0.05 \right) e \right] - B \tan^2 \psi + \delta_R \quad (7.34)$$

where T is the temperature in degrees Kelvin and the other symbols have the same meaning as in Eq. 7.32.

If the standard values of P and e are placed in Eq. 7.34 at the zenith distance $\psi = 0°$, the corresponding range correction is $\Delta S = 241.6$ cm, which is 2.6 cm smaller than the corresponding standard value in Table 7.9.

EIGHT

MAXIMUM MEASURABLE DISTANCE

In working with medium or long wave transmission the maximum measurable distance is not limited by geometric or atmospheric considerations but by the transmitter power. The ray path in this case follows approximately the curvature of earth. In the use of short waves, microwaves, or light waves, the maximum distance depends on the curvature of the earth, which is always greater than that of the ray path, and on obstructions along the ray path.

When planning to measure new lines it is always advisable to check that there is enough clearance for the ray path to propagate between the terminal points of the line. In Fig. 8.1 the earth is assumed to be a "rectified" plane, and the ray path then contains the effects of both refraction and the curvature of the earth. The problem of determining the maximum range can be divided into two categories with respect to the characteristics of the terrain—that is, cases for flat terrain or water surface and cases for mountainous areas. In the former case, the maximum distance D_{max}, is a function of the *minimum elevation* G of the ray path from sea level. In the latter case it is a function of the elevation h_{Dh} of a terrain obstruction above the sea level and of its distance D_h from the initial terminal of the line.

To solve the problem stated in the first case, the following expression can be used (*Wilson*, 1950; *Laurila*, 1960):

$$D_{max} = 4.0(\sqrt{K - G} + \sqrt{H - G})$$ (8.1)

where D_{max} is the maximum measurable distance in kilometers, K and H

164

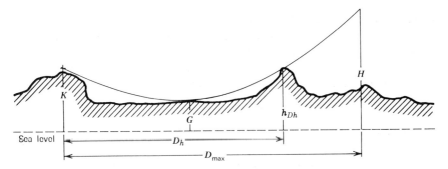

Figure 8.1   Ray path over earth, assumed to be a plane (Laurila, 1960).

are the elevations of the terminal points of a line in meters, and G is the minimum elevation in meters. The constant 4.0 is computed to represent National Advisory Committee for Aeronautics (NACA) moist atmosphere at lower troposphere. If measurements are made over water at sea level, G is zero. Otherwise, its value must be estimated or taken from an existing contour map.

To illustrate the computation of the measurable maximum distance over seawater where $G = 0$, let $K = 286$ m and $H = 3085$ m. Substituting these values into Eq. 8.1, we find the maximum distance to be $D_{max} = 289.6$ km.

In the case of mountainous terrain, the length of a line to be measured and the elevations of the terminal points are usually known accurately enough to permit a "reconnaissance" type of computation of the survey possibilities over that line. An atmospheric model at lower troposphere is known or may be determined to compute the average curvature of the ray path $\bar{Q}$, from Eq. 5.22. The usual problem is to learn whether an obstruction, such as a mountain top, is cutting off the free passage of the ray path between the terminal points. To determine the actual clearance between the ray path and the top of an obstruction, Eq. 5.17 can be used:

$$h = K + D_h \left[\frac{H - K}{D_A} - \frac{D_A}{2} \left(\frac{1}{R} - \bar{Q} \right) \right] + \frac{D_h^2}{2} \left(\frac{1}{R} - \bar{Q} \right)$$

where h is the sea level elevation of the ray path above the obstruction, K and H are the elevations of the terminal points of the line, D_A is the total length of the line, D_h is the distance from the initial terminal of the line to the obstruction, R is the curvature radius of earth, and $\bar{Q}$ is the average curvature of the ray path. All distances are in kilometers.

To illustrate this case, a reference line is chosen for the sample computation (Section 7.1.3), where $D_A = 65$ km, $K = 0.286$ km, $H = 3.085$ km, $R = 6371$ km, and $\bar{Q} = 0.0000247$. Let us assume that there is a mountain top along the line, having an elevation of $h_{D_h} = 1.640$ km and located at a distance of $D_h = 40$ km from the initial terminal point. The substitution of these values, except h_{D_n}, into Eq. 5.17 yields

$$h = 1.656 \text{ km}$$

which shows that there is a clearance between the ray path and the mountain top, $h - h_{D_h} = 16$ m.

A fast overall check to solve problems of the two classes mentioned, or any combination of possibilities, can be obtained by graphical means. The parabola, (Eq. 5.17) is constructed on a transparent sheet of paper, which is moved parallel to the rectangular axes of a graph, where positions and elevations of the terminal points and possible obstructions are plotted. If two of these elements are known, the third can be found when the parabola is made to pass through the two known points.

Chapter

NINE

INFLUENCE OF IONOSPHERIC REFLECTION ON THE LOW FREQUENCY PHASE–MEASURING TECHNIQUE

The propagation of electromagnetic radiation in low frequency bands, such as utilized by Decca with a wavelength of $\lambda = 3000$ m, is divided into two components. The component that propagates along the earth's surface or in the atmosphere close to it is called *ground wave* propagation. The other component propagates upward and is reflected back from a highly conductive layer of free electrons known as the *E-layer*. The altitude of this layer varies from 70 to 100 km, followed by the *F*-layer, which extends up to 400 km. The reflected waves from the *E*-layer are called the *sky waves*, and they interfere with the ground waves in the receiver at the point of location. This interference causes erroneous phase displacement in a phasemeter where phases of two ground wave signals are being observed. The erroneous phase displacement is directly proportional to the field strength of the sky waves and inversely proportional to the field strength of the ground waves when they reach the measuring receiver.

The field strength of the sky waves is dependent on three factors, namely, the distance of the measuring point from the transmitting stations, the height of the reflecting layer, and the reflection or absorption coefficient of the reflected waves. According to the studies of *Williams* (1951) and *Weekes* and *Stuart* (1952), the average height of the *E*-layer is 70 km in the daytime and 90 km at night. The reflection coefficient

167

varies greatly according to the distance from the ground transmitter to the receiver, thus is dependent on the incident angle that the wave path forms with the E-layer. The magnitude of the reflection coefficient increases when the distance increases; it is approximately 0.001–0.05 in the daytime and considerably more at night. To avoid the influence of sky wave interference, any precision survey should be made in the daytime and at distances of, say, less than 200 km. The erroneous phase displacement cannot be corrected, but its magnitude can be analyzed.

The field strength, stated in volts per unit length of a vertical antenna of the reflected waves as a function of the aforementioned factors, may be computed as follows (*Williams*, 1951):

$$E_r = \frac{2E_0}{D} \psi \cos^2 \left(\arctan \frac{2h}{D} \right) \tag{9.1}$$

where E_0 denotes the field strength of the radiation of the transmitter antenna, D is the distance between the transmitter and the receiver along the ellipsoid, ψ is the reflection coefficient, and h is the height of the reflecting layer.

The field strength of the ground waves, stated in volts per unit length of a vertical antenna, is obtained from the following standard formula:

$$E_g = 120\pi \frac{Ih_e}{\lambda D} k \tag{9.2}$$

where I is the antenna current, h_e is the effective height of the transmitting antenna, λ is the wavelength, D is the distance between the transmitter and the receiver along the ellipsoid, and k is an attenuation coefficient dependent on the electrical characteristics of the ground.

The phase difference between the direct and reflected waves (Fig. 9.1) obviously is

$$\alpha = 2\pi \frac{d}{\lambda} \tag{9.3}$$

where d denotes the difference in the distances traveled by the ground waves and the reflected waves and λ is the wavelength. Since the distance d, usually is longer than λ, α may yield any values between 0° and 360° but still has relatively small influence to the actual phase displacement. If we take E_r as the field strength component of the reflected waves at a distance D and E_g as the field strength component of the ground waves, it can be seen from Fig. 9.1 that the maximum phase displacement Θ_{max} between the ground and the resultant waves occurs when the resultant component E_g' is tangential to the circle whose radius is E_r.

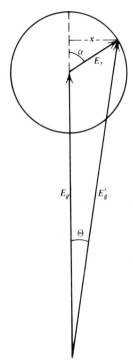

Figure 9.1 Phase relationship between ground waves and reflected waves (Laurila, 1960).

To obtain the phase displacement Θ as a function of any value of α, it may be stated, without making any significant error, that

$$\Theta = \frac{x}{E_g} \tag{9.4}$$

where x is obtained from Fig. 9.1 as follows:

$$x = E_r \sin \alpha \tag{9.5}$$

The phase displacement in Eq. 9.4 is given in radians, but to express it in terms of a fractional phase displacement equivalent to a hyperbolic coordinate, it must be divided by 2π. Thus when Eq. 9.5 is substituted into Eq. 9.4, we have

$$\Phi = \frac{E_r}{E_g} \frac{\sin \alpha}{2\pi} \tag{9.6}$$

The phase difference α is not known, except that it can have values between $0°$ and $360°$ with equal probability. Therefore, it may be replaced by the standard error of the periodic function of $\sin \alpha$, which is $m_{\sin \alpha} = \pm(1/\sqrt{2})$. With this substitution, Eq. 9.6 can be rewritten

$$m_{\Phi} = \pm \frac{E_r}{E_g} \frac{1}{2\pi \sqrt{2}} \tag{9.7}$$

or, approximately

$$m_{\Phi} = \pm \frac{E_r}{E_g} \frac{1}{9} \tag{9.8}$$

Equation 9.8 gives the standard error caused by the phase displacement of the *transmitted* ground waves when affected by reflected sky waves. In Decca surveying, the phasemeters or decometers indicate phase differences between so-called *comparison* waves, which are certain harmonics related to the transmitting frequencies by the factors $k_R = 3$, $k_{MR} = 4$, $k_G = 2$, and $k_{MG} = 3$ (see Section 20.2.2.1). Thus the standard errors caused by phase displacements of comparison waves at the receiver will be, for the red pattern

$$m_{\Phi_M} = \pm k_{MR} m_{\Phi}$$
$$m_{\Phi_R} = \pm k_R m_{\Phi} \tag{9.9}$$

and for the green pattern

$$m_{\Phi_M} = \pm k_{MG} m_{\Phi}$$
$$m_{\Phi_G} = \pm k_G m_{\Phi} \tag{9.10}$$

Since the hyperbolic lanes are determined by the phase differences of comparison waves, to obtain the final standard errors of phasemeter readings stated in lanes, values in Eq. 9.9 must be added together quadratically for the red phasemeter error and values in Eq. 9.10 must be added together in similar way for the green phasemeter error. The errors thus will be

$$m_{N_R} = \pm m_{\Phi} \sqrt{k_{MR}^2 + k_R^2} \tag{9.11}$$
$$m_{N_G} = \pm m_{\Phi} \sqrt{k_{MG}^2 + k_G^2}$$

or, if the numerical values of the coefficients k are substituted:

$$m_{N_R} = \pm 5.0 m_{\Phi}$$
$$m_{N_G} = \pm 3.6 m_{\Phi} \tag{9.12}$$

To give an idea of the magnitude of the errors caused by sky wave reflection, the following examples are presented. If 100 km is considered to be an average distance between the ground stations and the survey receiver, one might select $\psi = 0.005$ as the reflection coefficient to correspond to that distance (*Weekes* and *Stuart*, 1952). The substitution into Eq. 9.1 of this value together with $D = 100$ km, $h = 70$ km, and $E_0 = 300$ mV, corresponding to 1 kW output power yields the field strength of the reflected waves:

$$E_r = 10 \; \mu \text{V/m}$$

According to *Sanderson* (1952), the field strength of the ground waves at that distance will be $E_g = 3000 \; \mu \text{V/m}$. By substituting the values of E_r and E_g into Eq. 9.8, the standard error caused by the phase displacement of the transmitted waves is found to be

$$m_\Phi = \pm 0.00037$$

When applied to Eq. 9.12, this yields

$$m_{N_R} = \pm 0.002 \text{ lane} \quad \text{and} \quad m_{N_G} = \pm 0.001 \text{ lane} \quad (9.13)$$

It is to be emphasized that an exact evaluation of the sky wave effect is very difficult to make because its magnitude varies greatly as the elevation of the E-layer and the reflection coefficient vary. In particular, the sky wave effect increases greatly at night, making accurate survey operations almost impossible between sunset and sunrise. The rapid increase of the sky wave error as a function of increased distance has been illustrated by extensive field tests using an error elimination system (*Laurila*, 1953). The results, which showed an average value of $m_N = \pm 0.007$ lane at a distance of 220 km, were supported by the work of *Sanderson* (1952): $m_N = \pm 0.004$, and *Mendoza* (1947): $m_N = \pm 0.007$.

TEN

INFLUENCE OF SOIL CONDUCTIVITY ON THE LOW FREQUENCY PHASE–MEASURING TECHNIQUE

The effects of soil conductivity on the propagation velocity of electromagnetic radiation have been extensively studied by *Norton* (1937) over flat earth and by *Bremmer* (1949) over the spherical earth. Norton also investigated the effects of transmission frequency. In applications of the low frequency phase-measuring technique, as in the case of Decca, the influence of these phenomena must be considered when computing the propagation velocity over various soils of differing conductivities.

Based on the Norton-Bremmer theory, *Verstelle* (1957) developed a practical procedure for the computation. By taking $\delta_w = 5 \times 10^{-11}$ electromagnetic unit (emu) to represent the conductivity of average seawater and $\delta_1 = 10^{-14}$ emu to represent the conductivity of soil with dry sand, he computed ratios c_w/c_1 for intermediate values of δ. In these ratios c_w is the propagation velocity of low frequency transmission over seawater and c_1 is the velocity over land. When the δ value for a given type of ground is estimated, the ratio c_w/c_1 can be found from Fig. 10.1. Since the velocity over seawater, $c_w = 299,700$ km/sec, may be considered a constant, c_1 can easily be computed.

To achieve the best geometric configuration of the ground stations in hydrographic surveying, it is sometimes necessary to erect the transmitting stations inland, at some distance from the shoreline. In this case, the electromagnetic waves are transmitted over both land and sea. According

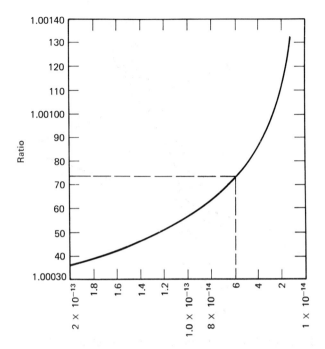

Figure 10.1   Ratio of propagation velocity over sea and land as function of soil conductivity, assuming c to be equal for master, red, and green stations. (Laurila, 1960).

to the theoretical work of *Verstelle* and practical test surveys made by *Pressey et al.* and *Laurila* (Section 7.2.2), the difference between the sea and land velocities may be as large as 466 km/sec. A good way to eliminate the effects of this discrepancy is to use the sea velocity in the computation of the lattice charts and to derive a correction factor as a function of land velocity and land distances. The correction factor permits one to draw correction lines on the chart and apply them to the observed Decca hyperbolic coordinates. *Larsson* (1954) derived such a method of corrections, which is presented next with respect to one master-slave pair of stations.

If the propagation velocity over land in Fig. 10.2 is c_l and c_w is the velocity over water, the comparison wavelengths will be

$$\lambda_l = \frac{c_l}{f} \quad \text{and} \quad \lambda_w = \frac{c_w}{f} \tag{10.1}$$

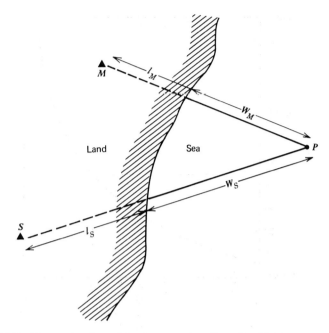

Figure 10.2   Determination of land correction (Laurila, 1960).

where f is the comparison frequency corresponding to the pattern concerned. Let us assume that the phase angles at the master and slave stations at a given time are Φ_M and Φ_S, respectively. The phase angle of the master transmission when it arrives at point P is thus

$$\Phi_{PM} = \Phi_M - \frac{2\pi l_M}{\lambda_l} - \frac{2\pi W_M}{\lambda_W} = \Phi_M - 2\pi f\left(\frac{l_M}{c_l} + \frac{W_M}{c_W}\right) \tag{10.2}$$

Similarly, the phase angle of the slave transmission at point P is

$$\Phi_{PS} = \Phi_S - \frac{2\pi l_S}{\lambda_l} - \frac{2\pi W_S}{\lambda_W} = \Phi_S - 2\pi f\left(\frac{l_S}{c_l} + \frac{W_S}{c_W}\right) \tag{10.3}$$

In these formulas, l_M and l_S, and W_M and W_S, are the distances the waves traverse over land and water, separately, from the master and slave stations, to point P.

The difference in phase angle at point P is thus

$$\Delta\Phi_P' = \Phi_M - \Phi_S - 2\pi f\left(\frac{l_M - l_S}{c_l} + \frac{W_M - W_S}{c_W}\right) \tag{10.4}$$

Since the lattice chart is computed by using the velocity over water, the corresponding difference in phase angle at point P when $c_l = c_W$ will be:

$$\Delta\Phi_P'' = \Phi_M - \Phi_S - 2\pi f\left(\frac{l_M - l_R}{c_W} + \frac{W_M - W_S}{c_W}\right) \tag{10.5}$$

The correction to be applied to the observed phase difference is obtained by the difference between values of Eqs. 10.4 and 10.5 as follows:

$$\Delta\Phi_P = 2\pi f\left(\frac{l_M - l_S}{c_W} - \frac{l_M - l_S}{c_l}\right) \tag{10.6}$$

If we accept the notations $c_l - c_W = \Delta c$ and $l_M - l_S = \Delta l$, Eq. 10.6 can be written as

$$\Delta\Phi_P = 2\pi f\frac{\Delta c\ \Delta l}{c_W(c_W + \Delta c)} \tag{10.7}$$

Since the term Δc is small compared to c_W, a good approximation of Eq. 10.7 is

$$\Delta\Phi_P = \frac{2\pi f\ \Delta l\ \Delta c}{c_W^2} \tag{10.8}$$

To transform Eq. 10.8 to be stated in units of lanewidth rather than phase difference, it is necessary to divide by 2π, and the so-called land correction finally yields

$$\Delta N = k\ \Delta l \tag{10.9}$$

where $k = f\ \Delta c/c_W^2$.

The following example illustrates the entire procedure of determining the land correction for a hypothetical pair of master and red slave stations.

The average soil conductivity is $\delta = 6 \times 10^{-14}$ emu. From Fig. 10.1, the ratio between the sea and land velocities is obtained as $c_W/c_l = 1.00073$. If the velocity over water is assigned the value $c_W = 299{,}700$ km/sec, the velocity over land will be $c_l = 299{,}448$ km/sec and the difference $\Delta c = 252$ km/sec. The comparison frequency in the Survey Decca System for the red pair of stations is $f_R = 354{,}065$ Hz. The master station is assumed to be 20 km inland in the direction of the point to be located, and the red slave station is 5 km in the corresponding direction. If the difference $\Delta l = 15$ km, together with the foregoing values of f, Δc, and c_W, is placed into Eq. 10.9, we find the land correction for the red hyperbola at the point to be located:

$$\Delta N_R = +0.015 \text{ lane}$$

ELEVEN

INFLUENCE OF GROUND REFLECTION ON THE MICROWAVE MEASURING TECHNIQUE

Under microwave conditions the ground reflection is similar to the ionospheric reflection described in Chapter 9 in that a part of the radiation reflects from ground and interferes with the direct beam traversing between two instruments. The erroneous phase displacement thus occurring cannot be corrected, but its magnitude can be estimated, and by certain types of repetition its influence can be lessened.

All ground-to-ground microwave instruments operate with the same common principle of comparing phases of modulations of the outgoing and incoming carrier waves. The carrier wavelength is usually of the order of 10 or 3 cm, and directive parabolic reflector antennas permit the production of relatively narrow beams of electromagnetic radiation; thus the most economic power consumption is achieved and the edges of the beam are prevented from reflecting from ground or from objects off the line. Even though the cone width of the radiation between 3 dB or half-power limits is about 10° or 6° depending on the wavelength used, reflections usually occur, distorting the readings of phase comparison.

This distortion is a function of the excess length between the direct and the reflected ray paths, that is, the length of the line and the line clearance from the ground, the carrier, and the modulation wavelengths, and the so-called *reflection coefficient*. A formula developed by *Fejer* and reported by *Wadley* (1957) has proved to be a good practical approximation for estimating the reflection error in terms of the above-mentioned factors

$$\Delta T = \frac{r\lambda_m}{2\pi} \cos \frac{2\pi \, \Delta d}{\lambda_c} \sin \frac{2\pi \, \Delta d}{\lambda_m} \qquad (11.1)$$

where ΔT is the reflection error in meters, r is the reflection coefficient (which can vary between 0.1 and 0.4 depending on the type of topography over which the electromagnetic radiation propagates), Δd is the difference in path lengths between the direct and reflected signals in meters, λ_c and λ_m are the carrier and modulation wavelengths, respectively, in meters.

The excess length Δd between the direct and the reflected paths can be computed by the standard formula

$$\Delta d = \frac{2h^2}{D} \qquad (11.2)$$

where h is the line clearance from the ground below and D is the length of the direct path of radiation, both in meters; Δd is also in meters. The reflection coefficient r reaches its minimum value of 0.1 over bushy, rugged terrain and its maximum value of 0.4 over calm or frozen water surface.

In work with all microwave modulation techinique instruments, the modulation wavelength λ_m can be termed a "yardstick" for the fine reading, or the reading giving the last fractional part of the length of the line. It is a constant for the particular instrument used. For a given line, Δd is also constant and therefore the entire sine term in Eq. 11.1 is constant and, determines along with the reflection coefficient r, the amplitude of the reflection error. To lessen the influence of the reflection error, ΔT is made to fluctuate sinusoidally several times, and usually an arithmetic mean of the readings forming the sine (cosine) curve is used as the most probable value of the fine reading. This fluctuation is made possible by including in each line measurement a set of observations, usually 12, and by changing the carrier frequency or cavity tune slightly from one observation to another. It is easy to see that the cosine term will cause one cycle of error if $2\pi \, \Delta d/\lambda_c$ is varied through 2π rad. To find the total number of cycles appearing in a set of observations made over a given line defined by Δd, the following expression can be written:

$$2\pi K = \frac{2\pi \, \Delta d}{\lambda_{c_H}} - \frac{2\pi \, \Delta d}{\lambda_{c_L}} \qquad (11.3)$$

where K is the number of cycles, λ_{c_H} refers to the highest carrier frequency within the cavity range, thus the shortest wavelength, and λ_{c_L}

refers to the lowest carrier frequency and thus longest wavelength. Equation 11.3 can now be rewritten

$$K = \frac{\lambda_{c_L} - \lambda_{c_H}}{\lambda_{c_H} \lambda_{c_L}} \Delta d \qquad (11.4)$$

To analyze the functioning of the sine term, note from Eq. 11.1 that the maximum amplitude of the cycling error occurs when $2\pi \Delta d / \lambda_m = \pi/2$, $3\pi/2, \ldots$ This is obtained when $\Delta d = \frac{1}{4}\lambda_m, \frac{3}{4}\lambda_m, \ldots$ The corresponding minimum or zero value of the sine term occurs when $2\pi \Delta d / \lambda_m = \pi$, $2\pi, \ldots$ This is obtained when $\Delta d = \frac{1}{2}\lambda_m, \lambda_m, \ldots$

The foregoing analysis makes it quite clear that before a survey is made of any preselected line, the expected ground reflection, which is sometimes called the swing error, can be predicted. Also, before selecting sites for new lines, the combinations of h and D, which introduce Δd, should be chosen to produce as many cycles as possible with a minimum amplitude.

To illustrate this procedure in practice, let us select two pairs of instruments representing two different modulation wavelengths and two different cavity ranges. The older model Tellurometers MRA-1 and MRA-2 utilize $\lambda_m = 30$ m, $\lambda_{c_L} = 0.105$ m, and $\lambda_{c_H} = 0.095$ m. The newer model MRA-101 Tellurometer and the Cubic Corporation Electrotape DM-20 utilize $\lambda_m = 20$ m, $\lambda_{c_L} = 0.0300$ m, and $\lambda_{c_H} = 0.0286$ m. Table 11.1 supplies examples of different excess line lengths, error amplitudes, and number of error cycles developed when r equals 0.4 (*Laurila*, 1965).

The self-explanatory Table 11.1 indicates that when Δd increases, the number of cycles increases but the amplitude fluctuates between zero and the maximum value. The maximum amplitude occurs when both the sine

TABLE 11.1

	MRA-1 and MRA-2			MRA-101 and Electrotape DM-20	
Δd (m)	Amplitude (m)	Number of Cycles	Δd (m)	Amplitude (m)	Number of Cycles
7.5	1.9	7.5	5.0	1.3	8.2
15.0	0.0	15.0	10.0	0.0	16.4
22.5	1.9	22.5	15.0	1.3	24.6
30.0	0.0	30.0	20.0	0.0	32.8
37.5	1.9	37.5	25.0	1.3	41.0
45.0	0.0	45.0	30.0	0.0	49.2

and cosine terms are 1; Eq. 11.1 then can be written

$$\Delta T_{max} = \frac{r\lambda_m}{2\pi} \qquad (11.5)$$

For a fast estimate of the ground swing error, especially useful for field reconnaissance purpose, *Gunn* (1964) constructed a graph (Fig. 11.1) based on Eqs. 11.1 and 11.2, where the maximum amplitude of the reflection error is readily obtainable as a function of the length of the line and the line clearance. The graph also gives a direct reading of the length difference Δd for the computation of the number of cycles formed. For simplicity's sake, the reflection coefficient is assigned a value of $r = 0.5$.

Example: line length $D = 8$ km
  line clearance $h = 208$ m

From the vertical axis, select the clearance and draw a horizontal line until it intersects the distance curve at point P_1. From P_1 draw a vertical line until it intersects the error curve at point P_2, and from P_2 draw a horizontal line back to the vertical axis. Using the second scale, read the maximum amplitude $\Delta T = 1.8$ m. To obtain the length difference, extend the vertical line from P_2 to the horizontal axis and read the length difference $\Delta d = 10.9$ m. This graph was designed for the older model Tellurometers, where the number of cycles is always

$$K = \Delta d$$

when λ_{c_L} and λ_{c_H} are substituted into Eq. 11.4. In our example, $K = 10.9$ cycles. If other carrier frequencies are used (Table 11.1), the number of cycles K differs from the length difference Δd.

Reflection occurs from solid objects such as walls, bridges, and moving vehicles, off the line, in addition to the ground. A test was carried on to reduce the side beams, preventing them from reaching the reflecting surfaces by placing specially designed shields on both sides of the parabolic antenna of the instrument: the aim was to eliminate the side reflection or, on the front to eliminate the ground reflection (*Laurila*, 1963, 1965). We used VHP-3 rubber absorbents, made by the B.F. Goodrich Rubber Company, for the experiment. This type of absorbent has the capability for about 1000-fold attenuation of the radiation even at very small incident angles. A simulated test was made in a parking lot at the Ohio State University campus, where a solid wall ran parallel to the test line. The length difference, $\Delta d = 7.5$ m, between the direct and reflected paths was selected to represent the worst possible situation.

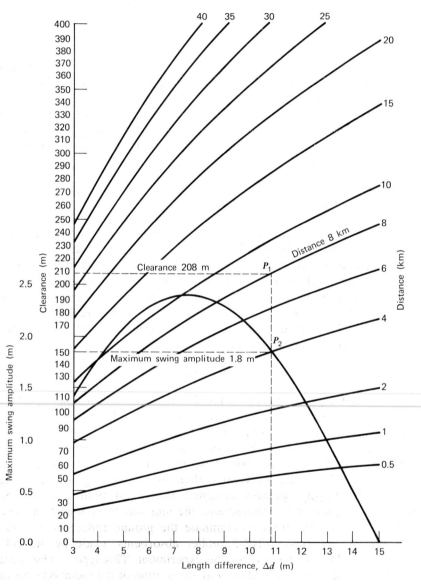

Figure 11.1 Error template for ground reflections (from *Gunn*, 1964, Graph VI).

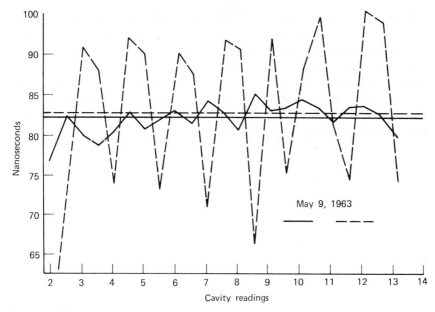

Figure 11.2 Ground reflection with (solid lines) and without (dashed lines) absorbents.

Also, the instruments were turned to make the antenna axes coincide with the bisector of the direct and reflecting lines; producing equal radiation power from antenna to antenna and from antenna to the reflection point.

A model MRA-1 Tellurometer was used for the experiment, and from the data presented previously, we expected 7.5 cycles to appear. The broken line in Fig. 11.2 represents the reflection pattern without the absorbents, and the solid line indicates the pattern with the absorbents. The figure shows that approximately 7 cycles appeared, having an average amplitude of 9 nsec, which in the case of model MRA-1 is equivalent to 1.35 m in distance. By assigning the value 1 for the sine and cosine terms in Eq. 11.1, the reflection coefficient r was computed as

$$r = \frac{1.35 \times 2\pi}{30} = 0.28 \qquad (11.6)$$

where 30 in the denominator is the modulation wavelength in meters. The value obtained for r represents fairly well the reflection coefficient of a concrete surface containing a few breaking objects (in this case, two doors in the wall, with small trees in front of it).

The effectiveness of the absorbent is evident, since the average amplitude of 9 nsec was reduced to about 2 nsec. Later the relatively heavy and bulky rubber absorbents were replaced by cardboard plates coated with aluminum foil and finally by ordinary window screens, to prevent the wind from moving the absorbent. The plates seemed to have capability for screening the unwanted radiation equal to that of the rubber absorbent, not through absorption but rather by bouncing the side beams in directions where there were no reflecting surfaces (*Gunn*, 1963).

Chapter

TWELVE

CYCLIC ZERO ERROR
IN MICROWAVE
DISTANCERS

In the late 1950s it was assumed that the zero error in the Tellurometer was a constant, about -3.0 cm, as estimated by Wadley, the inventor of the instrument. In the early 1960s it was reported that the Tellurometer zero error varies in a cyclic manner as a function of the A fine reading or as a function of the fractional part of the last half-wavelength of the A pattern (*Poder* and *Bedsted*, 1963). This cyclic variation was first discovered in the early model Tellurometers MRA-1 and MRA-2, but since these old workhorses are still in use, and the phenomenon is also reflected in the new model microwave instruments, the general trend of testing is presented here.

The first formula for total zero error was published by *Bedsted* (1962), and the error was assumed to propagate as

$$\Delta T = b_1 + b_2 \sin \frac{4\pi A}{100} \qquad (12.1)$$

where b_1 is the actual constant instrument error, b_2 is the amplitude of the cyclic variation, and A is the fine reading in which one total revolution equals 100 nsec. The second term in Eq. 12.1 is based on the discovery that if the instrument was moved over a total range of 15 m, which equals 100 nsec, and measurements were made at certain intervals, two complete error cycles appeared in the distance (time) readings. Since $4\pi = 720°$, Eq. 12.1 can also be written

$$\Delta T = b_1 + b_2 \sin (7.2° A) \qquad (12.2)$$

183

In this first formula no phase shifting of the cyclic function was anticipated, and the amplitude, by the use of MRA-1, was $b_2 = 6.1$ cm.

In Sweden a similar test was performed with MRA-2 and the cyclic variation was obvious with an amplitude of $b_2 = 6.5$ cm (*Brook*, 1962). In a test made in Norway, *Bakkelid* (1963) employed a formula similar to Eq. 12.1 but added a term dependent on the carrier frequency:

$$\Delta T = b_1 + b_2 \sin \frac{4\pi A}{100} + K \tag{12.3}$$

where K is a function of the initial and the mean cavity-set carrier frequency. The amplitude of the cyclic variation was approximately 8–9 cm, and the instrument was the MRA-1.

In tests made in Canada with the MRA-1, *Lilly* (1963) first brought the phase shift of the cyclic variation into the analysis, and the following formula was adapted:

$$\Delta T = b_1 + b_2 \sin (7.2°(A + b_3)) \tag{12.4}$$

where b_3 is the phase shift of the cyclic variation. Equation 12.4 can be written in the form of

$$\Delta T = x + y \sin (7.2°A) + z \cos (7.2°A) \tag{12.5}$$

where $y = b_2 \cos (7.2°b_3)$ and $z = b_2 \sin (7.2°b_3)$. After the solution of the normal equations, the coefficients can be computed as follows:

$$b_1 = x$$

$$b_3 = \frac{\tan^{-1}(z/y)}{7.2°} \tag{12.6}$$

$$b_2 = y \sec (7.2°b_3) = z \operatorname{cosec} (7.2°b_3)$$

The amplitude of the cyclic variation was approximately 3.5 cm and the phase shift was noticeable, varying between $b_3 = 0$ and $b_3 = -4.3$ nsec. Two different pairs of instruments were used for measurements over two separate baselines. Since both the constant errors b_1 and the amplitudes b_2 varied from line to line and from instrument to instrument, Lilly concluded that no prefabricated formula can be applied to measurements of lines of unknown lengths. This conclusion puts the cyclic zero error into the same category as the ionospheric and ground reflection errors discussed in previous chapters. They cannot be corrected except by certain repetitions, but their magnitude can be predicted.

Another significant series of tests was made at Ohio State University over a calibrated baseline with a length of 432.072 ± 0.005 m (*Hatch*,

1964). A procedure similar to that already described was performed, but only one pair of instruments, MRA-1, was used on the same baseline and the sets of measurements were made on seven different days: January 19, January 22, February 8, March 23, April 15, April 17, and April 25, 1964. This was done to determine the influence of the changing weather conditions to the total error. Given the shortness of the baseline, and because careful observations were made to compute the refractive index, the contribution of the refractive index to the total error was negligible.

In the adjustment of his observations, Hatch applied a form of Fourier series by adding terms of lower and higher order harmonics as follows:

$$\Delta T = b_1 + b_2 \sin\left(\frac{2\pi}{100}(A + b_5)\right) + b_3 \sin\left(\frac{4\pi}{100}(A + b_6)\right)$$
$$+ b_4 \sin\left(\frac{6\pi}{100}(A + b_7)\right) \quad (12.7)$$

The quadratic means of the standard errors for one observation before and after the adjustment by using all sets of measurements proved to be

$$m_B = \pm 7.4 \text{ cm} \quad \text{and} \quad m_A = \pm 4.3 \text{ cm} \quad (12.8)$$

respectively, strongly supporting the assumption of cyclic variation. A typical postadjustment curve appears in Fig. 12.1, where the standard errors of one measurement before and after adjustment were

$$m_B = \pm 10.2 \text{ cm} \quad \text{and} \quad m_A = \pm 4.8 \text{ cm}$$

Analyzing all the data of the series of tests, Hatch found a clear correlation between the total errors and the humidity expressed as the partial pressure of water vapor. This phenomenon was strongly backed by the work of *Laurila* (1963), who used the same baseline and instrument over one year without changing the position of the instrument between measurements. Especially strong correlation existed between the total error and the humidity, expressed as grams per cubic meter, when the reverse readings of the A pattern were used (Fig. 12.2). It must be noted, however, that different sets of instruments may react differently for changes in humidity. Also, the use of lower and higher order harmonics in Eq. 12.7 becomes relatively insignificant in practical error prediction, which converts Eq. 12.7 back to Eq. 12.4.

To achieve an ultimate accuracy with older model Tellurometers, the following conditions should be considered. At least one end of the line should be level, to allow the instrument to be moved 15 m for sets of observations, which are then reduced to the initial point and averaged. Two such sets should be completed within a day (or longer period) when

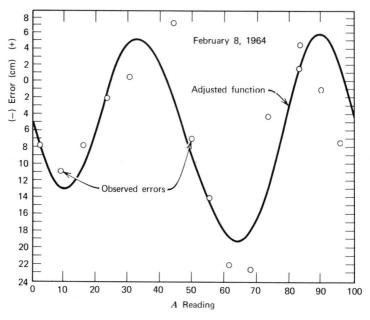

Figure 12.1 Cyclic zero error in MRA-1 fine readings.

the diurnal humidity reaches its minimum value at about 4:00 A.M. and its maximum value at about 12:00 noon (*Burger*, 1965). An example of such arrangements comes from measurements made by Hatch on February 8, 1964, and April 17, 1964. The absolute humidity values were 2.5 g/m^3 and 7.4 g/m^3, respectively, and the mean values of the measured sets were

$$D_{\text{Feb}} = 432.008 \text{ m}$$
$$D_{\text{Apr}} = 432.111 \text{ m}$$
$$\overline{D_{\text{mean}} = 432.060 \text{ m}}$$

The difference between this measured mean value and the true value of the line is $\Delta D = -1.2$ cm. This is a notable improvement, since the largest dispersion between two observations within both sets were -27.4 and $+16.5$ cm, and -12.0 and $+11.8$ cm, correspondingly.

The model MRA-2 Tellurometer differs little in design from the model MRA-1. However, there is no need to calibrate the crystals of the remote set in the field, and each instrument is interchangeable in such a way that it can be used as either the master or the remote unit in turn. Tests made by *Furry* (1965) over the baseline described in the previous paragraphs

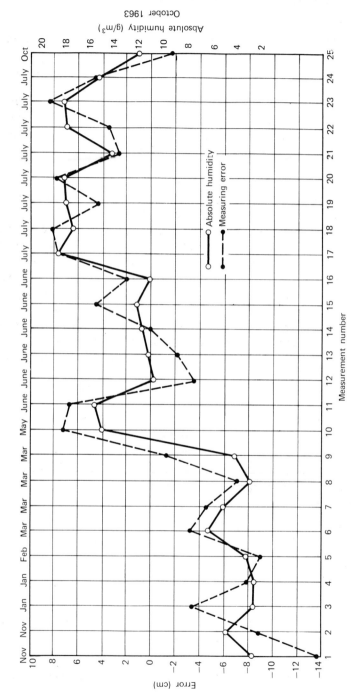

Figure 12.2 Absolute humidity compared to measuring error in MRA-1 reverse readings

187

showed that the magnitude of the errors and their cyclic characteristics were almost identical to those found when using model MRA-1.

The model MRA-3 Tellurometer is fully transistorized, uses 10 GHz carrier waves, and is also master/remote interchangeable. It was extensively tested over the same baseline at Ohio State University (*Bossler*, 1964; *Bossler* and *Laurila*, 1965). Because the modulation frequency was different from that of earlier models, the distance to cover the A pattern scale was 10 m. Two instruments, identified by numbers 349 and 350, were used for the test survey.

Unit 349, the master, introduced all-positive errors that followed very clearly a sinusoidal half-cycle pattern, repeating itself when a measurement was extended beyond the 10 m limit. Unit 350 introduced a similar but all-negative error curve when used as the master. The more complex patterns of sinusoidal curves with higher order harmonics, found by Hatch, apparently are not present in the model MRA-3 measurements. Figure 12.3 illustrates a typical error plot when both units are used as master and remote in turn at each position. To investigate a possible phase shift between measurements made on different days, a set of measurements made using one unit as master all the time was compared to another set of measurements made using the other set as master all the time. Although several days elapsed between the sets of measurements,

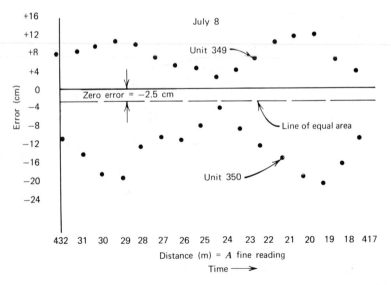

Figure 12.3   Typical error plot of MRA-3 readings. (Bossler and Laurila, 1965).

the error curves were rigidly phased together, with opposite signs. This suggests that the all-transistorized model MRA-3 is not vulnerable to external circumstances such as humidity. The standard error of one position measurement proved to be $m = \pm 1.5$ cm, and the standard error of the mean of the multiposition measurements was $M = \pm 0.5$ cm.

A similar but limited test was performed to find the cyclic zero error in the Electrotape model DM-20, manufactured by Cubic Corporation. This instrument is very similar to the MRA-3 Tellurometer by design, having the same modulation and carrier frequencies. Since the material available (only 22 measurements) is not sufficient to make any statistical analysis, caution must be exercised in drawing conclusions. The errors in measurements made when both instruments were used as the master in turn were highly correlated and no canceling effect could be seen. No cyclic variation seemed to appear in the plotted error curve. This finding was backed up by the observation that the error values at points 0 and 10 were quite different, although these are the same points in the A-pattern reading cycle. The standard error of one position measurement proved to be $m = \pm 0.7$ cm, and the standard error of the mean of the multiposition measurements was $M = \pm 0.3$ cm. Also, the length of the test line, in this case unknown, became the same $D = 435.107$ m when both instruments served as the master separately over the 10 m position shifting distance (*Laurila*, 1965).

THIRTEEN

ELLIPSOIDAL
PARAMETERS

For several of the reductions and applications in the following chapters, an ellipsoid of revolution is the mathematical reference surface. Readers lacking familiarity with the concepts of geometric geodesy may find the following summary of the most important ellipsoidal parameters helpful.

If a and b are the semimajor and semiminor axes of the ellipsoid, respectively, and f is the flattening, the following expressions can be written:

$$f = \frac{a-b}{a} \tag{13.1}$$

or

$$\frac{b^2}{a^2} = (1-f)^2 \tag{13.2}$$

The eccentricity e of the ellipsoid is usually given in squared form:

$$e^2 = \frac{a^2-b^2}{a^2} = (2-f)f \tag{13.3}$$

and

$$(1-e^2) = \frac{b^2}{a^2} = (1-f)^2 \tag{13.4}$$

In some cases the so-called second eccentricity e'^2 is used as

$$e'^2 = \frac{a^2-b^2}{b^2} \tag{13.5}$$

or

$$(1+e'^2) = \frac{a^2}{b^2} \tag{13.6}$$

190

which yields

$$(1 - e^2)(1 + e'^2) = 1 \tag{13.7}$$

By combining both eccentricity squares, we have

$$e'^2 = \frac{e^2}{1 - e^2} \tag{13.8}$$

In many geodetic applications, two expressions are constantly repeated. They are therefore called auxiliary terms having symbols W^2 and V^2 as follows:

$$W^2 = (1 - e^2 \sin^2 \phi) \tag{13.9}$$

$$V^2 = (1 + e'^2 \cos^2 \phi) \tag{13.10}$$

where ϕ is the latitude at the point at which the computations are to be made.

Since an ellipsoid of revolution is defined by two axes of different lengths, it is obvious that there cannot exist common radius for the entire ellipsoid. Radii are needed in all ellipsoidal computations; therefore the curvature radius of a very small element of arc going through a point can be derived in various forms depending on the task at hand. The direction of all curvature radii coincides with the normal to an infinitesimal plane that is tangential to the ellipsoid at the point concerned, but the lengths of the radii vary. The first basic radius of curvature is that one of the meridional arc at the point concerned:

$$M = \frac{a(1 - e^2)}{W^3} \tag{13.11}$$

The second basic radius of curvature, called the radius of curvature in the prime vertical, refers to the curve of intersection of the surface of the ellipsoid and a plane that is located perpendicular to the meridional plane and contains the normal to the ellipsoid at the point concerned:

$$N = \frac{a}{W} \tag{13.12}$$

The length of N always terminates at the polar axis of the ellipsoid, thus forms a closed triangle with the polar axis and the parallel radius.

Any short elemental line that is not a part of the meridian or a line perpendicular to the meridian has the radius of curvature at a given azimuth, which refers to the curve of intersection of the surface of the ellipsoid and a plane that is located at an angle A from the meridional plane and contains the normal to the ellipsoid at the point concerned:

$$R_A = \frac{NM}{M \sin^2 A + N \cos^2 A} \qquad (13.13)$$

If an average value for R_A at a given latitude is sufficient for the computations involved, the mean radius of curvature is obtained by taking an integrated mean of R_A when A goes through 0 to 2π:

$$R = \sqrt{MN} \qquad (13.14)$$

At the poles all radii of curvature are equal and are called the polar curvature radius:

$$c = \frac{a^2}{b} \qquad (13.15)$$

At the equator M, N, and R, have the following values:

$$M_0 = \frac{b^2}{a} \qquad (13.16)$$

$$N_0 = a \qquad (13.17)$$

$$R_0 = b \qquad (13.18)$$

The relation between W and V is

$$V = \frac{a}{b} W \qquad (13.19)$$

which, together with the polar curvature radius, yields

$$M = \frac{c}{V^3} \qquad (13.20)$$

$$N = \frac{c}{V} \qquad (13.21)$$

$$R = \frac{c}{V^2} \qquad (13.22)$$

In 1924 the International Geodetic Association accepted Hayford's 1910 ellipsoid as an "International Ellipsoid," whereas Clarke's 1866 ellipsoid is used in the United States as a standard reference ellipsoid. The numerical values of the parameters of both ellipsoids are as follows.

	International Ellipsoid	Clarke's 1866 Ellipsoid
a	6,378,388.0000 m	6,378,206.4000 m
b	6,356,911.9462 m	6,356,583.8000 m
f	0.00336700336	0.00339007530
$1/f$	297.0	294.9787
e	0.0819918898	0.0822718542
e^2	0.006722670	0.006768658
e'	0.0822688896	0.0825517104
e'^2	0.0067681702	0.0068147849
L	10,002,288 m	10,001,887 m
$\bar{R}$	6,371,228 m	6,370,997 m

$$(13.23)$$

where L is the length of one-quarter of the meridian and $\bar{R}$ is the radius of curvature for the earth, which is assumed to be a sphere having the same area as the reference ellipsoid. The $\bar{R}$-value in reference to the international ellipsoid is called the "radius of international sphere."

FOURTEEN

UNIVERSAL SPACE RECTANGULAR COORDINATE SYSTEM

In modern geometric geodesy and particularly in electronic surveying, long straight lines (chord distances) in space are used to locate points or are reduced onto the earth ellipsoid for further computations. For these purposes the *universal space rectangular* (USR) *coordinate system* has proved to be of great value.

In the universal space rectangular coordinate system there are three mutually perpendicular axes, located as follows. The origin of the system is at the center of the ellipsoid. The Z-axis coincides with the minor semiaxis or polar axis of the ellipsoid. The X- and Y-axes are situated in the equatorial plane, perpendicular to each other, the Y-axis going through the zero-meridian. In some applications of this method the X- and Y-axes are rotated $90°$, causing the X-axis to go through the zero-meridian and the Y-axis to have a negative sign.

The derivation of these coordinates is obtained from Fig. 14.1 in the following way. The projection N' of the curvature radius N in the prime vertical at point P, on the XY-plane, is

$$N' = N \cos \phi \qquad (14.1)$$

where ϕ is the latitude of point P. The components of N' along the X- and Y-axes are

$$X = N' \sin \lambda$$
$$Y = N' \cos \lambda \qquad (14.2)$$

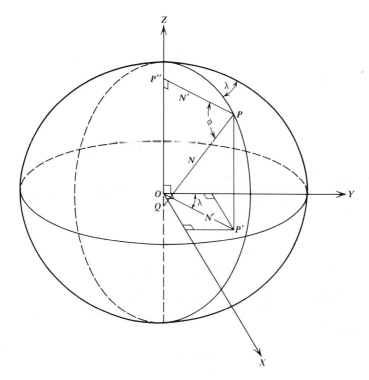

Figure 14.1   Derivation of USR coordinates, X, Y, and Z (Laurila, 1960).

where λ is the longitude of point P. Substituting Eq. 14.1 into Eq. 14.2 yields

$$X = N \cos \phi \sin \lambda$$
$$Y = N \cos \phi \cos \lambda \tag{14.3}$$

The derivation of the Z-coordinate is somewhat more complicated. If $P'Q$ in Fig. 14.1 is designated Z', $P'O$ is designated Z, and if the two parts to be separated from the curvature radius N by the XY-plane are designated a' and b', it is known from analytic geometry that

$$\frac{a'}{a'+b'} = \frac{b^2}{a^2} \tag{14.4}$$

and, from Fig. 14.1

$$\frac{Z}{Z'} = \frac{a'}{a'+b'}$$

Hence

$$Z = Z' \frac{b^2}{a^2} = Z'(1 - e^2) \tag{14.5}$$

From Fig. 14.1, it is seen that

$$Z' = N \sin \phi$$

which yields

$$Z = N(1 - e^2) \sin \phi \tag{14.6}$$

when substituted into Eq. 14.5.

If the point to be computed is situated at an elevation H above sea level, Eqs. 14.3 and 14.6 can be written as

$$
\begin{aligned}
X &= (N + H) \cos \phi \sin \lambda \\
Y &= (N + H) \cos \phi \cos \lambda \\
Z &= (N(1 - e^2) + H) \sin \phi
\end{aligned}
\tag{14.7}
$$

The transformation formula from the universal space rectangular coordinates X and Y into the geodetic longitude λ is obtained from Eq. 14.3 or 14.7 as

$$\frac{X}{Y} = \tan \lambda \tag{14.8}$$

hence

$$\lambda = \tan^{-1} \left(\frac{X}{Y} \right) \tag{14.9}$$

Figure 14.1 indicates that

$$\tan \phi = \frac{Z'}{N'} = \frac{Z'}{\sqrt{X^2 + Y^2}} \tag{14.10}$$

and substituting the value of Z' from Eq. 14.5 gives us

$$\tan \phi = \frac{a^2}{b^2} \frac{Z}{\sqrt{X^2 + Y^2}} \tag{14.11}$$

hence

$$\phi = \tan^{-1} \left(\frac{a^2}{b^2} \frac{Z}{\sqrt{X^2 + Y^2}} \right) \tag{14.12}$$

It was seen that the determination of the geodetic longitude λ from the space rectangular X- and Y-coordinates was not dependent on the elevation H. The geodetic latitude ϕ was obtained from space rectangular coordinates X, Y, and Z (Eq. 14.12) at sea level. If the point to be found is situated at an elevation H above sea level, a good approximation can be obtained by adding H to both the semimajor and semiminor axes of the ellipsoid, and Eq. 14.12 then can be written

$$\phi = \tan^{-1}\left[\left(\frac{a+H}{b+H}\right)^2 \frac{Z}{\sqrt{X^2+Y^2}}\right] \tag{14.13}$$

The error introduced into Eq. 14.13 occurs because the elevation of the point is no longer exactly H but a value close to it. If Eq. 14.13 is used for high precision survey work, the following correction can be applied (*Hirvonen*, 1961):

$$d\phi = +0.''0007H \sin\phi \cos\phi \tag{14.14}$$

where H is given in kilometers. For example, if $H = 10\,\text{km}$ and $\phi = 45°$, the correction will be $d\phi = +0.''0035$, which is approximately equivalent to 10 cm along the meridian on the ellipsoid.

An approach based on iteration can be derived by dividing Z in Eq. 14.7 by $N' = (N+H)\cos\phi$, where H is an extension of N over point P:

$$\frac{Z}{N'} = \frac{[(1-e^2)N+H]}{N+H}\tan\phi \tag{14.15}$$

Substituting $N' = \sqrt{X^2+Y^2}$ and solving for ϕ yields

$$\phi = \tan^{-1}\left[\frac{N+H}{[(1-e^2)N+H]}\frac{Z}{\sqrt{X^2+Y^2}}\right] \tag{14.16}$$

Since N in the right-hand side of Eq. 14.16 is a function of ϕ, the first approximation for ϕ can be obtained, for example, from Eq. 14.13; then the iteration should be continued until the required accuracy level of ϕ is achieved.

If the chord distance between two points either on or above the ellipsoid is required, the space rectangular coordinates X, Y, and Z are computed for both points and the following formula is used:

$$D_C = \sqrt{\Delta X^2 + \Delta Y^2 + \Delta Z^2} \tag{14.17}$$

where ΔX, ΔY, and ΔZ are the coordinate differences between the two points.

FIFTEEN

REDUCTION OF SPATIAL CHORD DISTANCE INTO ELLIPSOIDAL CHORD DISTANCE

In this chapter we analyze geometric reductions only; thus the input data consist of the spatial chord distance D_C, the elevation of the initial point of the line K, the elevation of the terminal point of the line H, and the curvature radius of earth R. The output data are the ellipsoidal chord distance d_C. In practical surveying, the combined ray path arc to spatial chord and ellipsoidal chord to ellipsoidal arc correction ΔD_2 (Eq. 5.4) is added to the computations, and the spatial chord distance D_C, which appears in the following equations, is replaced by the ray path arc distance D_A and the output data are the ellipsoidal arc distance d_A.

Two approaches can be derived to obtain the geometric reduction of spatial chord distance into ellipsoidal chord distance. One approach is based on step-by-step partial reductions using series expansions; the other is based on the corrected value of d_C obtained by a rigorous, closed formula. The advantage of the first approach is that especially when short lines are to be measured, good approximations of the partial reductions can be obtained immediately in the field by use of a slide rule. The second approach provides accurate corrected values for lines of any length within the limits of capability of the instrument chosen for the measurements.

In the first case, two major reductions—namely, the reductions to the horizontal and the ellipsoidal levels—must be derived, together with additional correction terms needed in high precision surveys. In the derivation of the horizontal reduction, the vertical side P_2L in the triangle P_1P_2L is assigned the value of the true elevation difference ΔH (Fig. 15.1). With

198

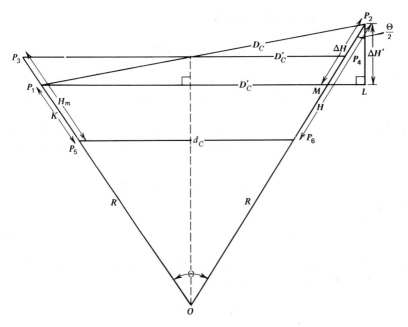

Figure 15.1 Determination of the partial reductions from spatial chord distance into ellipsoidal chord distance.

this substitution, it is seen from Fig. 15.1 that

$$D_C - D'_C = D_C - (D_C^2 - \Delta H^2)^{1/2}$$

$$= D_C - D_C\left[1 - \left(\frac{\Delta H}{D_C}\right)^2\right]^{1/2} \tag{15.1}$$

When the bracketed term has been developed into a series expansion, Eq. 15.1 yields

$$D_C - D'_C = D_C - D_C\left(1 - \frac{\Delta H^2}{2D_C^2} - \frac{\Delta H^4}{8D_C^4} - \cdots\right) \tag{15.2}$$

and the horizontal reduction C_L will be

$$C_L = D'_C - D_C = -\left(\frac{\Delta H^2}{2D_C} + \frac{\Delta H^4}{8D_C^3}\right) \tag{15.3}$$

With the assumption that $\Delta H = \Delta H'$ (Fig. 15.1), a small error was introduced in ΔH in the derivation of Eq. 15.3. To correct that error, the first term in Eq. 15.3 is differentiated with respect to ΔH (the error in the second term is insignificant) as follows:

$$dC_L = -\frac{\Delta H}{D_C} d \Delta H \tag{15.4}$$

where

$$d \Delta H = \Delta H' - \Delta H = \Delta H \cos\frac{\Theta}{2} - \Delta H \tag{15.5}$$

Without making any significant error, we can write $\Theta/2 = D_C/2R$ which, when substituted into Eq. 15.5 and developed into a series expansion, yields

$$d \Delta H = \Delta H\left(1 - \frac{D_C^2}{8R^2} + \cdots\right) - \Delta H$$

hence

$$d \Delta H = -\frac{D_C^2 \Delta H}{8R^2} \tag{15.6}$$

where R is the curvature radius of earth. The substitution of Eq. 15.6 into Eq. 15.4 gives the additional correction term

$$dC_L = +\frac{D_C \Delta H^2}{8R^2} \tag{15.7}$$

For the derivation of the reduction to ellipsoidal level, the line $P_3P_4 = D'_C$ parallel to the line P_5P_6 on the ellipsoidal level is used (Fig. 15.1). The elevation of this line H_m is the mean elevation of the terminal points of the line P_1P_2:

$$H_m = \frac{K + H}{2} \tag{15.8}$$

From the triangles OP_3P_4 and OP_5P_6, the following value for d_C is obtained:

$$d_C = D'_C \frac{R}{R + H_m} \tag{15.9}$$

or

$$d_C = D'_C \frac{1}{1 + (H_m/R)} \tag{15.10}$$

The development of Eq. 15.10 into a series expansion gives us

$$d_C = D'_C\left(1 - \frac{H_m}{R} + \frac{H_m^2}{R^2} - \cdots\right) \tag{15.11}$$

or

$$d_C = D'_C - \frac{D'_C H_m}{R} + \frac{D'_C H_m^2}{R^2} \tag{15.12}$$

and finally, the reduction to ellipsoidal level:

$$C'_E = d_C - D'_C = -\left(\frac{D'_C H_m}{R} - \frac{D'_C H_m^2}{R^2}\right) \tag{15.13}$$

In the derivation of Eq. 15.13, the line $P_3P_4 = D'_C$ was used as the reference line instead of the original input data, the line $P_1P_2 = D_C$. In the second term of Eq. 15.13, D'_C may be replaced by D_C without any significant error. If D'_C is replaced by D_C in the first term of Eq. 15.13, an additional correction must be obtained by differentiating that term with respect to D'_C

$$dC'_E = -\frac{H_m}{R}\,dD'_C \tag{15.14}$$

which leads to Eq. 15.15 when $dD'_C = D'_C - D_C = -(\Delta H^2/2D_C)$ from Eq. 15.3 has been substituted.

$$dC'_E = +\frac{H_m\,\Delta H^2}{2D_C R} \tag{15.15}$$

With this correction added to the total reduction, Eq. 15.13 can be rewritten

$$C_E = -\left(\frac{D_C H_m}{R} - \frac{D_C H_m^2}{R^2}\right) \tag{15.16}$$

and the total reduction will be the sum of Eqs. 15.3, 15.7, 15.15, and 15.16, as follows:

$$C_T = C_L + C_E + dC_L + dC'_E \tag{15.17}$$

The omission of the correction terms $(dC_L + dC'_E)$ in the computations of lines with lengths up to 500 km and with elevation differences between the terminals of 10 km causes an error of less than 10^{-6}. The figures given represent the approximate maximum distance that can be measured by modern ground-to-air survey systems. At distances less than 50 km both correction terms dC_L and dC'_E can be neglected without causing an error of more than 10^{-8}.

The derivation of the closed formula to compute the ellipsoidal chord distance as a function of the spatial chord distance is basically the same as presented by Ewing (1955) (see Laurila, 1960) but in the reverse direction. In Fig. 15.2 the line $P_1P_2 = D_C$ is the spatial chord distance, the line $P_5P_6 = d_C$ is the ellipsoidal chord distance, H and K are the elevations of the terminals of the line P_1P_2, and R is the curvature radius of earth. The line $P_1P_4 = s_1$ is drawn parallel to the line P_5P_6 at point P_1, and the line $P_3P_2 = s_2$ is drawn parallel to the line P_5P_6 at point P_3.

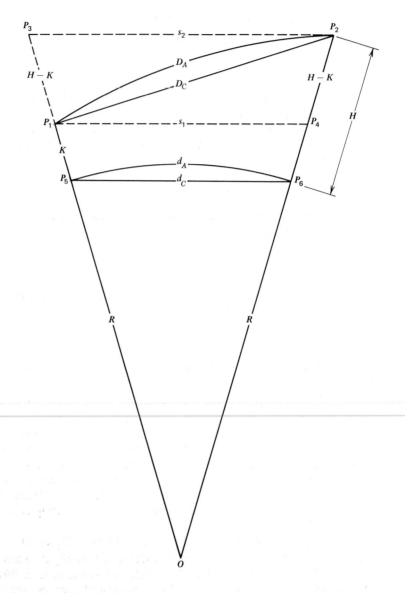

Figure 15.2 Determination of ellipsoidal chord distance as function of spatial chord distance (Laurila, 1960).

According to standard geometry, the square of the diagonal D_C of a trapezoid $P_1P_3P_2P_4$ is expressed as follows:

$$D_C^2 = s_1 s_2 + (H-K)^2 \tag{15.18}$$

From the triangles OP_5P_6, OP_1P_4, and OP_3P_2, the values of s_1 and s_2 can be found:

$$s_1 = \frac{R+K}{R} d_C \quad \text{and} \quad s_2 = \frac{R+H}{R} d_C \tag{15.19}$$

When substituted into Eq. 15.18, these values yield

$$D_C^2 = \frac{(R+K)(R+H)}{R^2} d_C^2 + (H-K)^2 \tag{15.20}$$

The ellipsoidal chord distance is obtained by solving Eq. 15.20 for d_C:

$$d_C = \left[\frac{(D_C^2 - (H-K)^2)R^2}{(R+K)(R+H)} \right]^{1/2} \tag{15.21}$$

To compare the two methods and to illustrate the magnitude of terms involved, consider the following example.

$$D_C = 400 \text{ km} \qquad H_m = 6 \text{ km}$$
$$K = 1 \text{ km} \qquad R = 6371 \text{ km}$$
$$H = 11 \text{ km} \qquad R+K = 6372 \text{ km}$$
$$\Delta H = 10 \text{ km} \qquad R+H = 6382 \text{ km}$$

Substituting the necessary data into Eq. 15.21 yields the value of d_C:

$$d_C = 399,498.836 \text{ m} \tag{15.22}$$

Substituting these data into Eq. 15.17 gives us the following total reduction of C_T:

$$C_T = -(125.000 + 0.020) - (376.707 - 0.355) + 0.123$$
$$+ 0.118 = -501.131 \text{ m} \tag{15.23}$$

When adding C_T to D_C, the corrected ellipsoidal chord distance will be

$$d_C = 399,498.869 \text{ m} \tag{15.24}$$

and the difference between these two approaches thus appears to be -0.033 m, or -8×10^{-8}. In shorter distances, the relative error will become still smaller.

A third method is also possible. Derived by *Wong*, (1949) and *Laurila* (1960), it gives the final ellipsoidal arc distance d_A as a function of the spatial chord distance D_C as follows:

$$d_A = \left[\frac{12R^2M}{12(R+H)(R+K)-M}\right]^{1/2} \tag{15.25}$$

where $M = D_C^2 - (H-K)^2$.

Substituting the necessary data from our example into Eq. 15.25 yields the following value of d_A:

$$d_A = 399,564.275 \text{ m} \tag{15.26}$$

To compare this value with the one obtained from Eq. 15.21, d_C in Eq. 15.22 must be given the ellipsoidal chord to ellipsoidal arc correction (Eq. 4.65), and the ellipsoidal arc d_A then becomes

$$d_A = 399,498.836 + 65.451 = 399,564.287 \text{ m} \tag{15.27}$$

the difference between the last two approaches being only $+0.012$ m, or $+3 \times 10^{-8}$.

To apply the *Ewing* method, Eqs. 5.6 and 15.21 must be worked to obtain the ellipsoidal arc distance as a function of the ray path arc distance. Similarly, for the *Wong* method, Eqs. 5.7 and 15.25 should be applied.

It must be noted that in all high precision surveys the curvature radius of earth must be computed with Eq. 13.13, in which the azimuth of the line is also taken into consideration.

Chapter

SIXTEEN

COMPUTATION OF
COORDINATES FROM
GROUND-TO-AIR DATA

Section 4.2 presents general outlines of the procedures to reduce ground-to-air data onto the ellipsoid. Also, a graphical method of establishing a circular grid for ground-to-ground and ground-to-sea surveying has been described. Reduction techniques in photogrammetric mappings are complicated because of the highly sloped lines from ground stations to the aircraft. Aerial mapping often involves high speed dynamic position-fixing techniques, and the next sections are devoted to a graphical method based on initial distance readings, thus providing instantaneous fixes, together with a numerical method.

16.1. GRAPHICAL METHOD

An aircraft carrying out an aerial mapping mission maintains its altitude with high precision. To provide a fast position fix, a circular grid may be constructed as a function of the noncorrected distance readings from the ground stations to that preselected mission altitude. Grid circles drawn on a projection plane such as the UTM thus represent integer distance readings having the desired intervals. The intervals of such circles on the projection plane, however, are not constant but vary according to the distance between the ground station and the aircraft.

In deriving the circle intervals on a plane, we assume that spatial chord distances serve as a basis for the grid. Since the grid construction is based

205

on a known reference point and on reference circle principles, distances within one map sheet are small enough to justify the omission of the curvature correction (Eq. 5.7), and in practice, spatial chord distances may be replaced by spatial arc distances.

To construct the grid, we must know plane coordinates X and Y and geographic coordinates ϕ and λ of the two ground stations. Within the survey area, a reference point P_0 with known X, Y, ϕ, and λ is selected for the construction of the reference circle (Fig. 16.1). Taking the space rectangular coordinates (Eq. 14.7), the spatial chord distance D_{C_0} from ground station A to point P_0' is computed (Eq. 14.17); P_0' is the point located above P_0 at the mission altitude H. The distance D_{C_0} is not an integer number, and if D_{C_1} is the closest integer number, the distance $P_0 P_1 = d_{A_0}$ on the projection plane can be found from the following expression (*Laurila*, 1954, 1960):

$$dA_0 = (D_{C_1} - D_{C_0})\left[1 + \frac{[\frac{1}{4}(D_{C_1} + D_{C_0})^2 + (R + H)^2 - (R + K)^2]^2}{8(R + H)^2 D_{C_1} D_{C_0}}\right] C_E C_P$$

$$(16.1)$$

where D_{C_1} and D_{C_0} are the spatial chord distances as indicated in Fig.

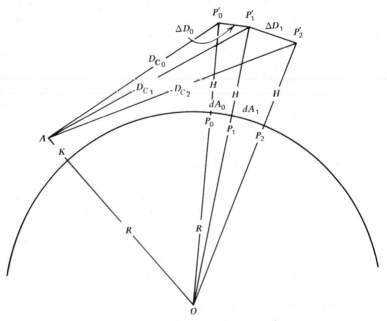

Figure 16.1   Determination of circle intervals in circular grid.

16.1, K is the elevation of the ground station, H is the flying height, and R is the curvature radius of the earth and can be taken as that of the radius of international sphere, $R = 6371$ km; C_E is the correction to reduce the spatial distances ΔD (Fig. 16.1), onto the ellipsoidal level

$$C_E = \frac{R}{R+H} \qquad (16.2)$$

and C_P is the correction to convert the ellipsoidal distances of the circle intervals onto the projection plane. Since the distances are small, the following simplified formula may be applied to the UTM projection:

$$C_P = 0.9996 + \frac{X_m^2}{2R^2} \qquad (16.3)$$

where X_m is the distance from the midpoint of the line (circle interval along the radius AP_0) to the nearest central meridian. The intervals of each consecutive circle are now computed with Eq. 16.1 by placing into it the desired chord distance D_{C_2}, D_{C_3}, and so on; in turn, and noting that C_E, C_P, R, H, and K remain constant.

The circles are constructed as described in Section 4.2, starting at point P_1 and noting that the radius will be $D_{G_1} = (\Delta X^2 + \Delta Y^2)^{1/2} + d_{A_0}$, where ΔX and ΔY are the coordinate differences between points A and P_0. The radii of the consecutive circles are then correspondingly:

$$D_{G_2} = D_{G_1} + d_{A_1}$$
$$D_{G_3} = D_{G_2} + d_{A_2}$$
$$\begin{array}{cc} \cdot & \cdot \\ \cdot & \cdot \\ \cdot & \cdot \end{array}$$

To illustrate the magnitude and variation of the circle intervals, a hypothetical case is presented where

$$H = 5 \text{ km}$$
$$K = 0.5 \text{ km}$$
$$R = 6371 \text{ km}$$
$$D_C = 30 - 150 \text{ km at 10-km intervals}$$

The survey area is assumed to locate at an average distance of 100 km from the central meridian. The circle intervals and their variations are given in Table 16.1 which indicates that the increments of the circle

TABLE 16.1

Distance, D (km)	Circle Interval, d_A (km)	Increment, Δd_A (m)
30		
	10.0772	
40		33.7
	10.0435	
50		16.7
	10.0268	
60		9.6
	10.0172	
70		6.0
	10.0112	
80		3.9
	10.0073	
90		2.8
	10.0045	
100		2.0
	10.0025	
110		1.5
	10.0010	
120		1.1
	9.9999	
130		0.8
	9.9991	
140		0.7
	9.9984	
150		

intervals are very small when the flying height is small compared to the total distance but increase rapidly when the distances decrease. Taking into account practical drafting accuracy, at long distances the circle intervals should be checked only, say, at every 30 to 50 km and the intermediate circles within those limits be drawn with equal intervals.

16.2. NUMERICAL METHOD

The method presented in this section is a slightly modified version of the original procedure developed and used by the *U.S. Geological Survey* (1950). It is slower than the graphical method, of course, but it meets the accuracy demands of any precision survey. Since the measurements are made from ground to air, several reductions must first be employed. These have been presented in previous chapters. Referring to Fig. 4.10, a summary of these preliminary computations is given as follows.

1. The measured arc distances D_A and D_B from the ground antennas A' and B' to the airborne antenna P' must be corrected to correspond to the true propagation velocity of electromagnetic energy. The procedure used for this reduction depends on the length of the line and the instrumentation chosen (see Section 7.3).
2. The corrected arc distances must be reduced to the chord distances D'_A and D'_B using Eq. 5.7.
3. The chord distances must be reduced to the ellipsoidal arc distances d_A and d_B by using Eq. 15.25. The total reduction from the corrected spatial arc distances to the ellipsoidal arc distances also can be made by combining Eqs. 5.6 and 15.21.

After these computations have been made, the sides d_A, d_B, and b of the spherical triangle ABP on the ellipsoid are known. Since the coordinates of the ground stations A and B are known on the UTM projection, the length of the base m_b on that projection is obtained from Fig. 16.2 as follows:

$$m_b = (\Delta X^2 + \Delta Y^2)^{1/2} \tag{16.4}$$

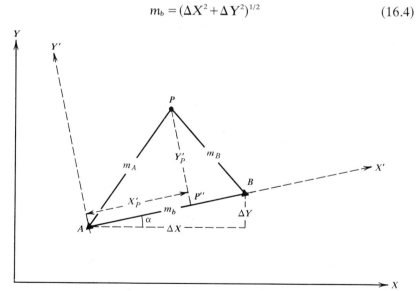

Figure 16.2 Computation of coordinates on plane (Laurila, 1960).

The scale factors k_A and k_B to reduce the distances d_A and d_B onto the UTM projection can be found by the following method. On any existing small-scale map, or on graph paper, the approximate location of point P is plotted by the intersection of two circles whose radii are the ellipsoidal distances d_A and d_B and whose centers are situated at the points A and B, respectively. The approximate coordinates $\bar{X}_P$, $\bar{Y}_P$, together with the true coordinates of the ground stations X_A, Y_A, and X_B, Y_B, permit us to obtain the scale factors for the ellipsoidal distances from the following formulas:

$$k_A = 1 + \frac{X_A^2 + X_A \bar{X}_P + \bar{X}_P^2}{6R^2}$$

$$k_B = 1 + \frac{X_B^2 + X_B \bar{X}_P + \bar{X}_P^2}{6R^2}$$

(16.5)

where R is the radius of the international sphere. The map distances between points AP and BP are now

$$m_A = k_A d_A$$

$$m_B = k_B d_B$$

(16.6)

From this stage on, the following derivation for control point determination applies to any circular method on the ground or at sea.

Before the final plane coordinates of point P in the XY-coordinate system can be computed, an auxiliary coordinate system $X'Y'$ is established, having its origin situated at ground station A and the X'-axis going through ground station B. The coordinates of point P in this new system are obtained directly from Fig. 16.2 as follows:

$$Y_P'^2 = m_A^2 - X_P'^2$$

$$Y_P'^2 = m_B^2 - (m_b - X_P')^2$$

(16.7)

When solved for X_P' Eqs. 16.7 yield

$$X_P' = \frac{m_A^2 - m_B^2 + m_b^2}{2m_b}$$

(16.8)

After X_P' has been computed from Eq. 16.8, Y_P' is obtained from Eq. 16.7 as follows:

$$Y_P' = \sqrt{m_A^2 - X_P'^2}$$

(16.9)

The last step in the computation procedure is to transform the coordinates of point P computed in the $X'Y'$-system into the original XY-system. This can be done by the well-known transformation formulas

$$X_P = X_A + X'_P \cos \alpha - Y'_P \sin \alpha$$
$$Y_P = Y_A + X'_P \sin \alpha + Y'_P \cos \alpha$$

(16.10)

Trigonometrical tables are not needed in the computation of the final coordinates, however, since the values of $\sin \alpha$ and $\cos \alpha$ are directly obtainable as functions of the already known and used elements m_b, ΔX, and ΔY, as follows:

$$\sin \alpha = \frac{\Delta Y}{m_b} \quad \text{and} \quad \cos \alpha = \frac{\Delta X}{m_b}$$

(16.11)

Substituting Eqs. 16.11 into Eqs. 16.10 yields the following simple transformation formulas:

$$X_P = X_A + X'_P \frac{\Delta X}{m_b} - Y'_P \frac{\Delta Y}{m_b}$$
$$Y_P = Y_A + X'_P \frac{\Delta Y}{m_b} + Y'_P \frac{\Delta X}{m_b}$$

(16.12)

SEVENTEEN

SOLVING THE INVERSE PROBLEM IN GEODESY BY THE CHORD METHOD

Numerous high precision formulas are available to compute superlong distances between two points on the ellipsoid. Many are based on iterative procedures requiring long and cumbersome computations on high-speed computers, which are not always available, especially in the field or in shipboard operations. We now describe a method that makes possible the rapid calculation of long distances for navigational purposes with an aid of small pocket calculator and seven-decimal-place trigonometric tables. The approach is based on an exact chord distance between two points on an ellipsoid and an approximate arc distance derived from the chord distance.

17.1. DETERMINATION OF THE ELLIPSOIDAL CHORD DISTANCE

Two methods are presented to compute the rigorous chord distance between two points on the reference ellipsoid. If the *universal space rectangular (USR) coordinates* are utilized, the first step is to compute the coordinates X, Y, and Z of the two points in that coordinate system as functions of the known geographic coordinates ϕ and λ. This is done by Eqs. 14.3 and 14.6. The ellipsoidal chord distance d_C is then obtained as a function of the differences in the USR coordinates of the points from Eq. 14.17.

For the purpose of further analysis, let us write Eq. 14.17 in a complete form:

$$d_C = [(N_1 \cos \phi_1 \sin \lambda_1 - N_2 \cos \phi_2 \sin \lambda_2)^2$$
$$+ (N_1 \cos \phi_1 \cos \lambda_1 - N_2 \cos \phi_2 \cos \lambda_2)^2 \qquad (17.1)$$
$$+ (1 - e^2)^2 (N_1 \sin \phi_1 - N_2 \sin \phi_2)^2]^{1/2}$$

where ϕ_1, λ_1, ϕ_2, and λ_2 are the latitude and longitude values of points P_1 and P_2; N_1 and N_2 are the curvature radii in prime vertical at points P_1 and P_2, and e^2 is the eccentricity square of the reference ellipsoid.

Another rigorous solution can be obtained through a method developed by *Ewing* (1955). In Fig. 17.1, P_1 and P_2 are two known points on the ellipsoid, and the ellipsoidal chord distance d_C is to be computed between them. The center of the parallel of latitude through P_1 lies on the polar axis and is designated O_1. The center of the parallel of latitude through P_2 is also on the polar axis and is shown as O_2. The intersection of the meridian through P_1 with the parallel of latitude through P_2 is called P'_2, and the intersection of the meridian through P_2 and the parallel of latitude through P_1 is designated P'_1.

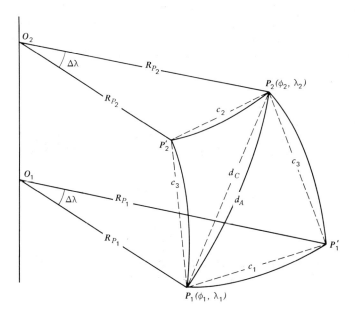

Figure 17.1 Trapezoid on the ellipsoid (Laurila, 1960).

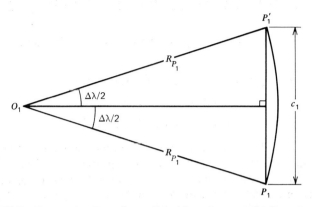

Figure 17.2 Determination of parallel sides of trapezoid (Laurila, 1960).

Let c_1 represent the chord joining points P_1 and P'_1, and let c_2 represent the chord joining points P'_2 and P_2. These two lines are parallel to each other but of differing lengths. The chords joining points P_1 and P'_2 and points P'_1 and P_2 are equal in length. They are not parallel but symmetrically tilted inward. It is now obvious that the four chords c_1, c_2, c_3, and c_3 form a trapezoid on a plane within the ellipsoid, and the unknown chord d_C is the diagonal of this trapezoid. According to standard geometry and as expressed in Eq. 15.18, the diagonal will be

$$d_C = (c_1 c_2 + c_3^2)^{1/2} \qquad (17.2)$$

Now it is necessary to indicate the method of obtaining the four sides of the trapezoid. As Fig. 17.1 shows, c_1 is a chord of the parallel of latitude through P_1 and R_{P_1} is the radius of that parallel. The sector $O_1 P_1 P'_1$ is removed from Fig. 17.1 and presented in Fig. 17.2, which indicates that

$$c_1 = 2R_{P_1} \sin \frac{\Delta\lambda}{2}$$

and similarly   (17.3)

$$c_2 = 2R_{P_2} \sin \frac{\Delta\lambda}{2}$$

where $\Delta\lambda$ is the longitude difference between points P_1 and P_2.

The parallel radius values are obtained from Eq. 14.1 as follows:

$$R_P = N \cos \phi$$

where N is the curvature radius in prime vertical.

The two equal sides c_3 of the trapezoid must now be determined.

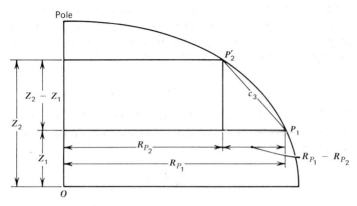

Figure 17.3 Determination of equal sides of trapezoid (Laurila, 1960).

Figure 17.3 shows a meridian plane in which c_3 is the chord joining points P_1 and P'_2. It is the hypotenuse of the right triangle whose sides are $(R_{P_1} - R_{P_2})$ and $(Z_1 - Z_2)$. The values of Z are obtained from Eq. 14.6 as follows:

$$Z = N(1 - e^2) \sin \phi$$

Consequently, the length of the chord c_3 is

$$c_3 = [(R_{P_1} - R_{P_2})^2 + (Z_1 - Z_2)^2]^{1/2} \tag{17.4}$$

and the chord distance d_C, which we want to learn, is obtained by substituting Eqs. 17.3 and 17.4 into Eq. 17.2:

$$d_C = \left[4N_1 N_2 \cos \phi_1 \cos \phi_2 \sin^2 \frac{\Delta\lambda}{2} + (N_1 \cos \phi_1 - N_2 \cos \phi_2)^2 \right.$$
$$\left. + (1 - e^2)^2 (N_1 \sin \phi_1 - N_2 \sin \phi_2)^2 \right]^{1/2} \tag{17.5}$$

Both approaches are based on rigorous, closed formulas. The times required to use the formulas are about the same when pocket or desk calculators are employed. It should be noted, however, that eight trigonometric functions must be obtained from trigonometric tables when Eq. 17.1 is applied, but only five are needed when Eq. 17.5 is used.

17.2. DETERMINATION OF THE ELLIPSOIDAL ARC DISTANCE

The determination of the ellipsoidal arc distance d_A consists of two principal operations, as pointed out by *Macomber* (1957). First, find the average curvature radius of the ellipsoid along the line to be determined; and use this curvature radius to replace the ellipsoid by a sphere. Second, use this curvature radius and the precisely computed chord d_C to determine the geocentric angle Θ representing the unknown arc distance d_A.

The average curvature radius is obtained by applying the *Simpson rule* as follows:

$$R_m = \tfrac{1}{6}(R_{A_1} + 4R_{A_m} + R_{A_2}) \tag{17.6}$$

where R_{A_1}, R_{A_m}, and R_{A_2} are the curvature radii at the azimuths A_1, A_m, and A_2 at the beginning, at the midpoint, and at the end of the line P_1P_2. These curvature radii are computed by Eq. 13.13 as follows:

$$R_A = \frac{NM}{M \sin^2 A + N \cos^2 A}$$

where M is the meridian curvature radius and N is the curvature radius in prime vertical.

To compute the curvature radii R_A at these three points, we need the latitudes ϕ and the azimuths A. The latitudes of the terminal points P_1 and P_2 are known, and the azimuths A_1, A_m, A_2, and the latitude ϕ_m are obtained from the following auxiliary spherical triangle (Fig. 17.4):

$$\cos \delta = \sin \phi_1 \sin \phi_2 + \cos \phi_1 \cos \phi_2 \cos \Delta\lambda$$

$$\sin A_1 = \frac{\cos \phi_2 \sin \Delta\lambda}{\sin \delta}$$

$$\sin A_2 = \frac{\cos \phi_1 \sin \Delta\lambda}{\sin \delta} \tag{17.7}$$

$$\sin \phi_m = \sin \phi_1 \cos \rho + \cos \phi_1 \sin \rho \cos A_1$$

$$\sin A_m = \frac{\cos \phi_2 \sin A_2}{\cos \phi_m}$$

where $\Delta\lambda = \lambda_2 - \lambda_1$ and $\rho = \delta/2$.

The trigonometric functions (Eq. 17.7) can be solved with the aid of a good slide rule; alternatively, the needed parameters A_1, A_m, A_2, and ϕ_m can be determined graphically by the simultaneous use of a *gnomonic* projection grid, where all great circles are straight lines, and any conformal projection grid, where all angles are correct.

The geocentric angle Θ is now obtained as

$$\Theta = 2 \sin^{-1} \frac{d_C}{2R_m} \tag{17.8}$$

by a procedure similar to that described in connection with Fig. 17.2, and

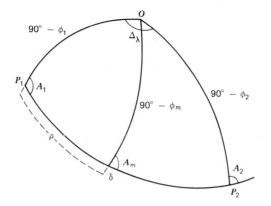

Figure 17.4   Auxiliary spherical triangle.

the unknown arc distance thus is

$$d_A = R_m \Theta \qquad (17.9)$$

If Eq. 17.8 is developed into a series expansion from which the first three terms are used, Eq. 17.9 becomes

$$d_A = d_C + \frac{d_C^3}{24R_m^2} + \frac{3d_C^5}{640R_m^4} \qquad (17.10)$$

When d_C is given the value of 1000 km and R_m is given the value of 6371 km, the difference between the two arc distances d_A obtained from Eqs. 17.9 and 17.10 is only 1 cm.

The weakness of this method obviously lies in the determination of the average curvature radius R_m. If d_A in Eq. 17.10 is differentiated with respect to R_m in the second term (the third term is insignificant in the error analysis), we have

$$dd_A = -\frac{d_C^3}{12R_m^3} dR_m \qquad (17.11)$$

The substitution into Eq. 17.11 of the foregoing values of d_C and R_m, together with $dR_m = 1$ km, yields

$$dd_A = -0.32 \text{ m}$$

which means that to maintain an accuracy of 1 ppm at the distance named, an error of ± 3 km in R_m is allowed.

By using the procedure just described, the average curvature radius R_m can be determined at that accuracy even along lines longer than 1000 km. A hypothetical test line was computed by the arc method between two points $P_1(\phi_1 = 5°N, \lambda_1 = 5°E)$ and $P_2(\phi_2 = 5°S, \lambda_2 = 5°W)$ along the surface of the International Ellipsoid. When applying the chord method, this type of line, crossing the equator at an angle of approximately 45°, is the most vulnerable to the error in R_m. The result was compared to the computation made by using *Sodano's* (1958) method and is presented in the following:

Arc method: $d_A = 1,568,065.78$ m

Sodano method: $d_A = 1,568,065.34$ m

The difference was $+0.44$ m or $+2.8 \times 10^{-7}$.

EIGHTEEN

REVERSE LEVELING

In a classical sense, leveling means carrying elevation information from a reference surface called the mean sea level (MSL) to the earth's surface or topography. The information may be transferred by means of differential (precision) leveling, where a leveled telescope is aimed at two stadia rods located at opposite sides of the leveling instrument. The vertical distance between the level surfaces going through the points at which the rods are located is found from the differences of the observed readings in the engraved stadia rods. Elevation differences between two points may also be obtained by observing the vertical angle formed between the line of sight and the horizontal. If the distance between points is known, the elevation difference can be worked out by trigonometric functions. This type of leveling is called trigonometric leveling. The third type of classical leveling is based on observations of barometric pressures at each point at which the elevation is to be determined. By using the laws of the interrelation between the pressure, altitude, and temperature, the elevation differences between points can be determined. These three methods were presented in descending order of measuring accuracy, but one characteristic is common to all of them—namely, all measurements are made "upward" from the common reference surface, the MSL.

The term "reverse leveling" is used in this chapter because the observations initially are made from a reference surface within the atmosphere or from a known point in space "downward" to the point whose elevation is to be determined. The first case refers to aircraft altimetry, where the reference surface is an accurately defined *isobaric surface* in the atmosphere. The second case refers to a satellite altimetry, where the point of reference is located at a known altitude in a predetermined *satellite orbit*. In both these cases the final reference surface is the MSL, the point from which the elevations of points on earth's surface are computed.

18.1. AIRCRAFT ALTIMETRY

In the exploitation of vast, unsurveyed land areas for natural resources, a map is one of the most vitally important tools in the hands of explorers. Based on minimum or no ground control, maps must be compiled and elevations determined indirectly from the air. This is done photogrammetrically by taking advantage of the stereoscopic principle, which requires basic planimetric and elevation control points. Aircraft altimetry provides the answer by establishing the elevations. It also represents a fast means for determining elevation control for reconnaissance mapping, where accuracy is not at a premium.

To achieve in aircraft altimetry the ultimate accuracy that accords with the precision provided by modern altimeters and variometers, attention must be devoted to the study of certain atmospheric phenomena, including the behavior of the equal-pressure or isobaric surface, where the same atmospheric pressure prevails simultaneously at every point. In reality such a surface is extensively curved; a small part of it can be considered to be a plane, however, even though it is hardly ever wholly horizontal. Besides the inclination of this plane, diurnal variations occur in its altitude.

The principle of utilizing aircraft altimetry is presented in Fig. 18.1. At the beginning of the air leveling mission, the aircraft settles at the initial absolute altitude H_0 using the radioaltimeter over the sea level surface or over a terrain with known elevation. During the mission, the pilot tries to maintain the initial altitude with the aid of a precision barometer. However, variations occur in the aircraft's path from that constant-pressure or isobaric surface, and these variations ΔS are recorded automatically on a sensitive variometer. If the point P_n is considered to be a sample point on the earth's surface, the altitude of the aircraft H_n above that point is measured with the radioaltimeter. Finally, if the elevation difference ΔI between the isobaric surface and a surface parallel to MSL at an altitude H_0 can be computed, the desired topographic elevation Z_n of the point P_n is obtained as follows:

$$Z_n = H_0 - H_n + \Delta S + \Delta I \qquad (18.1)$$

To achieve as narrow a beamwidth as possible, microwave instruments

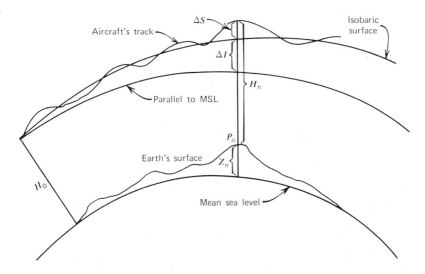

Figure 18.1 Principle of aircraft altimetry (Laurila, 1960).

operating on the X-band and provided with large parabolic antennas serve as radioaltimeters. The variometers may be of different design. Earlier models were based on very sensitive liquid barometers or *stato-scopes*, capable of detecting small variations in altitude. The liquid levels in the U-tube of the statoscope were recorded photographically on the main aerial film, or the difference in the levels was detected electronically. In the newer models, an aneroid capsule is directly linked to a parallel-plate condenser and the change in pressure is recorded from the change in capacitance.

18.1.1. Determination of the Isobaric Surface

There are two ways of determining the inclination of the isobaric surface over the survey area at the mission altitude. One approach is based on observation of atmospheric pressures and temperatures at three points around the survey area at approximately the mission altitude. These values are usually obtained from meteorological soundings if aerological stations are available. In the absence of such stations, the inclination may be determined from observations made in the aircraft carrying out the mission, of the velocity and direction of wind at the mission altitude.

In the first approach the interdependency between altitude, pressure, and temperature is obtained in a slightly modified form (*Humphreys*,

1940; *Schönholzer*, 1938) as follows:

$$H = 18,400 \log \frac{P_0}{P_H}(1 + 0.0037 t_m) \qquad (18.2)$$

or

$$\log P_H = \log P_0 - \frac{H}{18,400(1 + 0.0037 t_m)} \qquad (18.3)$$

where H is the geopotential altitude in meters, P_0 is the barometric pressure at sea level, P_H is the barometric pressure at the altitude H given in the same units as P_0, and t_m is the mean-altitude temperature in degrees centigrade.

To determine the inclination of the isobaric surface, there is no need to know the elevations H to the three points but rather, the elevation differences ΔH between them. Therefore, by differentiating Eq. 18.3 with respect to H and P_H, an important expression is obtained for the dependence of the elevation difference ΔH with the pressure difference dP_H at a given pressure level and temperature:

$$\Delta H = 7991(1 + 0.0037 t_m) \frac{dP_H}{P_H} \qquad (18.4)$$

In reverse air leveling, Eq. 18.4 has practical value only when the pressure and temperature can be recorded *at the mission altitude* over the three observation sites. This is possible if the actual geometric height H of a radiosonde used for sounding can be observed independently of the barometric pressure (radar tracking). In barometric ground surveys Eq. 18.4 rapidly yields an accuracy high enough for reconnaissance or other lower order leveling. In this case t_m is the mean temperature between the two points to be compared, P_H is the mean pressure between the same points, and dP_H is their pressure difference.

In meteorological soundings, recordings are usually made at intervals of even 50 mb of pressure through the troposphere and at shorter intervals within the stratosphere. From these pressure values, together with the recorded temperature values, the geopotential heights are computed and tabulated to correspond to the even-numbered pressure levels. If the isobaric surface is inclined, these heights are not equal and do not represent the mission altitude. In practical air-leveling procedures, the isobaric surface, or level that is located closest to the mission altitude, is selected as the level of reference. This is best illustrated by the following actual example (Fig. 18.2).

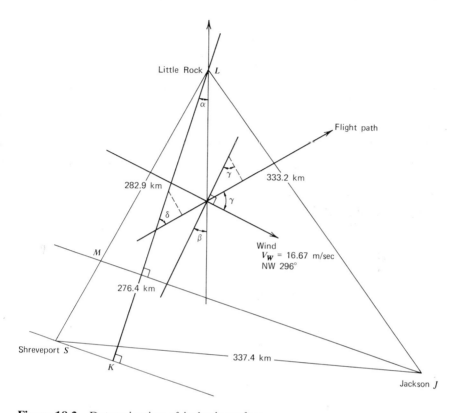

Figure 18.2 Determination of isobaric surface.

Date of meteorological soundings: October 2, 1966
Time: 00.00 Greenwich time
Sounding sites: Little Rock, Arkansas; Shreveport, Louisiana;
Jackson, Mississippi
Isobaric level: 600 mb
Geopotential heights: Little Rock, $H_L = 4341$ m
                       Jackson, $H_J = 4365\ m$
                       Shreveport, $H_S = 4375$ m
Distances between sites: Little Rock–Shreveport, $D = 282.9$ km
                          Shreveport–Jackson, $D = 337.4$ km
                          Jackson–Little Rock, $D = 333.2$ km

The lowest altitude, which is the one at Little Rock, is selected as the reference altitude, and the altitude differences are

$$\Delta H_L = 0 \text{ m}$$
$$\Delta H_J = +24 \text{ m}$$
$$\Delta H_S = +34 \text{ m}$$

The length of line LS is divided proportionally to the altitude differences, hence

$$LM = \frac{24}{34} LS$$

and line JM is now the tilt axis of the inclined isobaric surface. Line LK, which is drawn perpendicular to line JM, is the direction of the maximum slope of the isobaric surface and forms an angle α with the meridian going through L. In this example $\alpha = 18°$. The magnitude of the slope is obtained by drawing line SK parallel to JM and measuring the distance LK, which at the scale of the drawing is $LK = 276.4$ km. The maximum inclination of the isobaric surface is then

$$\Delta H_{max} = \frac{34}{276.4} = 0.123 \text{ m/km} \tag{18.5}$$

If the flight path makes an angle δ with the direction of the maximum slope (Fig. 18.2), the actual correction ΔI between the isobaric surface and a surface parallel to the MSL will be

$$\Delta I = D_F \Delta H_{max} \cos \delta \tag{18.6}$$

where D_F is the distance flown.

If the mission altitude differs greatly from the altitude of the closest even-numbered pressure level, more accurate results are obtained if the closest pressure level below and the closest pressure level above the mission altitude are selected as levels of reference. At both levels the altitude differences ΔH are computed and their values are then reduced or interpolated exactly to the mission altitude. One altitude (e.g., H_L in our example) serves as reference for the interpolation.

To determine the isobaric surface completely independently of any ground observation, and to permit use of modern electronic equipment, we resort to the theory of the *geostrophic wind*. According to the Buys-Ballot law, the movements of an air parcel in the atmosphere are determined by three different forces affecting it. An air parcel moving freely in the atmosphere is first affected by the pressure gradient force, which strives to carry the air parcel in a straight line from an anticyclone to a cyclone. Once it is in motion, the air parcel is affected by the Coriolis force resulting from the rotation of the earth, and finally by the centrifugal force resulting from the curvature of the air parcel's field of motion. The Coriolis and centrifugal forces resist the movement of the air parcel in the direction of the pressure gradient, and after having turned the field of motion of the air parcel in the northern hemisphere approximately 90° to the right of the original course, ultimately achieve a magnitude equal to the pressure gradient force. In addition, the movement of the air parcel is affected in lower air layers by the frictional force caused by topographic variations of the terrain.

At all practical flying altitudes the frictional force can be neglected, and applying the expressions of the three forces, we obtain

$$\frac{1}{\rho}\frac{dB}{\Delta D'} = 2\omega V \sin \phi \pm \frac{V^2}{r} \tag{18.7}$$

where ρ is the density of the air at the measuring altitude, $dB/\Delta D'$ is the pressure gradient, ω is the angular velocity of the earth, V is the wind velocity, ϕ is the degree of latitude, and r is the curvature radius of the field of motion of the air parcel. The sign of the second term of the right-hand member of Eq. 18.7 is positive in cyclonic and negative in anticyclonic conditions.

If in Eq. 18.7 the term $(1/\rho)\,dB$ is replaced by the equivalent quantity $\Delta H \times g$, we obtain the so-called *gradient wind equation*

$$\Delta H = \frac{2\omega}{g} V \sin \phi\, \Delta D' \pm \frac{\Delta D' V^2}{gr} \tag{18.8}$$

where g represent the acceleration of the gravitational force and ΔH is the variations of the altitude in the isobaric surface. From the terms in the right-hand side of Eq. 18.8 it is seen that the influence of the Coriolis force increases with any increase in the degree of latitude.

Since photogrammetric flights and air-leveling work are usually performed in stationary anticyclonic conditions, the curvature radius r of the field of motion of an air parcel is relatively large. Therefore, especially at mid or high latitudes, the second term of Eq. 18.8 can be neglected. To investigate the justification of this deletion, a statistical study was based on 99 photogrammetric flights flown between latitudes 60° and 67° during 5 years (*Laurila*, 1953). The influence of the centrifugal term in Eq. 18.8 was found to be only 5.8% of the slope of the isobaric surface as computed solely by means of the Coriolis term. The expression thus obtained is called the *geostrophic wind equation*

$$\Delta I = \frac{2\omega}{g} V \sin \phi D_F \sin \gamma \qquad (18.9)$$

where D_F is the distance flown and γ is the angle between the direction of flight and the direction of wind.

When Eq. 18.9 is used, the direction and velocity of wind must be known at the mission altitude, which means that at least one meteorological sounding station must be situated within the survey area or at proximity to it. To illustrate the computation procedure and to compare the results to those obtained by way of the first method described, we use the same sounding data.

Sounding Site	Wind Velocity (m/sec)	Wind Direction	Pressure Level (mb)
Little Rock	19	NW 295°	600
Shreveport	15	NW 295°	600
Jackson	16	NW 298°	600

The mean wind velocity was $V_W = 16.67$ m/sec and the mean wind direction was NW 296°. The latitude at the middle of triangle *LSJ* is $\phi = 33°$N, the acceleration of gravity at that latitude is $g = 9.796$ m/sec^2, and the angular velocity of earth is a constant $\omega = 0.0007272$ rad/sec. Placing the last two values, together with $V_W = 16.67$ m/sec, $\sin 33° = 0.5446$, and $\gamma = 90°$ into Eq. 18.9, we find the maximum slope of the isobaric surface

$$\Delta H_{max} = 0.135 \text{ m/km} \qquad (18.10)$$

The direction of the maximum slope of the isobaric surface in this case forms an angle $\beta = 26°$ with the meridian that goes through L. The results of both approaches agree well within this randomly selected area at a random observation date. The difference in direction of the maximum slope is 8°, and the difference in magnitude of the maximum slope is 0.012 m/km. The results obtained must be considered as examples of the execution of these approaches only, not as guidelines for accuracies. Based on studies made by *Möller-Sieber* (1937), *Dines* (1938), *Pettersen* (1940) and *Franssila* (1944), a summary of accuracies obtainable from the geostrophic wind theory at mid or high latitudes and under relatively stationary anticyclonic conditions proved to be (*Laurila*, 1953)

$$m_{\Delta H} = \pm 0.020 \text{ m/km} \qquad (18.11)$$

averaged through 0° to 360° between the directions of wind and the flight path.

If air leveling is to be performed in areas where no meteorological sounding stations are available, a "self-sufficient" application of the geostrophic wind equation may be employed. Since the movement of an aircraft through the air is affected by the wind, Eq. 18.9 can be derived again as follows (*McCaffrey*, 1950; *Laurila*, 1953):

$$\Delta I = \frac{2\omega}{g} V_g \tan \delta \sin \phi D_F \qquad (18.12)$$

where V_g is the ground speed of the aircraft in meters per second and δ is the drift angle of the aircraft caused by the wind. These elements can be determined by photogrammetric means when photomissions are involved. Otherwise, and with even better results, Doppler radar as described in Chapter 27, may be chosen.

The diurnal fluctuation of the height of the isobaric surface in the first approach can be taken into account by computing the tilt of the surface from two different sounding times, 00.00 and 12.00 Greenwich time, and reducing them to the proper time of the mission. In the latter case, because the aircraft is continuously observing its ground velocity and drift angle, the time is not critical.

18.2. SATELLITE ALTIMETRY

Although aircraft altimetry has been used effectively in local mapping for decades, the first serious thought of applying satellite altimetry in the various geosciences became public in 1966, when an ad hoc Scientific Advisory Group was formed. Its task was to advise the Geodetic Satellite Office of NASA of the ability to make geodetic measurements with an accuracy of ± 1 m and ± 10 cm, and the resulting report led to funding for the development and research on feasible satellite altimeters. The National Academy of Sciences Space Applications Seminar Study at Woods Hole emphasized in 1967 the importance of satellite altimetry in oceanographic studies by measuring such dynamic features of the earth's surface as the departure of the sea surface from the perfect geoid due to tidal action, ocean currents, and wind forces. A multidisciplinary group was organized and met in Williamstown, Massachusetts, in 1969; among other recommendations it reemphasized the long-term need for altimeters capable of ± 10 cm accuracy. Currently the NASA Earth and Ocean Physics Applications Program (EOPAP) is projecting a long-range program relying on altimetry as one of the principal sensor techniques

(*McGoogan et al.*, 1974a). The first satellite altimetry was tested and data were collected for analysis on May 14, 1973, when the S-193 radar altimeter was riding in Skylab, which was placed into earth's orbit at a mean altitude of 440 km and an inclination of 50°. The radar altimeter was built by the General Electric Company, Ithaca, New York. The frequency is 13.9 GHz and the pulse repetition rate is 250 pulses/sec. The pulse widths are nominally 10 and 100 nsec, but the actual data indicated that the pulse widths equivalent to 3 dB, or half-power limits, are 18 and 72 nsec, correspondingly. The altitude data are recorded at a rate of eight samples per 1.04 sec. The antenna is a 1.1 m diameter paraboloid, and with the frequency used it yields a total 3 dB beamwidth of about 1°. The S-193 altimeter is a nadir-pointed radar system, which means that the axis of parabolic antenna, or the direction of the radar beam, coincides with the line connecting the radar antenna to the earth's center of mass, thus passes through a point on the spheroid called the nadir point.

One of the main tasks of the S-193 altimeter in Skylab was to collect information and gain experience for use in designing future satellite altimeters. In addition, scientists wanted more information on such geodetic, geophysical, and oceanographic phenomena as geoid undulations, correlations with ocean floor topography, current detections, wave heights, and land topography.

The basic theory of satellite altimetry is analogous to that of aircraft altimetry in that from the stable platform provided by a satellite, vertical measurements are made to the surface below (ocean, lake, solid land). In measurements over ocean surfaces, the altimeter measures to the instantaneous electromagnetic mean sea level (IEMSL), and this measurement is averaged over a special "footprint" or *spot size* (*McGoogan*, 1974). In Fig. 18.3 the accurate positioning of the satellite is done by spatial intersection from known ground tracking stations with reference to the ellipsoid. By observing a priori such instantaneous and long-range ocean surface dynamics as wave heights, currents, tides, and winds, the relationship between the IEMSL, MSL, and ultimately the geoid may be determined. If, however, the geoid is accurately defined, valuable data become available for ocean dynamics investigation. The beamwidth and spot size are of great importance in satellite altimetry. Narrow beamwidth results in small spot size and limits the freedom of verticality stabilization, whereas larger beamwidth serves to control spot size by the pulsewidth, and in moderate conditions the nadir point is illuminated by the beam cone. Figure 18.4 illustrates the impingement of a radar pulse on a hypothetically flat and smooth plane (sea surface) within the beam cone. At the moment the pulse touches the surface there are no footprints (Fig.

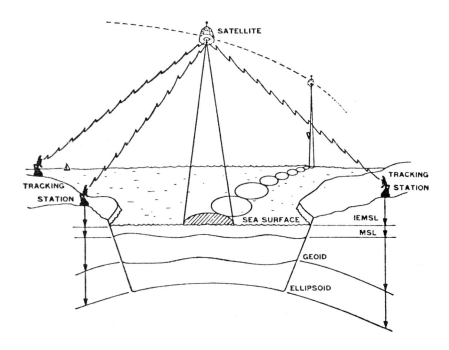

Figure 18.3 Geometry of satellite altimetry (from *McGoogan*, 1974).

18.4*a*). After the pulse has "sunk" into the surface with its total width the footprint has rapidly reached the maximum spot size, a solid circle with radius

$$r = \sqrt{hcT} \qquad (18.13)$$

where h is the satellite altitude, c is the velocity of light, and T is the pulsewidth (*Pierson* and *Mehr*, 1970; *Miller et al.*, 1973; *McGoogan et al.*, 1974b). Beyond this moment the reflecting circle becomes a reflecting ring with no illumination at the center circle (Fig. 18.4*c*). This nearly constant plateau stage continues until the illuminated ring starts falling outside the beam cone, and the return signal strength gradually fades out (Fig. 18.4*d*). Of course, the beamwidth is not an absolute boundary at which the signal is "switched off" but rather a reference value. The footprint may be considered to be a spatial filter in the analysis of ocean surface dynamics. In data analysis the wave height features are obtained from the front slope characteristics of the return wave form, whereas the tailing slope of the wave form is related to the nadir pointing.

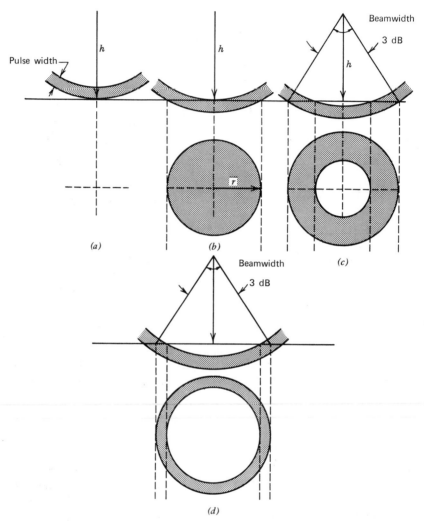

Figure 18.4   Surface scattering of impinging pulse.

The calibration of a satellite altimetry system is very interesting in its multipurpose complexity. First, an altimeter placed in so massive and bulky a body as Skylab, orbiting at a relatively low altitude, is bound to be affected by radial errors caused by variations in the vessel's orbit. The effective area of Skylab is $293 \, m^2$, and its mass is 87,441 kg; thus it is vulnerable to atmospheric drag. Also, thrusting used to maneuver the spacecraft before and after the altimeter data are obtained can cause errors in the orbit computation. Therefore ground-based tracking stations are needed to determine the altimeter's altitude during the mission. Assuming that the spacecraft tracking takes place, we now have the following data at hand to be combined in various ways: the true position of the satellite altimeter in space, thus the satellite altitude (h_s) above the reference spheroid; the satellite altitude (h_a) above the contact area on the ocean surface (the altitude being measured by the satellite altimeter); the dynamic ocean effect Δ_h, and the geoid height or undulation h_g. From Fig. 18.5 it is clear that if h_s is known, any of the remaining three parameters, h_g, h_a, or Δ_h, can be determined as a function of the others, and a general expression can be written as follows:

$$h_g = h_s - h_a - \Delta_h \tag{18.14}$$

in which case the geoid height can be obtained if the dynamic ocean effect Δ_h has been predicted and the satellite altitude h_a has been measured.

The main systematic error sources in the Skylab S-193 altimetry are the instrument bias, which is of the order of 2–10 m; the pointing error, which can vary between 1 and 30 m depending on whether the trailing edge of the wave form is used to generate off-nadir position; and the orbit determination error, which can vary between 5 and 20 m depending on the availability of tracking stations. The resolution in terms of standard error is about ±1 m. The correction used for pulse retardation through the troposphere is −2.79 m, which is significantly larger than the value of −2.42 m obtained from Saastamoinen's Eq. 7.34 by using standard sea level temperature and pressure values, together with an average value of $e = 10.00$ mb, or the value −2.39 m obtained by Laurila (Table 7.8) and based on four widely separated meteorological soundings. An important factor in determining the geoid by satellite altimetry is the knowledge of

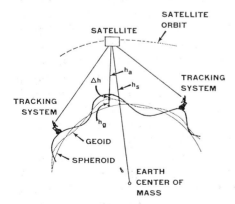

Figure 18.5 18.5 Determination of geoid height (from *McGoogan*, 1974).

the ground truth or sea state (ocean surface dynamics). To develop a method for such data collection for present and future satellite altimeters, a joint effort was made by the Naval Research Laboratory (NRL) and NASA Wallops Flight Center (*Walsh*, 1974). An X-band radar was mounted in a NASA Wallops Flight Center C-54 aircraft, and experiments were carried out over calm water in Chincoteague Bay. The average aircraft altitude was about 3600 m, and during the mission the vertical motion of the aircraft was tracked by the Wallops Island FPQ-6 radar. A Naval Oceanographic office laser profilometer supplied ground truth data on the sea surface for the NRL radar altimeter. After a bias of 180 m was removed from the NRL radar data (which had not been calibrated a priori), the standard deviation of the NRL data from the smoothed altitude determined by the FPQ-6 was only ±6 cm. To use NRL radar for collection of ground truth data to the satellite altimeter, the radar should be more sensitive to the sea state than the satellite altimeter for which it provides the information. By analyzing the leading edge characteristics of the NRL radar, it was proved that under the worst conditions the radar could detect a change of 38 cm in the significant wave height (SWH). Since the strictest specifications for sea state measurements applied to satellite altimeters to date and in the near future are 50 cm or 10% of the wave height, the tests prove the feasibility of using NRL-type radar sets as ground truth instruments.

The various applications of satellite altimetry in geodesy, geophysics, and oceanography are well summarized in the report of *McGoogan et al.* (1974b). In geoid determination the great advantage in using the satellite altimetry results from its ability to detect short wavelength features such as trenches, seamounts, ridges, and gaps. In the Marianas and Puerto Rico Trenches, depressions in the ocean surface of 15–20 m were detected, although these anomalies did not show in the Marsh-Vincent geoid, which is a detailed gravimetric geoid. Abrupt changes in ocean floor topography have been detected as significant changes in mean sea level at the Blake Escarpment region in the southeast coast of the United States. The rise of sea level due to persistent currents such as the Gulf Stream, which has a rise of 1 m over distances of 50 to 100 km, should be detectable by instruments available in the near future. The prospects of satellite altimetry over solid land with rapidly changing height features cause some misgivings. However, the comparison of elevations between lake surfaces and between lake and ocean surfaces provide good height reference points. The elevation of Lake Erie with respect to the Atlantic Ocean has been determined within a few meters of the published data.

The current status of NASA satellite altimetry program can be divided into three stages (*McGoogan*, 1974). Presently the data reduction of the Skylab S-193 altimeter results are being done with the primary goal of analyzing the scattering characteristics of the ocean surface, to be used for designing future altimeters. Although Skylab itself and its orbit were not well suited for long-arc geodetic experiments, the spacecraft could carry large antenna and power well adaptable for the intended research. In late 1974 the GEOS-C spacecraft was scheduled to be launched into an orbit suitable for geodetic experiments. The mean satellite altitude was to be 843 km, which meant that the spacecraft was less affected by local gravity anomalies and was visible from more tracking stations. The determination of the orbit was planned to be very precise with the addition of laser retroreflectors, *C*-band and *S*-band transponders, Doppler beacons and satellite-to-satellite tracking (SST) facilities. The launch occurred in early 1975, but because of the unexpected spin of the satellite, the GEOS–C has not been used for altimetry purpose. In 1978 the SEASAT–A is scheduled for launching into orbit primarily to supply oceanographic research data relating to topography, surface temperature, directional winds, directional waves, and wave heights. Table 18.1 gives the main features of the three satellite altimeters.

An interesting feasibility study has been done by *Yionoulis* and *Black* (1974) at the Johns Hopkins University Applied Physics Laboratory (APL) to determine the deflection of the vertical (DOV) at sea. It was proposed that two satellites be separated by about 200 km in the same, near-polar orbit. Each carries a radar altimeter. A 3 GHz satellite-to-satellite Doppler link connects the two satellites, providing necessary

Table 18.1[a]

Feature	Skylab	GEOS–C	SEASAT–A
Mean altitude	435 km	843 km	725 km
Antenna beamwidth	1.5°	2.6°	1.5°
Attitude control	0.5°	1.0°	0.5°
Frequency	13.9 GHz	13.9 GHz	13.9 GHz
Peak power	2.0 kW	2.5 kW	2.5 kW
DC power	0.250 kW	0.125 kW	0.125 kW
Nominal pulse width	10 nsec	12.5 nsec	3.0 nsec
Spot size	2.3 km	3.6 km	1.6 km
Resolution	±1.0 m	±0.6 m	±0.1 m
Weight	45 kg	68 kg	45 kg

[a] From "Precision Satellite Altimetry," *J. T. McGoogan*, NASA Wallops Station, Wallops Island, Va., 1974.

information to remove biases. By subtracting consecutive readings of the altimeter in one satellite, the DOV component along the satellite's ground track is determined. The earth's rotation causes the satellite ground tracks to be separated by about 14 km. Thus by subtracting the altimeter readings in one satellite from those in the other, the DOV component across the ground tracks is found. Since all altimeter readings are differenced, no absolute accuracies of altitudes are required. From any perpendicular track system, the north-south and east-west components of the DOVs then can be derived. The authors have estimated that the accuracy in determining the DOVs at points on a 10×15 grid will be of the order of ± 2–4 arc seconds.

MULTIWAVELENGTH
DISTANCE MEASUREMENTS

In this chapter we discuss systems that use several (two or three) wavelengths simultaneously in distance measurements. As we know from Section 6.2.1, two wavelengths in the light spectrum have different group refractive indexes under standard atmospheric conditions; these indexes are constants and can be computed by the *Barrell-Sears* formula (Eq. 6.13) as functions of wavelengths. When propagating through variable atmosphere, these waves are retarded differently depending on the atmospheric parameters, and the ambient refractive index can be computed either by Eq. 6.14 or by Eq. 6.17—the two are equivalent. By comparing the optical distances obtained by measurements with these two wavelengths, the effect of the density of dry air for the measurements, thus the correct geometric distance, can be found directly from the measurements and without auxiliary meteorological observations. The mathematics involved in two-wavelength measurements is discussed by *Thompson* and *Woods* (1965).

If the two wavelengths are in the red visible region λ_R and in the blue region λ_B, the measured optical distances will be

$$R_B = \bar{n}_B L$$
$$R_R = \bar{n}_R L \qquad (19.1)$$

where $\bar{n}_B$ and $\bar{n}_R$ are the average refractive indexes of blue light and red light propagations over the line to be measured, and L is the true geometric length of the line. From Eqs. 6.14 and 6.17, we see that the

235

average refractive indexes for red and blue transmissions are obtained by taking integrated averages of these equations; thus we write

$$(\bar{n}_B - 1) = (n_{g_B} - 1)M \qquad \text{and} \qquad (\bar{n}_R - 1) = (n_{g_R} - 1)M \qquad (19.2)$$

where n_{g_B} and n_{g_R} are the group refractive indexes of blue and red light, respectively, and M is the dry air density coefficient in Eqs. 6.14 and 6.17; note that this coefficient is the same for red and blue light.

Dividing the first expression in Eq. 19.2 by the second expression and solving for $\bar{n}_B$ yields

$$\bar{n}_B = \frac{(n_{g_B} - 1)(\bar{n}_R - 1)}{n_{g_R} - 1} + 1 \qquad (19.3)$$

From Eq. 19.1 we obtain

$$R_B - R_R = (\bar{n}_B - \bar{n}_R)L \qquad (19.4)$$

which yields

$$R_B - R_R = \left[\frac{(n_{g_B} - 1)(\bar{n}_R - 1)}{n_{g_R} - 1} - \frac{(n_{g_R} - 1)(\bar{n}_R - 1)}{n_{g_R} - 1}\right]L \qquad (19.5)$$

when $\bar{n}_B$ from Eq. 19.3 has been substituted. Furthermore,

$$R_B - R_R = \alpha_R(\bar{n}_R - 1)L \qquad (19.6)$$

where $\alpha_R = (n_{g_B} - n_{g_R})/(n_{g_R} - 1)$, which is a known constant. Since we can say from Eq. 19.1 that $\bar{n}_R = R_R/L$, Eq. 19.6 can be written as

$$R_B - R_R = \alpha_R(R_R - L) \qquad (19.7)$$

from which L can be solved as follows:

$$L = R_R - A_R(R_B - R_R) \qquad (19.8)$$

where $A_R = 1/\alpha_R$. If the refractivity numbers $N = (n-1)10^6$ are used, $A_R = N_R/(N_B - N_R)$.

At this stage of multiwave two-wavelength distance measurements we are facing two crucial sources of errors. First, the coefficient A_R in Eq. 19.8 is a large number. If light waves at $\lambda_Y = 5791$ Å yellow and $\lambda_V = 4047$ Å violet from mercury arc light sources are used (*Thompson* and *Wood*, 1965), then A_Y is 17.4. If light sources are the helium-neon and helium-cadmium (He-Cd) lasers having wavelengths $\lambda_R = 6328$ Å red and $\lambda_B = 4416$ Å blue, correspondingly, A_R is approximately 20.0. Since in Eq. 19.8 the atmospheric correction is the product of A_R and the difference between the measured optical paths R_B and R_R, the difference must be measured with an accuracy 20 times higher than the desired system accuracy. If we can assume that the two distance measurements are partially independent, the accuracy of each measurement must be even higher than 20 times the desired system accuracy. This requirement sets forth a "tall order" in instrument design. The second source of errors is due to the neglect of the influence of the average water vapor density in air. As was pointed out in Chapter 6 (Eqs. 6.20 and 6.28), the refractive indexes of light and microwave transmissions were about equal in sensitivity to temperature and pressure, but the refractive index of microwave propagation was about 100 times more sensitive to the partial pressure of water vapor than the refractive index of the light wave propagation. Therefore, a third wavelength has been added to the multiwavelength system, namely, a microwave whose frequency is about 10 GHz (*Thompson*, 1968; *Huggett* and *Slater*, 1947b); also the modified *Barrell* and *Sears* formula has been used to include the effects of water vapor in light wave transmission.

The derivation of the distance equation for a three-wavelength system is long and cumbersome; therefore, only a summary is given in the following (*Huggett* and *Slater*, 1974b).

By designating the average index of refraction for red light transmission as $\bar{n}_R$, we can write $\bar{n}_R = 1 + (N_R 10^{-6})$, where N_R is the refractivity number. If water vapor density is included, the following set of equations can be written:

$$N_R 10^{-6} = \alpha_R D + \beta_R D_W$$

$$N_B 10^{-6} = \alpha_B D + \beta_B D_W \qquad (19.9)$$

$$N_M 10^{-6} = \alpha_M D + \beta_M D_W$$

where D is the dry air density, D_W is the water vapor density, and α_i and β_i are constants of proportionality. The refractivity number for the microwave transmission can refer to any of the formulas (6.23–6.25) in Chapter 6. Then

$$R_R = (1 + N_R 10^{-6})L \qquad \text{and} \qquad R_B = (1 + N_B 10^{-6})L \qquad (19.10)$$
and
$$R_B - R_R = 10^{-6}(N_B - N_R)L \qquad (19.11)$$

which, after substitution of two first expressions from Eq. 19.9 yields

$$R_B - R_R = (\alpha_B - \alpha_R)DL + (\beta_B - \beta_R)D_W L \qquad (19.12)$$

In a similar way, we can take the first and third expression from Eq. 19.9 to derive

$$R_M - R_R = (\alpha_M - \alpha_R)DL + (\beta_M - \beta_R)D_W L \qquad (19.13)$$

By eliminating D and D_W first from Eqs. 19.12 and 19.13, the geometric path length is found as follows:

$$L = R_R - A(R_B - R_R) - B(R_M - R_R) \qquad (19.14)$$

where

$$A = \frac{\alpha_R \beta_M - \alpha_M \beta_R}{(\alpha_B - \alpha_R)(\beta_M - \beta_R) - (\alpha_M - \alpha_R)(\beta_B - \beta_R)}$$

$$B = \frac{\alpha_B \beta_R - \alpha_R \beta_B}{(\alpha_B - \alpha_R)(\beta_M - \beta_R) - (\alpha_M - \alpha_R)(\beta_B - \beta_R)}$$

Given blue light with $\lambda_B = 4416$ Å, red light with $\lambda_R = 6328$ Å, and microwave with $\lambda_M = 3$ cm, A is approximately 20 and B is approximately 0.02. The accuracy requirement associated with the light-emitting instrument is still the primary concern. With reference to the foregoing example, the blue and red light-emitting components in the system must be accurate to about $\pm 5.0 \times 10^{-9}$ if the atmospheric correction (second term in Eq. 19.14 is to be kept within $\pm 10^{-7}$. In this case, of course, the total system error will be entirely dominated by that term only, because the coefficient B is only 0.02; thus there is no difficulty in keeping the third term well within this limit by using any existing microwave distance measuring equipment.

A very promising instrument design featuring two light wave transmissions and one microwave transmission has been reported by G. R. *Huggett* from the Applied Physics Laboratory and L. E. *Slater* from the Geophysics Program, University of Washington (1974a, 1974b). The optical sources are a 5-mW He-Ne gas laser operating in the red region of the spectrum at 632.8 nm and a 5 mW He-Cd metal vapor laser operating in the blue region of the spectrum at 441.6 nm. The two laser beams enter a Wollaston prism at the proper angle and polarization to ensure that the outgoing beams are collinear (Fig. 19.1). The light passes through a Pockels modulator that varies the ellipticity of the polarized light at 3 GHz. The combined beams are expanded to a diameter of 20 cm by a Cassegrainian telescope in which a concave paraboloid mirror is the reflector and a convex mirror serves as an interceptor. The light traverses the path being measured and is returned by a cat's eye retroreflector, a 20-cm parabolic reflector with a plane mirror at its focal point. Received by the same optics used for transmission, the beam passes a second time through the modulator, where the ellipticity of the polarization is increased or canceled depending on the phase of the modulator excitation. The second pass through the modulator and the Wollaston prism results in optical signals that are incident on photodetectors. The frequency that

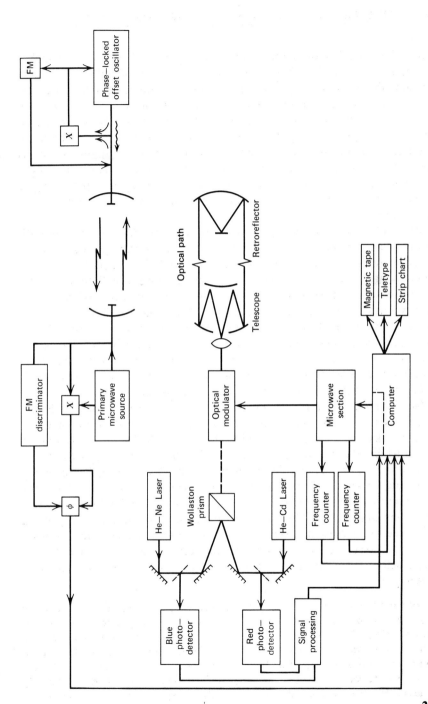

Figure 19.1 Functional diagram of APL three-wavelength distance meter (from *Hugget and Slater*, 1974a).

239

drives the Pockels modulator is controlled by a small computer in such a way that the optical path length is always an integral number of half-wavelengths of the modulation. The period T of the modulation is about 3×10^{-10} sec, which corresponds to a modulation wavelength of only 10 cm. The analog outputs of the photodetectors are converted to digital information and used by the computer to determine the modulation frequency correction. The transmitted light in the APL instrument thus is polarization modulated (as in the Kern Mekometer); the half-wavelength unit distance is not maintained by changing the optical path length, however, but by changing the modulation frequency (as in the prototype Geodimeter).

Simultaneously with the optical measurements, the microwave path length is determined by standard direct phase measurement of a 9.6 GHz carrier frequency. The phase data and the modulation frequencies for the optical system are input to the computer for distance calculations. At the conclusion of each calculation (every 10 sec), the corrected distance and all raw data are output to magnetic tape and other peripheral devices. The APL instrument has two complete servo systems. One controls the modulation frequency on the red laser beam, ensuring that it is an integral number of half-wavelengths; the other system controls the modulation on the blue beam. The servos are identical, and the common RF elements and the optical modulator are time shared.

The RF modulation frequency at 3.005 GHz is derived from 5 MHz by an offset RF phase-locked oscillator. To secure a stable RF frequency, the reference oscillator is phase locked to the system clock and the computer-controlled 5 MHz frequency is added to it. The system master clock is a high precision crystal oscillator having a stability of $\pm 10^{-11}$/sec and a long-term stability of 5×10^{-11}/day. The clock is the time base for frequency counters and is set with respect to a color television network's *rubidium standard* (used like the cesium standard) to within a few parts in 10^{10}. The crystal oscillator is set with respect to the rubidium-derived reference every two days to maintain an accuracy of better than 10^{-9}.

According to G. R. Huggett and L. E. Slater, of the University of Washington:

The microwave transmission system was designed by the ESSA *Institute for Telecommunication Sciences.* It is a direct phase measuring system operating at about 9.6 GHz. Two CW microwave signals, at slightly different frequencies, are transmitted in opposite directions over the distance to be measured. At each terminal a strong local signal and a weak received signal are heterodyned. This gives two intermediate frequency (IF) signals, whose relative phase is linearly related to the sum of the individual phase shifts experienced by the microwave signals in traveling over the path. At the remote terminal the IF signal is used to phase-lock a voltage-controlled crystal oscillator for the return transmission and to frequency modulate a second microwave signal that is also transmitted back over the path. These signals are received, separated by filtering and amplified. The IF signal goes to the phase detector and the FM signal is fed into an FM discriminator. The reference phase from the discriminator is compared in a phase detector with the IF signal, thus producing an output signal that is proportional to twice the propagation time (T) and hence to the microwave refractivity. This signal goes to an analog-to-digital converter being input to the computer.

The long-term stability of the APL instrument was tested over a known line 5.3 km long, based on data taken from June through August, 1974. The maximum fluctuations of the atmospheric parameters were: temperature 15°C, pressure 14 mb, and relative humidity 65%. The standard deviation proved to be $m = \pm 1.3 \times 10^{-7}$. Over a line of 10.1 km, the short-term standard deviation was about 5×10^{-8} (*Huggett* and *Slater*, 1974a). The three-wavelength electronic distance measuring instrument is undergoing electronic modifications to improve the long-term stability to several parts in 10^{-8}. There is no doubt that such an instrument will play an important role in geophysical fault measurements and earthquake predictions.

TWENTY

INSTRUMENTATION

World War II introduced radar as a weapon for warfare. Radar measured distances electronically with fairly high accuracy, thus could be used to guide aircraft to their targets. Therefore it was not pure accident that after the war geodesists took a hard look at the possibility of applying this principle for geodetic and navigational purposes. Decca and Loran had been developed for commercial use in the field of navigation in the early 1940s. At the same time Shoran was being converted from a wartime air navigation system to a system capable of measuring long lines. Then in the late 1940s, after intensive testing, the Geodimeter was ready for mass production. This instrument, which uses light as a carrier, is primarily for high precision baseline surveying. About 10 years later the Tellurometer entered the market as the first instrument to use microwave modulation for precise geodetic surveying. The foregoing examples of different early systems indicate the kinds of devices that helped mark the breakthrough of electronics into geodesy and allied fields.

After the pioneering period, the development of electronic distance measuring (EDM) instruments was phenomenal. New equipment also was being developed and manufactured by leading electronic and optical companies around the world. New models and new brands are entering competition so rapidly that the author assumes responsibility in describing only samples of systems in common use at the time of writing.

20.1. LAND-TO-LAND SURVEY

Because of the multitude of systems and instrumentation presently available for various survey operations, it is difficult to make logical groupings.

242

To keep the presentation consistent with the previous analysis of velocity application in actual surveying (Chapter 7), grouping is carried out according to the purpose of the survey.

20.1.1. Long Distances

Included in this section are types of instruments capable of measuring distances at least 10 or even 50 km or more. Grouping is according to whether the carrier is light or microwaves, and each group is presented in chronological order of system development.

20.1.1.1. *Geodimeters*

All Geodimeters employ visible light as the carrier. The measuring set consists of an active transmitter and receiver at one end of the line to be measured and a passive retrodirective prism reflector at the other end. Continuous light emission in the transmitter is intensity (amplitude) modulated, using a precision radio frequency generator and an electro-optical shutter to form sinusoidal light intensity waves. The distance is obtained by comparing the phases of the outgoing modulation waves with those received by the receiving component after reflection from the distant reflector.

This principle is, in effect, the same as used by Fizeau in his terrestrial experiment to measure the velocity of light (see Chapter 6). *Bergstrand* (1950) developed the basic concept of the modern Geodimeter in 1941–1943 and used the prototype in 1947–1949 to determine the velocity of light. The Swedish AGA Company (Svenska Aktiebolaget Gasac-cumulator) was licensed to redesign the prototype equipment for commercial production, and the NASM-1 entered into market in 1949, followed by NASM-2 and -2A in 1952, NASM-3 in 1956, and NASM-4 in 1959. These models are no longer produced and have been replaced by models M6 (1964), M6A (1968), M6B (1971), M6BL (1972), and M8 (1967).

OPERATION AND MEASURING PRINCIPLES. The period of development from the first NASM-1 model to the latest models M 6BL and M 8 saw several modifications, but the basic principle remained the same. To illustrate the operation principle, we present Fig. 20.1, which refers to models NASM-3 and NASM-4; changes in the later models are explained in the next section.

A steady output of light from a tungsten or mercury lamp L is focused onto the modulation system, consisting of two polaroids P_1 and P_2 and the

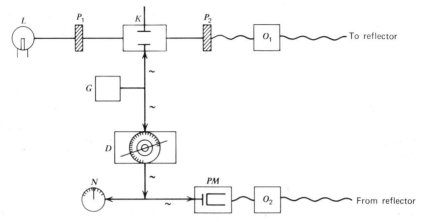

Figure 20.1 Block diagram of NASM-3 and NASM-4 Geodimeters.

Kerr cell K. The polarization planes of P_1 and P_2 are perpendicular to each other, and no light can pass through the second polaroid. When an electric field is introduced between the electrodes of the Kerr cell, it becomes doubly refractive and reverses the polarization of the incoming light by 90°, thus allowing the light to emerge through the second polaroid. The electric field in the Kerr cell is generated by a crystal-controlled generator G that oscillates at a frequency of about 30 MHz. The intensity of light that passes the second polaroid varies according to the output voltage of the generator. The optical transmitting system O_1 transmits this modulated light beam to the distant reflector in nearly parallel rays. Light returning to the receiver from the distant reflector is collected by the optical receiving system O_2 and is focused onto the cathode of the photomultiplier PM, causing photoemission of electrons. In the photomultiplier the photoemission of electrons is converted to an electrical signal whose strength varies according to the incoming beam modulation. The same generator output applied to the Kerr cell is also applied to the photomultiplier. Consequently, the sensitivity of the photomultiplier varies according to the incoming beam modulation. The phase difference between the strength of the incoming signal (photoemission) and the sensitivity of the photomultiplier depends on the distance measured, or rather, on the fractional part of the modulation wavelength included in that distance.

If the modulation wavelength is designated λ and the distance to be measured is assumed to include an integer amount of $\lambda/2$, the incoming signal and the sensitivity of the photomultiplier are in phase at that time and also thereafter, whenever the distance increases or decreases by an amount of $\lambda/2$. This distance interval is called the unit length U. If the 180° out-of-phase condition is observed also, as is the case in older model Geodimeters, the unit length will be $U' = \lambda/4$.

The fractional value L of the last noninteger unit length is obtained from the adjustable delay line D, which is placed between the oscillating generator and the photomultiplier. The delay line has variable impedance that can be controlled manually at the front panel of the instrument. By controlling the impedance of the delay line, the phase angle of the signal from the generator can be changed or delayed until the strength of the photoemission and the sensitivity of the photomultiplier are in phase. The in-phase condition can be observed visually from the null detector N, which is provided by a galvanometer needle. Since the modulation wavelength is known, the unit length U also is known and the fractional part L or U is obtained from the ratio between the delayed phase angle and 180°.

From the preceding data we can conclude that the basic expression for the length of an unknown line will be

$$D = nU + L + k \qquad (20.1)$$

where n is an integer multiple indicating how many unit lengths are included in the line, L is the fraction of the last noninteger unit length, and k is the combined instrument and reflector constant. If only one modulation frequency were used, the length of the line would be known within $U = \lambda/2$, and n thus would be known. To extend the length of the unambiguous known distance, we take three slightly different modulation frequencies f_1, f_2, and f_3, and Eq. 20.1 can be written as follows:

$$D = n_1 U_1 + L_1 + k$$
$$D = n_2 U_2 + L_2 + k \qquad (20.2)$$
$$D = n_3 U_3 + L_3 + k$$

where the unit lengths U_1, U_2, and U_3 in the NASM-4 and in the newer models are related in the following way:

$$400 U_1 = 401 U_2 = 2000 \qquad \text{and} \qquad 20 U_1 = 21 U_3 = 100 \qquad (20.3)$$

In Eq. 20.2 the values of k, L, and U are known, but the n-values and D are unknown; thus Eq. 20.2 as such cannot be solved.

To obtain a valid solution, differences of the expressions in Eq. 20.2 are used. First, however, two aspects must be noted. When applying the difference of the first two expressions in Eq. 20.2, $n_1 = n_2 = m_{12}$ within

2000 m, as Eq. 20.3 indicates, and the total distance to be measured must be known to an accuracy of 2000 m. In other words, we must know how many times 2000 is included in the total distance. Correspondingly, when applying the difference of the first and the third expression in Eq. 20.2, $n_1 = n_3 = m_{13}$ within 100 m, and again we must know how many times 100 m is included in the total distance. The integer x of the multiple of 2000 m must be obtained by reconnaissance survey, from any existing map, or by using a fourth modulation frequency f_4 and the integer y of the multiple of 100 m is obtained from the solution of the first difference.

In explaining the mathematics involved, we take advantage of the difference of the first two expressions of Eq. 20.2 as a sample. By applying the first aspect noted previously to these expressions we obtain

$$D = x2000 + m_{12}U_1 + L_1 + k$$
$$D = x2000 + m_{12}U_2 + L_2 + k \tag{20.4}$$

and from their difference we have

$$m_{12} = \frac{L_2 - L_1}{U_1 - U_2} \tag{20.5}$$

From Eq. 20.3, $U_2 = (400/401)\,U_1$ which yields

$$m_{12} = \frac{L_2 - L_1}{U_1 - (400/401)U_1} = \frac{401(L_2 - L_1)}{U_1} \tag{20.6}$$

when substituted into Eq. 20.5.

The substitution of m_{12} from Eq. 20.6 into the first expression of Eq. 20.4 yields

$$D = x2000 + 401(L_2 - L_1) + L_1 + k \tag{20.7}$$

The known relationship between U_1 and U_2 permits us to write Eq. 20.7 in the following identical form:

$$D = x2000 + 400(L_2 - L_1) + L_2 + k \tag{20.8}$$

With similar analysis it can be shown that when the difference between the first and third expressions in Eq. 20.2 is utilized, the following equations for the total distance D are obtained:

$$D = y100 + 21(L_3 - L_1) + L_1 + k \quad \text{and}$$
$$D = y100 + 20(L_3 - L_1) + L_3 + k \tag{20.9}$$

Thus two independent measurements of the distance are made by using three different frequencies and observing three L-values. To visualize the physical meaning of Eqs. 20.7 to 20.9, let us consider them to represent a type of electronic vernier system; or, imagine measuring a distance using two rulers of slightly different scale with graduations at one end only.

GEODIMETER MODELS.   The early models *NASM-1* and *NASM-2* were large and boxlike, weighing about 125 kg including the power generator. They were primarily built for use in high precision baseline measurement and for checking sides of first-order triangulations. Only nighttime operation was possible, and distances of 50 and about 35 km could be reached with the NASM-1 and the NASM-2, respectively. The basic modulation frequency in both instruments was 10 MHz, corresponding to a modulation wavelength of approximately 30 m. The exact wavelength, thus the unit length U, was determined for each measurement by calibration. NASM-1 used two modulation frequencies, and the line was measured in units of 750 m. NASM-2 used three modulation frequencies, and the line was measured in units of 1500 and 250 m. The reflector in NASM-1 was a plane mirror, whereas NASM-2 first had a concave mirror, which was later replaced by retrodirective prisms. The accuracy of NASM-2 was about ± 20 mm $\pm 1.2 \times 10^{-6}$.

The model *NASM-3* was a lightweight instrument designed principally for geodesists and surveyors for use in various tasks such as trilaterations and supertraverses. The total weight of the instrument including the power generator was 60 kg. NASM-3 could be used only at night, and a distance of 25 km could be reached. The modulation frequency f_1 was selected such that at given atmospheric conditions, and by assuming $c_0 = 299,792.5$ km/sec, the unit length U_1 would be exactly 50 m. The atmospheric parameters used as reference were $T = -6°C$, $P = 760$ mm Hg, and $e = 50\%$ relative humidity. This condition yields the refractive index $n = 1.0003104$ and the "built-in" velocity $c = 299,699.5$ km/sec. As a result, $f_1 = 1.4985$ MHz. Only two modulation frequencies were used with $f_2 = 1.025 f_1$ and the lines were measured in units of 2000 m. Because of the increase in the unit length U from the 7.5 m of previous models to 50.0 m, and because of some modifications in the delay line and calibration methods, the accuracy in NASM-3 was reduced to about ± 50 mm $\pm 2 \times 10^{-6}$.

Several improvements were introduced with the *NASM-4* model. The weight of the instrument remained approximately the same as that of NASM-3, but the newer model was smaller. The basic modulation frequency f_1 was increased to 29.97 MHz, and when the velocity of light was assigned the value recommended by *Bergstrand*, namely, $c_0 = 299,792.9$ km/sec, it yielded the exact value for the unit length $U_1 = 5.000$ m. In NASM-4 the null meter is read to define the unit length to be

$U = \lambda/2$. Two other modulation frequencies f_2 and f_3 are also used to relate U_2 and U_3 with U_1 according to Eq. 20.3. In models NASM-4 and 4B, which differ only slightly from each other, the light source is a tungsten lamp. Depending on the number of reflectors, nighttime distances to 15 km may be measured. Even during daylight, distances to 1.0 to 1.5 km are measurable. A big improvement was achieved in model *NASM-4D*, in which the tungsten lamp was replaced by a mercury arc lamp. With this new source of light the daylight distance was extended to about 5 km, and it is possible to measure distances of 25 to 30 km at night. The NASM-4 is a rugged instrument, relatively easy to operate. Because of the reduction of the unit length U_1 to 5.000 m, accuracy was improved to about ± 10 mm $\pm 2 \times 10^{-6}$.

In model *M6* the optical principle, thus also the external appearance of the instrument, was changed drastically. In the previous models the transmitting and receiving optics were located side by side horizontally, giving the instrument a boxlike look. In model M6 the optics are coaxial, and the axes of the transmitting and receiving optical beam coincide. In the middle of the telescope a prism directs the light beam from the Kerr cell to the distant reflector, and after return it directs the beam to the photomultiplier for measurement purposes. For calibration the prism is removed from the light path, allowing the beam to propagate directly from the Kerr cell to the photomultiplier. This new optical principle makes the Geodimeter resemble a theodolite with a U-shaped fork and horizontal axis holding the telescope (Fig. 20.2). The instrument shown in Fig. 20.2 is the model M6B, which has the same external appearance as the M6.

The light source can be either a tungsten lamp or a mercury arc lamp. The tungsten lamp is powered by a conventional 12 V lead-acid battery or a lightweight and compact 12 V nickel-cadmium battery; the mercury arc lamp needs a power generator. The M6 utilizes the same modulation frequencies as the NASM-4, thus the same unit lengths. The electronic components in both instruments are almost the same except for small differences in the delay settings. In the M6 the tungsten and mercury lamps can be easily interchanged in the field. The total weight of the M6 is 43 kg, which is less than the NASM-4. The maximum obtainable distances for day and night are about the same in both models, but the accuracy of M6 is slightly higher.

In model *M6A* the optics are the same as in M6, but an essential change and improvement is made in the electronics, especially in the demodulation system. Instead of using only one generator feeding both the Kerr cell and the photomultiplier with the same modulation frequency

Figure 20.2   Model M6B Geodimeter (courtesy AGA Co.).

f_T through a variable delay line, there are two generators. One feeds the Kerr cell with a modulation frequency f_T (about 30 MHz) and the other, a reference generator, feeds the photomultiplier with a reference frequency f_R which is 1.5 kHz below f_T. The photomultiplier thus varies in sensitivity with the reference frequency f_R and, according to the heterodyne principle, the photomultiplier output signal has a frequency of $f_T - f_R =$ 1.5 kHz. The phase of this signal is identical with the phase of the received f_T signal and is measured by comparison with the reference signal $f_T - f_R$ derived directly from the frequencies of the transmitter and the reference generators. The variable delay line between the Kerr cell and the photomultiplier is replaced by a rotatable phase resolver between the frequency mixer and the phase detector. In this resolver the phase of the reference signal can be delayed until, in effect, the phases of the transmitted and received modulations are equal. This condition is indicated by a zero reading in the null meter, and the delay in the resolver can be read in meters. The receiving and demodulation system just explained can be better comprehended by studying the block diagram in Fig. 20.3, which actually refers to model M8 with identical receiving electronics.

The modulation frequencies in model M6A are the same as in model M 6 with an addition of the fourth frequency f_4, which is related to f_3 by

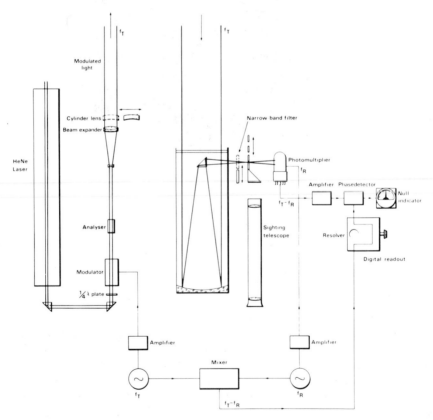

Figure 20.3 Block diagram of M8 Geodimeter (courtesy AGA Co.).

$f_4 = 10,000 f_3/10,001$. This yields the third equation of distance:

$$D = z\, 50,000 + 10,000(L_3 - L_4) + L_4 + k \qquad (20.10)$$

The total weight of the M6A is about 43 kg. The maximum distances obtained in daylight are 5 and 10 km, depending on whether tungsten or mercury lamps are used. The corresponding distances at night are 15 and 25 km. The improvements in the receiving electronics, the replacement of the delay line by the resolver, and the comparison of phases of low frequency heterodyned signals rather than the high frequency modulation signals increased the over-all accuracy to about $\pm 5\,\text{mm} \pm 10^{-6}$.

In essence, model M6B, is very similar to model M6A except that the modulation frequencies have been partially changed and rearranged to make the instrument more suitable for measurements of short lines. The frequencies have been renumbered, and f_1 has been added as $f_1 = 0.74925$ MHz; f_2 has the same basic value as f_1 in the NASM-4 and the earlier M6 models, $f_2 = 29.970$ MHz. The two other frequencies, f_3 and f_4, are related to f_2 and to each other in the following way: $f_3 = 1001 f_2/1000$ and $f_4 = 10,000\, f_3/10,001$ yielding the transmitting modulation frequencies $f_3 = 30.000$ MHz and $f_4 = 29.997$ MHz.

The frequency f_1 now determines the unit length U_1 as follows:

$$U_1 = \frac{c_0}{n\,2f_1} = 200.000 \text{ m}$$

and the frequency f_2 determines the unit length U_2 as follows

$$U_2 = \frac{c_0}{n\,2f_2} = 5.000 \text{ m}$$

where the refractive index is $n = 1.000309$ and the velocity of light is $c_0 = 299,792.5$ km/sec. Frequencies f_3 and f_4 are not used alone but their differences are employed, as before, to determine the higher order unit lengths which, when rounded up will be

$$U_{3-2} = \frac{c_0}{n\,2(f_3 - f_2)} = 5000 \text{ m} \qquad \text{and} \qquad U_{3-4} = \frac{c_0}{n\,2(f_3 - f_4)} = 50,000 \text{ m}$$

The distance equations can then be written as follows:

$$D = x\, 50,000 + 10,000(L_3 - L_4) + L_3 + k$$
$$D = y\, 5000 + 1000(L_3 - L_2) + L_3 + k \qquad (20.11)$$
$$D = n_1 200 + L_1 + k$$
$$D = n_2 5 + L_2 + k$$

The weight of the instrument, the maximum measurable distances, and the accuracy are the same as in model M 6A.

The latest model in the M 6 series is the *M 6BL*, which is identical to the M 6B in all respects except that the light source is a 1 mW He–Ne laser and the electronic circuitry is designed as plug-in modules for easier and more economical servicing. Because of the lower power consumption, a lightweight nickel-cadmium battery may be used as the power source. Also, by replacing the relatively heavy mercury lamp system with the laser, the total instrument weight is reduced to about 27 kg. The measurable distance, which is about 25 km is the same for daylight and at night. The accuracy is the same as that of models M 6A and M 6B.

The model *M 8* Geodimeter was designed to be used in high precision, long-distance applications in geodesy and geophysics. Its light source is a He–Ne laser similar to the one in M 6BL. Externally it resembles the NASM-4, having a boxlike housing and noncoaxial optics, which are not needed in long-line measurements (Fig. 20.4). The total weight of the instrument is about 32 kg. The measurable distance is double or more than that obtained by model M 6BL, and distances to 65 km have been measured (Wood, 1973). The improvement in the maximum distance is caused by several modifications in the transmitting optics (see the block diagram, Fig. 20.3).

The He–Ne laser for the M 8 is a Spectra-Physics type 120 with an output power of 5 mW. The radiated emission has wavelength $\lambda = 6328$ Å and a beam diameter of 0.65 mm. The beam divergence is 1.7 mrad, which at the distance of 1 km causes an illuminated spot 1.7 m in diameter. To make the instrument more compact and easier to service, the beam from the laser is deflected 180° by two prisms and at the same time is displaced horizontally. The beam then passes a quarter-wavelength plate in front of the modulator to stabilize the properties of the modulator with respect to temperature. The Kerr cell that served as the modulator in earlier models is replaced by a Pockels modulator in the M 8. After leaving the Pockels modulator, the beam is polarization modulated and is fed to an analyzer that has a polarization plane parallel to the rotated polarization; thus the beam passes through the analyzer.

To concentrate the highest possible light power at the reflector without changing the characteristics of the laser, after the beam leaves the analyzer it is fed to a beam expander, where a combination of a concave and a convex lens increases its diameter to 2 cm and the divergence is reduced to about 0.1 mrad. The beam now illuminates a spot with 1 m diameter at the distance of 10 km. After return from the distant reflector, the beam is fed to the photomultiplier through a narrow-band optical

Figure 20.4 Model M8 Geodimeter (courtesy AGA Co.).

filter that is highly transparent for the laser beam but almost opaque for light of other wavelengths within the range of sensitivity to the photomultiplier used.

With such a narrow beam and at long distances, the reflectors are difficult to locate. The M 8 is provided with a special lens that can be inserted in the beam. This lens widens or polarizes the beam vertically to about $0.°25$ without affecting it horizontally. Sweeping the target area horizontally, one can catch a glimpse of reflected light in the receiver eyepiece. Perfect reflection can be achieved after removing the special lens by vertical fine adjustment of the Geodimeter.

The modulation frequencies f_1, f_2, and f_3, thus also the unit lengths U_1, U_2, and U_3 are related according to Eq. 20.3 yielding the values of

$$f_1 = 29.970000 \text{ MHz} \qquad U_1 = 5.000000 \text{ m}$$
$$f_2 = 30.044925 \text{ MHz} \qquad U_2 = 4.987531 \text{ m}$$
$$f_3 = 31.468500 \text{ MHz} \qquad U_3 = 4.761904 \text{ m}$$

when $c_0 = 299{,}792.5$ km/sec and $n = 1.0003086$. As stated before, the reference frequencies are exactly 1.5 kHz below the modulation frequencies and are thus

$$f_1 = 29.968500 \text{ MHz}$$
$$f_2 = 30.043425 \text{ MHz}$$
$$f_3 = 31.467000 \text{ MHz}$$

The fourth modulation frequency f_4 needed to determine the number of 2000-m intervals is placed exactly 3.0 kHz below the modulation frequency f_3, which means that it is exactly 1.5 kHz *below* the reference frequency f_3. In this case, the modulation frequency f_4 and the corresponding unit length U_4 are

$$f_4 = 31.465500 \text{ MHz}$$
$$U_4 = 4.762358 \text{ m}$$

By this arrangement the reference frequencies f_3 and f_4 are the same. The determination of the number of 2000 m intervals is very simple. The difference between the unit lengths U_4 and U_3 is 0.000454. In a 2000 m interval there are 420 unit lengths of U_4, and the difference between L_3 and L_4 at the end of this interval will be $420 \times 0.000454 = 0.19068$. The number of 2000 m intervals x is now obtained as follows:

$$x = \frac{L_3 - L_4}{0.19068} \tag{20.12}$$

The accuracy of model M 8 with standard observation procedure is about $\pm 5 \text{ mm} \pm 10^{-6}$. The error proportional to distance in all models is primarily caused by the drift of the frequencies and by the uncertainty in determining the refractive index. Errors of the former type may be reduced by checking the crystals before and after the measurements with a high precision frequency counter (*Parm*, 1973); errors of the other type may respond to a least squares principle as described in Chapter 7.

REFLECTORS.   All reflectors in the modern instruments are based on the retrodirectivity principle. Plane mirrors used in early prototypes were extremely sensitive to even the smallest misalignment. Each unit in the reflector is a retrodirective prism made of three mutually perpendicular reflecting surfaces. The prism is often called a corner reflector because for stability, it is made by cutting a corner from a solid glass cube. Light entering the prism reflects from each of the three surfaces, and afterward this triple reflection returns to the instrument parallel to the incident beam. With respect to illumination power, the alignment of the reflector with respect to the Geodimeter instrument is not critical and alignment errors of the order of $10°$ may be tolerated.

Where distances are concerned, the alignment of a single prism is not critical, but great care must be taken to avoid vertical inclination of a housing where several prisms are stacked in a pile. This inclination causes measured distance to become too short or too long, depending on whether the inclination occurs toward the Geodimeter or away from it. An inclination of 1° of the vertical axis of a housing containing three prisms results in a centering error of approximately 4 mm. Small horizontal alignment errors can be tolerated, since all prisms in a multiprism reflector are symmetrically located with respect to the vertical axis of the reflector system. The effective reflecting plane of the prisms does not coincide exactly with the plumb point of the housing. This, together with the fact that the speed of light is slower in glass than in air, will cause the Geodimeter to give a false reading. For AGA prisms mounted in an AGA housing, users apply a correction called the reflector constant, which has a magnitude of −30 mm.

In all NASM models the transmitting and receiving optics were located side by side horizontally. To aim the returning beam from the reflector to the receiving optics, a small parallactic angle between the incident and reflecting beams at the reflector had to be established. This was done during manufacture by deviating the three reflecting surfaces of the prism slightly from the right angle condition. In addition, depending on the distance to be measured, glass wedges were placed in front of the prisms to introduce the necessary beam split. In all M 6 models having a coaxial optical system, the introduction of the parallax became unnecessary; thus the operation was simplified and the illumination power was increased. The present reflectors are available as a single unit or in groups of 3, 9, and 16. All reflectors have a target rod for angle observations and an adapter to connect them to regular theodolite tribrachs with optical plummets. A complete set of 18 prisms with a target rod and an optical plummet is illustrated in Fig. 20.5.

20.1.1.2. *Geodolite 3G and 3A*

Spectra-Physics, in Mountain View, California, has been engaged in the development of modulated CW laser distance measuring instruments since 1963. The Geodolite 3G, a ground-to-ground electro-optical laser distancer, was put on the market in 1966, and even though production has been discontinued, it still is the most accurate electro-optical *long-range* distancer. Its light source is a He–Ne laser. Five modulation frequencies are available, and a typical frequency accuracy is 10^{-7}. The highest modulation frequency is $f_1 = 49.16471$ MHz, with a phase measurement accuracy of 0.003 ft or better. The phase measurement between

Figure 20.5 Set of 18 prisms with target rod (courtesy AGA Co.).

the transmitted and received modulation signals is made by a beat frequency of 4.916 kHz. The modulation frequency patterns refer to the English system of distances in feet, and if the coarse readings are used, an unambiguous distance of 100,000 ft can be measured. The instrument is relatively heavy and bulky. The telescope assembly weighs about 45 kg, the control unit about 18 kg, and the readout unit about 14 kg. The maximum measurable distance at night is about 80 km, with an accuracy (resolution) of $\pm 1 \, \text{mm} \pm 10^{-6}$, whichever is greater. A simplified block diagram of the model 3G Geodolite appears in Fig. 20.6. At present, Spectra-Physics is manufacturing special lasers for AGA's Geodimeters.

Another member of the Geodolite series is the model 3A, which was designed for high precision aircraft or helicopter altimetry. The beam divergency was less than 0.1 mrad, permitting resolution of very small features on the ground. The maximum measurable altitude at night is about $11 \, \text{km} \pm 10^{-4}$. Because of the very narrow beamwidth, model 3A is especially suited to terrain profiling. Production of this model also has been discontinued.

20.1.1.3. *Rangemaster and Ranger III*

The production of *Laser Rangers* was initiated in 1969 by Laser Systems and Electronics, Inc., in Tullahoma, Tennessee. Later the Ranger series was expanded and built exclusively for the Keuffel & Esser Company (K&E). The *Rangemaster* (Fig. 20.7) features a 10 mW He–Ne laser as the light source. It is amplitude modulated, and there are two measuring frequencies. Phase angle measurement between the outgoing and incoming modulation waves is made by the pulse-counter method. The pulse-counting frequency has been selected to ensure that the range measurement is a direct result of the count. Both the counting frequency and the modulating frequency of the transmitted light beam are generated by the same precision oscillator, eliminating the need for calibration in connection with the phase measurement. At modulation frequency f_1, 1000 phase measurements are performed, averaged, and stored in a built-in computer. At modulation frequency f_2, another 1000 phase measurements are performed and averaged. The resulting values of the electronic counter are displayed in the range display window as one eight-digit numerical distance. This type of phase measuring principle is utilized in most of the short-range electro-optical EDM systems. A circular slide-rule type of calculator comes with the instrument, and it allows the user to obtain the atmospheric correction in parts per million and then to set the corrections in the instrument, which automatically includes it in the

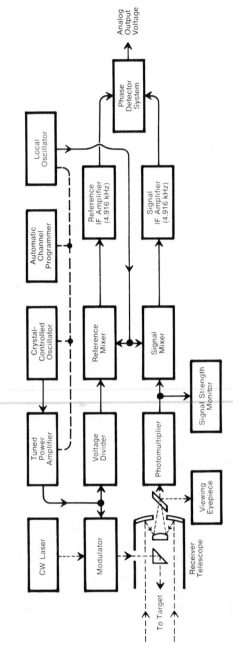

Figure 20.6 Simplified block diagram of model 3G Geodolite (courtesy Spectra-Physics).

Figure 20.7 Rangemaster (courtesy Keuffel & Esser Co.).

displayed distance. Also, any instrument and target offset constant can be set in the instrument and will be included in the final reading. An ft/m switch permits the readout to be shown in feet or meters. The Rangemaster cannot be used in connection with theodolites or transits, but optical plummets are available as optional accessories. Momentary interruptions of the beam by moving objects such as traffic will not affect the accuracy of the readout. After a 30 sec warmup, the measuring time is only a few seconds. The instrument's dimensions are $28.5 \times 28.5 \times 66.0$ cm, and it weighs 27 kg. The power consumption is 10 A at 12 V DC, and the operational temperature range is -28 to $+54°C$. The maximum measurable distance is about 60 km, with an accuracy of $\pm 3 \, \text{mm} \pm 10^{-6}$.

The *Ranger III* is almost identical to the Rangemaster; because of its lower power requirements and smaller laser, however, its range is about 12 km and its accuracy $\pm 5 \, \text{mm} \pm 2 \times 10^{-6}$. The light source is a 3 mW He–Ne laser, and the power consumption is 7 A at 12 V DC. It weighs 17 kg and has dimensions of $31.7 \times 22.8 \times 43.4$ cm.

20.1.1.4. *Tellurometers*

The Tellurometers use microwaves at about 3, 10, or 35 GHz as the carrier. The measuring set consists of two active units with a transmitter and a receiver; one is called the master and the other the remote unit. The carrier frequencies of the two units differ slightly, making it possible to utilize intermediate frequency (IF) amplification. The microwave carriers are frequency modulated by measurement or pattern signals that are slightly different at the master and remote units, to create beat frequencies in those units. Also, several other nearby pattern frequencies build up the total unambiguous distance. The range of the modulation frequencies varies according to the model between 5 and 25 MHz.

Because the carriers are microwaves, the beamwidths are narrow—between 2 and 20°. Measuring can be carried on either at night or daytime, through haze or light rain, although heavy rainfall may reduce the working range. The bare outlines of the measuring principle are as follows. A frequency-modulated carrier wave from the master station is sent to the remote station, where it is received and retransmitted to the master station. There the phase difference between the transmitted and received modulation or pattern waves is compared. The distance can be determined by knowing the average velocity of the radio waves along the ray path and also the master modulation wavelength. The development of the Tellurometer system of measurements, initiated by *Wadley* (1957) at the National Institute of Telecommunications Research of South Africa, is now the basis of all microwave distance measuring instruments. Commercial manufacturing was begun by the Tellurometer (Pty.) Ltd. in South Africa, and now the instrument is also manufactured and serviced in the United States, England, Canada, Australia, Japan, and Singapore.

The first commercial model, the MRA-1, appeared in 1957; it was followed by MRA-2 (1960), MRA-3 (1963), MRA-4 (1965), MRA-101 (1965), MRA-301 (1967), MRA-5, and the CA-1000 (1973). All models have duplex speech facilities, and the master and remote units are interchangeable in most.

OPERATION AND MEASURING PRINCIPLE. The following simplified presentation gives the principle of the Tellurometer system of phase propagation between the master unit and the remote units. Let us assume that at any instant the phase angle of the master modulation signal at the master station is Φ_M and the corresponding phase angle of the remote modulation signal at the remote station is Φ_R. Both signals travel an equal distance D, and when the remote signal arrives at the master antenna, the phase angle due to the distance covered will be

$$\Phi'_R = \Phi_R + \frac{2\pi f_R D}{c}$$

where f_R is the modulation frequency of the remote signal and c is the velocity of electromagnetic propagation over the line. At the master antenna both the local and the arriving modulation signals are mixed, yielding the low frequency beat signal $(f_M - f_R)$, where f_M is the modulation frequency of the master signal. The phase angle of the beat signal is then

$$\Phi''_R = \Phi_M - \Phi'_R = \Phi_M - \Phi_R - \frac{2\pi f_R D}{c} \qquad (20.13)$$

Equation 20.13 cannot be used to determine D because of the presence of the unknown initial phase angles Φ_M and Φ_R. Therefore, let the master signal travel to the remote antenna, where its phase angle will be $\Phi'_M = \Phi_M + 2\pi f_M D/c$. After the modulation signals are mixed at the remote antenna, the phase angle of the resulting beat signal becomes

$$\Phi''_M = \Phi'_M - \Phi_R = \Phi_M + \frac{2\pi f_M D}{c} - \Phi_R \qquad (20.14)$$

This phase "information" is now sent back to the master receiver as an additional modulation of the remote carrier. The same distance D will be traversed again, causing further phase lag but in this case with respect to the low frequency, remote beat signal $(f_M - f_R)$. The phase angle of the information signal at the master station then will be

$$\Phi'''_R = \Phi_M + \frac{2\pi f_M D}{c} - \phi_R + \frac{2\pi (f_M - f_R) D}{c} \qquad (20.15)$$

The phase angles Φ''_R (Eq. 20.13) and Φ'''_R (Eq. 20.15) are now compared, resulting in the phase difference $\Delta\Phi$ between the transmitted and received master modulations as follows:

$$\Delta\Phi = \Phi'''_R - \Phi''_R = \frac{4\pi f_M D}{c} \qquad (20.16)$$

Since in the phasemeter the partial cycle $\Delta\Phi'$ is read, both sides of Eq. 20.16 must be divided by 2π, which yields

$$\frac{\Delta\Phi}{2\pi} = \frac{2 f_M D}{c} = \Delta\Phi'$$

and, finally,

$$D = \frac{c}{2f_M} \Delta\Phi' = \frac{\lambda_M}{2} \Delta\Phi' \qquad (20.17)$$

where λ_M is the master modulation wavelength.

The arbitrary sequence of events in the phase propagation just presented was selected for the sake of convenience only. Equation 20.17 proves that the distance D is obtained as multiples and fractions of the modulation half-wavelengths, as was the case in the Geodimeter. The general distance equation then can be written as

$$D = (n + \Delta\Phi') \frac{\lambda_M}{2} \qquad (20.18)$$

where n is the integer amount of half-wavelengths included in the distance to be measured. Equation 20.18 also can be written in the following form, which is more suitable for further analyses:

$$D = \left(n + \frac{L}{100}\right) \frac{c}{2f_M} \qquad (20.19)$$

where L is the reading of the phasemeter dial, which is divided into 100 graduations.

The Tellurometer uses five modulation or pattern frequencies (with some exceptions, described later); these are designated f_A, f_B, f_C, f_D, and f_E. The basic frequency f_A determines the measuring unit and its fraction, and the corresponding reading L_A is called the "fine reading." The other frequencies are needed to solve the ambiguity of Eq. 20.19 with respect to n in such a way that decimals are introduced between frequencies. This is not done by separating the modulation frequencies from each other by multiples of 10 but rather by assigning them values that form differences in the decimal system when they are subtracted from the basic frequency f_A. This means that $(f_A - f_E) = 10(f_A - f_D) = 100(f_A - f_C) = 1000(f_A - f_B)$ and the corresponding readings L_B through L_E are called the "coarse readings." In this frequency-difference system Eq. 20.19 can be derived as follows with regard to $(f_A - f_B)$ as an example

$$D = \left(n_A + \frac{L_A}{100}\right) \frac{c}{2f_A}$$

$$D = \left(n_B + \frac{L_B}{100}\right) \frac{c}{2f_B}$$

First, the readings L_A and L_B are solved:

$$L_A = 100D \frac{2f_A}{c} - 100n_A \qquad \text{and} \qquad L_B = 100D \frac{2f_B}{c} - 100n_B$$

and their difference is taken as

$$L_A - L_B = 100D \frac{2(f_A - f_B)}{c} - 100(n_A - n_B)$$

From this point D can be solved as follows:

$$D = \left[(n_A - n_B) + \frac{L_A - L_B}{100} \right] \frac{c}{2(f_A - f_B)} \qquad (20.20)$$

Corresponding equations can be written for frequency differences ($f_A - f_C$), ($f_A - f_D$), ($f_A - f_E$), and for f_A itself with (f_A-). Based on the built-in velocity c, the modulation frequencies are related in such a way that the simulated wavelengths derived from the frequency differences are whole numbers of meters. Because of the decimal system adapted, the simulated wavelengths are then related to each other, so that $\lambda_B = 10\lambda_C = 100\lambda_D = 1000\lambda_E = 10{,}000\lambda_A$. Under these conditions, the second term of Eq. 20.20 determines the first approximation of D. To that approximation each consecutive simulated wavelength adds one significant digit, and the total distance can be expressed as a digital readout in the following way:

$$
\begin{array}{cc}
B & \text{X}\lfloor\text{X,000.00} \\
C_. & \text{X,}\lfloor\text{X00.00} \\
D & \text{X}\lfloor\text{X0.00} \qquad (20.21) \\
E & \text{X}\lfloor\text{X.00} \\
A & \text{X.XX} \\
\hline
 & \text{XX,XXX.XX}
\end{array}
$$

For checking purpose it should be noted that the second digit of each line from left will be the same as the first digit of the line below. The fine readings A are assumed to be made with an accuracy one decimal higher than the coarse readings. When the decimal system is employed, the first term of Eq. 20.20 becomes meaningless in practical computations except in the coarse pattern B, where it indicates the limit at which the system starts repeating itself.

To understand better the operation yielding the distance measurement in the Tellurometer system (Eq. 20.17), let us analyse a simplified block diagram of the MRA-3 (Fig. 20.8) with reference to the derivation of the phase propagation. In both the master and remote units the carrier wave is generated by a klystron oscillator operating at a frequency range between 10.025 and 10.450 GHz. During the observations of fine readings in the A patterns, the cavity settings of the klystrons are changed to eliminate the effect of the ground swing (this is described in detail in

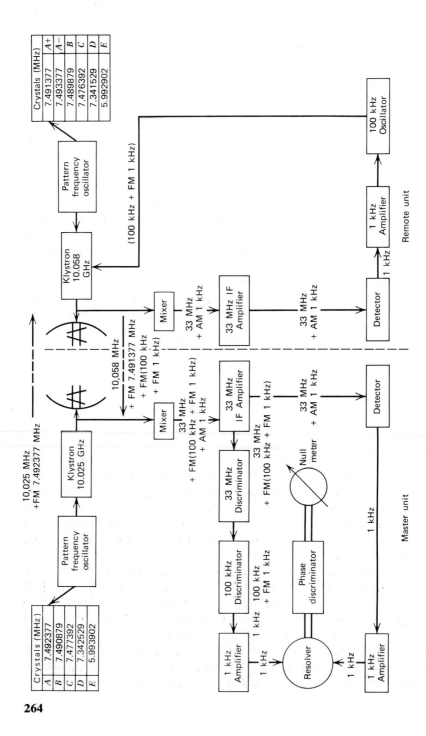

Figure 20.8 Block diagram of Tellurometer MRA-3.

Chapter 11). The modulation or pattern signals are generated in an oscillator that is controlled by temperature-stabilized crystals referring to each of the pattern frequencies A through E. The remote modulation frequencies are 1 kHz lower than the corresponding master frequencies, except an additional (A-) frequency in the remote unit, which is 1 kHz higher. The (A-) frequency is used to provide the measuring unit itself and its decimals. The pattern signals are applied to the klystron reflectors in both units, and they produce frequency modulation of the carriers, which are tuned in such a way that the remote frequency is 33 MHz higher than the master frequency.

The carrier wave from the master antenna modulated by the A pattern frequency arrives as 10,025 MHz + FM 7.492377 MHz to the remote antenna, where it is mixed with the remote carrier wave of 10,058 MHz modulated by the A+ pattern frequency of 7.491377 MHz. The two signals are mixed in the remote antenna, producing an intermediate frequency of 33 MHz with an amplitude modulation of 1 kHz (Eq. 20.14). The IF is amplified in the IF amplifier and is fed into a detector, where it is demodulated to the 1 kHz information signal. This signal, after passing the 1 kHz amplifier, frequency modulates the output of a local 100 kHz oscillator (100 kHz was selected to reduce the frequency separation between the carrier and the modulation values). The 100 kHz signal serves as a subcarrier, and the information signal arriving to the master antenna is of the following form: 10,058 MHz + FM(100 kHz + FM 1 kHz) (Eq. 20.15). Simultaneously, however, the remote carrier is frequency modulated by the A+ pattern frequency, and this component arrives at the master station as 10,058 MHz + FM 7.491377 MHz. After the modulated carrier waves have been mixed at the master antenna, the signal may be considered to be composed of two components, the information component from the remote station as 33 MHz + FM(100 kHz + FM 1 kHz) and the local component as 33 kHz + AM 1 kHz (Eq. 20.13). The signal is amplified in the IF amplifier and the two 1 kHz components are processed separately, to be compared by phases in the resolver.

The modulated IF with the AM 1 kHz local signal is applied to a detector circuit that passes the 1 kHz modulation but rejects the 33 MHz IF. The 1 kHz output from the detector is amplified and then fed into the rotor windings of the resolver. The modulated IF with the FM(100 kHz + FM 1 kHz) information signal is fed into a frequency discriminator circuit that rejects the 33 MHz IF but passes the 100 kHz with its FM 1 kHz. A further discriminator circuit rejects the 100 kHz frequency but passes the 1 kHz information signal, which, after amplification, is fed into the stator windings of the resolver. Finally, zeroing the null meter allows the phase comparison to be made for the distance determination (Eq. 20.17). It must be kept in mind in deriving the phase propagation (Eqs. 20.13–20.17) that once the instrument is in operation over a fixed line, the phasing in both instruments is stationary except for small variations caused by human errors, instrument errors, and variations in atmospheric conditions.

TELLUROMETER MODELS. The first Tellurometer, model *MRA-1*, was the pioneering design in ground-to-ground surveys using microwave modulation. The general operational principle is essentially the same as that described in the previous section, but there are several design differences. The measuring unit is time, and the display is presented in the form of a circular time base in the cathode ray tube (CRT), which is divided into 100 graduations. Four pattern frequencies are used—namely, $f_A = 10.00$ MHz, $f_B = 9.99$ MHz, $f_C = 9.90$ MHz, and $f_D = 9.00$ MHz—so that their differences from the f_A frequency form a decimal system. If in Eq. 20.20 the distance D is expressed in terms of velocity c and time T needed for the signal to travel from the master to the remote unit, we can now write

$$T = \left[(n_A - n_B) + \frac{L_A - L_B}{100} \right] \frac{1}{f_A - f_B} \text{ nsec} \qquad (20.22)$$

and the actual values of T for the entire pattern sequence, when only the second terms of Eq. 20.22 are used, are

$$T = 1000(L_A - L_B) \text{ nsec}$$
$$T = 100(L_A - L_C) \text{ nsec}$$
$$T = 10(L_A - L_D) \text{ nsec} \qquad (20.23)$$
$$T = L_A \text{ nsec}$$

which can then be applied to Eq. 20.21.

The complete rotation of the time base represents 100 nsec in the basic *A-* pattern frequency; thus each division equals approximately 15 cm in the measured direct distance. For the consecutive frequency-difference patterns, the corresponding figures are multiples of tens of the previous values. The highest coarse reading $(L_A - L_B)$ thus defines the length of the line with an accuracy of 15 km. To obtain the exact value of the distance with respect to the observed time, the observed time must be corrected to represent the velocity of light $c_0 = 299,792.5$ km/sec by dividing it by the ambient refractive index.

The carrier frequencies of both master and remote units are about 3.000 GHz with 33 MHz IF. The corresponding pattern frequencies of the master and the remote units differ from each other by 1 KHz. The 1 kHz local signal in the master unit forms the circular time base of 0.001 sec duration, and the 1 kHz information signal in the remote unit is pulse modulated, which causes a break in the time base after the signal arrives at the master unit. The break in the time base indicates the fractional time reading with reference to the total time represented by one revolution of the time base.

To remove the zero error of the reading system and to eliminate any nonlinearity of the time base, the direction of the sweep in the time base is reversed and the information signal in the remote unit is placed 180° out of phase. Thus one set of line readings is composed of $A+$, $A-$, $A+_{rev}$, and $A-_{rev}$ components.

The weight of either master or remote instrument is about 12 kg, battery excluded. The parabolic reflector radiating with a beamwidth of about 20° within the half-power limits is assembled outside the casing. A duplex telephone has the same channel and circuits used for measurements. The maximum measurable distance is about 100 km; with an accuracy of $\pm 5 \, cm \pm 4 \times 10^{-6}$. The master and the remote units are not interchangeable.

Tellurometer model *MRA-2* is very similar to model MRA-1 in external appearance, except that the instrument panel is on the side of the more recent device, the antenna being on the left end. The specifications are almost identical, but one big improvement in the design allows the master and the remote units to be interchangeable. This means that each unit can operate as the master and as the remote alternatively, thus providing checks for the measurements and strengthening the final mean values. Also, if one unit becomes partly defective in the field, it is usually possible to continue operations in reverse order.

Tellurometer model *MRA-3* (Fig. 20.9), was originally designed to meet U.S. Military Specifications and uses rather expensive missile-tested components. Several essential changes and improvements have been made compared to the earlier models (which are still widely used but are no longer in production). Except for the klystron, MRA-3 is an all-transistorized instrument. It houses a built-in nickel-cadmium battery that can be replaced by an outside 12 V DC source. The distance measurement is based on the resolver-null meter combination as represented in the block diagram (Fig. 20.8), and the readouts are obtainable directly in meters. Two presentation systems are available: namely, a direct digital readout and a dial readout featuring a vernier that allows the fraction of the graduation be read (in centimeters). The subtraction of two readings when two different modulation frequencies are used is done automatically

Figure 20.9 Tellurometer MRA-3 (courtesy Tellurometer–USA).

by a zero locking procedure. The MRA-3 units are interchangeable with each other and with the MRA-101 units but not with the MRA-2 units. The carrier frequency lies in the 10.025–10.450 GHz band, which with the built-in antenna introduces a beamwidth of 9° within the half-power limits. The instrument can be operated in the ambient temperature range of −40 to +50°C.

The basic pattern frequency f_A must be so chosen that in Eq. 20.19 a complete rotation of the reading counter is a whole number, in tens of meters. Since the speed of light $c_0 = 299,792.5$ km/sec and the refractive index $n = 1.000325$ are accepted as the built-in values, the basic master pattern frequency is $f_A = 7.492377$ MHz. With this frequency, one complete rotation of the reading counter represents a change of exactly 40 m

in the length of the double path and 20 m in the measured distance. The
difference $(f_A - f_{A-})$ then represents the measuring unit of 10 m.

To satisfy the condition for the decimal system, Eqs. 20.20 and 20.21,
the other master pattern frequencies are chosen as

$$f_B = 7.490879 \text{ MHz}$$
$$f_C = 7.477392 \text{ MHz}$$
$$f_D = 7.342529 \text{ MHz}$$
$$f_E = 5.993902 \text{ MHz}$$

(20.24)

To keep the IF exactly 33 MHz, the master klystron is controlled by an
automatic frequency control (AFC) system that modifies the carrier
frequency to match with any drift in the IF.

The instrument weighs about 15 kg without the battery and the carrying haversack. The measurable distance is about 80 km, with an accuracy
of $\pm 1.5 \text{ cm} \pm 3 \times 10^{-6}$. Model MRA-301 is also designed to military
specifications and like the MRA-3 has a digital readout.

Tellurometer model *MRA-101* was developed from the MRA-3 and is
primarily meant for civilian use. It is constructed with commercial components, which essentially lowers its price. The instrument fits in a compact
fiberglass transit case, and since it uses printed circuits, the weight is
reduced to about 12 kg. With a built-in reflector of 33-cm diameter, the
beamwidth is 6° between 3 dB limits (half-power). Its operational temperature range is slightly more limited than that of the MRA-3, namely,
−23 to +55°C. The readout is given in meters (and centimeters) directly,
and the subtraction of the two readings is done automatically. To keep
the 1 kHz beat frequency unchanged, a 1 kHz reference oscillator modifies the remote crystal frequencies correspondingly. Each unit is interchangeable with the others and with units of the MRA-3. The maximum
measurable distance is about 50 km, with an accuracy of $\pm 1.5 \text{ cm} \pm 3 \times 10^{-6}$.

Tellurometer *MRA-4* is the most accurate long-distance microwave
instrument in use today. There are two main reasons for this advance in
technology. First, the carrier frequency has been increased to 35 GHz, as
against the conventional 10 GHz used in most of the modern instruments.
This, together with the 33 cm diameter reflector, produces a narrow beam
of 2° between 3 dB limits, thus all but eliminating the ground swing
effect. Second, the modulation or pattern frequency has been increased to
75 MHz, which is 10 times larger than that used in most of the microwave
distancers, yielding a measurement resolution of 1 mm. The operational
principle is equivalent to that of the MRA-3 and MRA-101. Automatic

frequency control of the klystron and modified remote crystal frequencies ensure stability in phase comparison. The instrument is capable of operating within a temperature range of −40 to +55°C. The maximum measurable distance is about 60 km, with an accuracy of ±3 mm ±3 × 10⁻⁶.

Tellurometer *MRA-5* has been designed to meet the military standards of Royal Radar Establishment Specifications. It is very rugged and features a type of antenna and control unit separation similar to the Fairchild Microchain. There are three antenna positions. The antenna may be fitted in the receptacle on the front of the instrument, where it is controlled by the movement of the control unit. The antenna may be fitted to the azimuth mount on top of the control unit, where it can be rotated horizontally and vertically without moving the instrument, or the antenna may be separated completely from the control unit, fitted on the tripod, and connected by cable to the control unit. The MRA-5 has modular construction for easy maintenance. It has battery voltage tolerance between 10.8 and 34 V DC. The antenna has a beamwidth of 12° between 3 dB limits and weighs about 6 kg; the control unit weighs 12 kg. The operational temperature range is −32 to +44°C, and the maximum measurable distance is about 50 km. Voice communication is arranged by a loudspeaker on the front panel of the control unit and a handset. The carrier frequency range is 10.00–10.50 GHz. The readout is a fully automatic digital display, giving a resolution of 1 cm in fine readings and 10 cm in coarse readings.

The design of the Tellurometer *CA-1000* represents a drastic change in compactness, size, weight, and price compared to any other long-distance microwave instrument on the market. Figure 20.10 displays the master unit with the extended horn antenna but without the battery tray. The overall dimensions of the master and remote units are 305 × 146 × 76 mm, and the battery tray (identical in both units) is 299 × 121 × 38 mm. The total weight of each unit is 4 kg with the battery tray and 2 kg without it. The reduced size and weight result from the use of fiberglass casing and solid state miniature components. If an external 12 V DC power source such as a car battery is used, it is possible to operate the instrument and charge the internal nickel-cadmium batteries simultaneously. The current consumption is 380–600 mA at 12 V compared to 3.2 A at 12 V in an MRA-101; the carrier output power is about 15–30 mW compared to 50 mW in the MRA-101. This enables the nickel-cadmium batteries to be used for about 5 hours continuously. Duplex speech facilities between the non-interchangeable master and remote units are available at all times during measurements. The operational temperature range is from −55 to +60°C, and the maximum measurable distance is about 10 km using the

Figure 20.10   Tellurometer CA-1000 (courtesy Tellurometer–USA).

standard horn antenna (beamwidth 20°). The standard antenna can be replaced easily in the field by an extended range horn antenna that yields a maximum distance of 30 km.

In principle the CA-1000 is similar to the MRA-3, with a carrier frequency range of 10.00–10.45 GHz. Its five modulation frequencies A through E range from 19 to 25 MHz with a built-in refractive index $n = 1.000325$. A zero phase procedure is employed, and the main meter on the right-hand side of the front panel serves such multipurpose functions as battery testing, indication of microwave radiation, signal strength monitoring, phase null setting, and ultimately the distance measurements, together with the readout switch bezel. The readouts are in feet, but instruments that measure in metric units (bearing serial numbers with suffix E) are also available. The accuracy, according to the manufacturer's specification, is ± 1.7 cm $\pm 5 \times 10^{-6}$. The slightly poorer proportional accuracy compared to the earlier models may result from the weaker stabilization of the modulation frequencies. The light-weight CA-1000, also can be mounted on top of a theodolite.

20.1.1.5. *Electrotape DM-20*

The *Electrotape DM-20* (Fig. 20.11) is manufactured by the Cubic Corporation, San Diego, California, and was first put on market in 1962. Except for the klystron, it is an all-transistorized instrument operating with the carrier frequency band of 10.00–10.50 GHz which, with a built-in antenna, forms a beamwidth of 6° between 3 dB limits. The master klystron is provided with the automatic frequency control. The interchangeable master and remote units are housed in a fiberglass case and weigh 15 kg each. The operational temperature range is from −40 to +51°C, and the maximum measurable distance is about 50 km, with an accuracy of ±1.0 cm ±3×10⁻⁶. Power consumption is 4 A maximum at 12 V, which allows 2 hours of continuous use with built-in nickel-cadmium

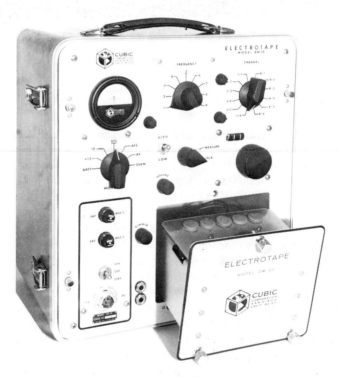

Figure 20.11 Electrotape DM-20 (courtesy Cubic Industrial Corp.).

batteries. The readout is a three-digit digital display on the front panel, and since the three digits are read in all pattern readings, the term $(L_A - L_B)/100$ in Eq. 20.20 becomes $(L_1 - L_i)/1000$. Also, by accepting the standards of $c_0 = 299.792.5$ km/sec and $n = 1.00320$, the five modulation frequencies used are assigned values to give the coefficient $V2(f_1 - f_i)$ (Eq. 20.20) values of whole numbers of meters 10, 100, 1000, 10,000, and 100,000.
These values are:

	Master (MHz)	Remote (MHz)	
f_6	7.490929	7.492427	
f_5	7.477443	7.478941	(20.25)
f_4	7.342579	7.344077	
f_3	5.993941	5.995439	
$f_2 = f_1$	7.492427	7.493925	

The general distance equations then become

$$D = 100(L_1 - L_6) \text{ m}$$
$$D = 10(L_1 - L_5) \text{ m}$$
$$D = 1.0(L_1 - L_4) \text{ m} \qquad (20.26)$$
$$D = 0.10(L_1 - L_3) \text{ m}$$
$$D = 0.01(L_1 - L_2) \text{ m}$$

It must be noted that $f_1 - (-f_2) = 2f_1$, obtained by reversing the modulation frequency f_1 into a negative f_2.

With some exceptions the measuring principle follows rather closely that of the Tellurometer MRA-3. The carrier frequencies are tuned in such a way that after the IF is mixed in both units it is exactly 46.5 MHz. This is amplified and remixed to 10.7 MHz IF and amplified again. The difference between the modulation frequencies in the master and remote units is $f_R - f_M = \Delta f = 0.001498$ MHz or approximately 1500 Hz. Before the information signal in the remote unit is fed into the klystron to frequency modulate the carrier wave, its frequency is divided into $\Delta f/2 = 750$ Hz. After the double mixing procedure has been completed in the master unit, there are two components to be compared, namely, the information signal $\Delta f/2$ and the local reference signal $f_R - f_M = \Delta f$. The voice communication signals have been omitted here. To simplify the conclusion, the frequency $\Delta f/2$ of the information signal is doubled back to Δf, and the signal is fed to one terminal of the phase measuring system and the local or reference signal is fed to the other terminal for phase comparison.

20.1.1.6. *Distomat DI-50 and DI-60*

The *Distomat DI-50* was originally developed by Albiswerke Zürich, Ltd, and later manufactured and distributed by Wild Heerbrugg, Ltd., in Switzerland. It entered the market in 1962 and was the first microwave instrument utilizing the pulse counter for phase comparison between the local and information signals. The interchangeable master and remote units have a separate antenna component connected to the measuring component by a cable 30 m long. The antenna assembly weighs 7.8 kg, and with the carrier frequency used, $f = 10.20$–10.50 GHz, forms a beam width of 6° between 3 dB limits. The measuring component weighs 15 kg. Once in operation, the remote unit automatically tunes its carrier frequency 37.5 MHz higher than that of the master unit, producing a 37.5 MHz IF. The power transmitted is 40 to 60 mW and the maximum measurable distance is 150 km with an accuracy of ± 20 mm $\pm 5 \times 10^{-6}$.

The operation and measuring principles can be described (*Strasser*, 1969; *Tikka*, 1973) as follows. The zero phase of the local reference signal in the master unit opens the pulse counter, and the zero phase of the returning information signal from the remote unit closes it. If the measured number of pulses is L and the total amount of pulses included in the modulation wavelength is k, the distance is obtained from the expression

$$D = \frac{1}{2}\left(n + \frac{L}{k}\right)\lambda = \left(n + \frac{L}{k}\right)\frac{c}{2f} \qquad (20.27)$$

where n is the total number of one-half modulation wavelengths included in the distance and f is the modulation frequency. Distomat DI-50 uses four modulation frequencies, f_1 through f_4, and f_4 is the basic frequency. The distance based on the sample difference $f_4 - f_1$ is then

$$D = (n_4 - n_1)\frac{c}{2(f_4 - f_1)} + (L_4 - L_1)\frac{c}{2k(f_4 - f_1)} \qquad (20.28)$$

where the first term indicates the limit at which the system starts repeating itself. In the rest of the distance equations the first term becomes meaningless.

The modulation frequencies f and the pulse constant k have been selected in a way that ensures that the scale factors $c/[2k(f_4 - f_i)]$ will be even multiple integer numbers. With $c_0 = 299,792.5$ km/sec and $n = 1.000320$, $c/2 = 149,848.3$ km/sec, the following parameters are obtained:

Modulation Frequencies (MHz)	Frequency Differences (MHz)	$c/[2(f_4 - f_i)]$ (m)	k	Unit of One Pulse
$f_1 = 14.977340$	$f_4 - f_1 = 0.007490$	20,000	20,000	1 m
$f_2 = 14.834980$	$f_4 - f_2 = 0.149850$	1,000	10,000	1 dm
$f_3 = 13.486350$	$f_4 - f_3 = 1.498480$	100	10,000	1 cm
$f_4 = 14.984830$	$f_4 = 14.984830$	10	10,000	1 mm

The general distance equations then will be:

$$D = 1.000(L_4 - L_1) \text{ m}$$
$$D = 0.100(L_4 - L_2) \text{ m}$$
$$D = 0.010(L_4 - L_3) \text{ m} \tag{20.29}$$
$$D = 0.001 L_4 \text{ m}$$

As was the case with Tellurometers and the Electrotape, where the phase comparison was made by way of low frequency 1.0 or 1.5 kHz signals instead of the high frequency modulation signals, a similar arrangement is established in the Distomat DI-50. In addition to the four modulation frequencies, a fifth crystal-controlled reference frequency $f_0 = 14.984580$ MHz is utilized. When this is mixed with the f_4 frequency, a low frequency $\Delta f = f_4 - f_0 = 0.00025$ MHz $= 250$ Hz is introduced, to make better use of the pulse counter. In short, the procedure is as follows, using f_4 and f_1 as an example. The modulation signal f_1 transmitted by the master unit is received in the remote unit and is modulated with f_4, and this difference modulation signal $f_4 - f_1$ is sent back to the master unit. Meanwhile, in the master unit both f_4 and f_0 are modulated with f_1. Mixing these two difference modulations $(f_4 - f_1) - (f_0 - f_1)$ yields the AM $f_L = f_4 - f_0 = 250$ Hz local reference signal, which is fed to the pulse-counting system. After returning from the remote unit the $f_4 - f_1$ signal is mixed in the master unit with the $f_0 - f_1$ signal, and the resulting FM $f_I = f_4 - f_0 = 250$ Hz information signal is also fed to the pulse-counting system for distance measurement.

Since only $2 \times 250 = 500$ Hz "pulses" are needed to count the 20,000 m scale unit and, in effect, 20,000 pulses in the pulse counter are utilized, the $(L_4 - L_1)$ readings are actually made 40 times and their mean value is automatically recorded. Similarly, the readings corresponding to the 1000, 100, and 10 m scale units are made 20 times and the mean values are used.

When the modulation frequencies f_1 and f_4 are plucked in, the pulse number $L_4 - L_1$ appears in a mechanical counter showing kilometers with their decimals up to 20 km. After a short interval, modulation frequencies f_2 and f_4 are applied and the pulse number $L_4 - L_2$ appearing in the counter indicates the hundreds of meters, whose decimals automatically correct the previous approximate decimals. This procedure is continued until f_4 itself is applied, when the pulse number L_4 indicates the meters, with decimals down to millimeters. This fine reading is repeated several times, and the mean value is used. Because of the pulse-counting system in the phase measurement, the modulation crystals are identical in the master and remote units. The voice connection is established with a built-in loudspeaker and microphone on the front panel, thus making a headset unnecessary. Production of the Distomat DI-50 has been discontinued.

Distomat DI-60 differs essentially from DI-50 in external appearance (Fig. 20.12). The separation of the antenna is omitted, the antenna is built into the measuring unit, and together they weigh only 14 kg. The voice connection retains a built-in loudspeaker, but the microphone is in a handset. Two main modifications have been made to the electronics. In addition to the four modulation frequencies featured in DI-50, a fifth modulation frequency $f_5 = 10 \times f_4 = 149.848300$ MHz is employed. This high modulation frequency is used for two reasons. First, it can serve for fine readings with a 1 m scale unit; second, it almost eliminates the effect of the ground swing error. As was seen in Eq. 11.1, the maximum amplitude of the ground swing is directly proportional to the modulation wavelength λ_m. In most of the modern microwave distancers; the effective modulation wavelength is about 15 MHz. If we place the value $r = 0.4$ in Eq. 11.1, the maximum swing amplitude will be 1.3 m. In the Distomat DI-60 the modulation frequency is 10 times higher, and the maximum swing amplitude is reduced to about one-tenth the previously given value, which appears in the fine readings of other instruments. Because there is of no need to rely on variable carrier frequencies to eliminate the ground swing error, the master and the remote carrier frequencies, after a few seconds of automatic tuning, are locked into fixed values, $f_M = 10.324300$ GHz with AFC and $f_R = 10.335000$ GHz (without AFC), introducing 10.7 MHz IF. Except for the extra frequency f_5, all modulation frequencies are identical in models DI-50 and DI-60.

Crystals are heated in a 75°C oven, and the heating can be done during

Figure 20.12 Distomat DI-60 (courtesy Wild Heerbrugg Ltd.).

transportation. Operational temperature ranges are −30 to +50°C in the standard model and −55 to +72°C in the special model. Power consumption at 20°C is 3.2 A at 12 V, which includes use of the quartz oven. Radiated power is about 30 mW, and the beamwidth at the half-power limit is 6°. Maximum measurable distance is about 150 km, with an accuracy of $\pm 10\,\text{mm} \pm 3 \times 10^{-6}$.

The Distomat DI-60 is no longer sold by Wild Heerbrugg but is still available from Siemens-Albiswerke, Zürich, and is known as the *SIAL MD 60*. Its operational and measuring principles are the same as in the DI-60 and it has the same high fine measurement frequency of 149.848300 MHz. There are two models: SIAL MD 60C, is a compact one-unit instrument, and the split model, SIAL MD 60S, is designed for operation on masts, towers, or in helicopters when a removable antenna is required. The control unit contains all operating elements as well as the circuitry of the measuring functions and is connected to the antenna unit by means of a plug-in cable 4 to 25 m long. The antenna unit accommodates the transmitter and the high frequency circuits, which means that the cable carries low frequency signals and supply voltages only. Its specifications are the same as the DI-60, except that the units have different weights. The total weight of SIAL MD 60C is 13.5 kg, whereas the antenna and the control units together in the SIAL MD 60S weigh 18.5 kg, not including the 25 m cable and a cable drum of approximately 10 kg.

20.1.1.7. *Summary of Long-Distance EDM Instruments*

In this section an attempt is made to summarize the ground-to-ground EDM instruments capable of measuring distances greater than 10 km. Instruments described in the previous sections are included, together with some instruments added for informational purposes only. References such as *Tikka* (1973), *Denison* (1963–1967), and *Romaniello* (1970) are cited, and the author makes no claims regarding completeness of the list in Table 20.1.

20.12. Short Distances

The instruments described in this section are designed for measurements from a few meters to a maximum of 10 km. They all operate on the electro-optical principle, the light source being either the He–Ne laser, the gallium-arsenide luminescent diode (Ga-As), or the xenon flash tube, and the average power consumption is about 15 W. They are presented in alphabetical order by manufacturer.

20.1.2.1. *Geodimeter M 700 and M 76*

Geodimeter M 700 is the most versatile electro-optical instrument manufactured by AGA, who put it on the market in 1971. It uses the He–Ne

TABLE 20.1

Manufacturer	Model	Emission Source	Range (km)	Accuracy (±mm ±ppm)		Remarks
AGA, Sweden	Geodimeter NASM-1	Tungsten	50			Obsolete
	NASM-2	Tungsten	35	21	1.2	First retrodirective reflector
	NASM-3	Tungsten	25	50	2.0	Lightweight
	NASM-4	Tungsten or mercury vapor	25	10	2.0	Daytime distance 5 km
	M 6	Tungsten or mercury vapor	25	8	2.0	Coaxial optics
	M 6A	Tungsten or mercury vapor	25	6	1.0	Delay line replaced by resolver
	M 6B	Tungsten or mercury vapor	25	5	1.0	Extra modulation frequency for short lines
	M 6BL	Laser	25	5	1.0	Daytime distance 25 km
	M 8	Laser	60	5	1.0	Kerr cell replaced by Pockels modulator
Anritsu, Japan	Autosurveyor ADM-4	Microwave	40	30	2.0	
Cubic Corp. United States	Electrotape DM-20	Microwave	50	10	3.0	Similar to MRA-3 Tellurometer
Electrim, Poland	Telemetre OG-2	Microwave	20	30	5.0	Similar to MRA-1 Tellurometer; carrier frequency 9 GHz

TABLE 20.1 Continued

Manufacturer	Model	Emission Source	Range (km)	Accuracy (±mm ± ppm)		Remarks
Ertel-Grundig, West Germany	Distameter III	Microwave	100	20	5.0	Similar to Electrotape DM-20
Fairchild, United States	Microchain MC8	Microwave	50	15	4.0	Military specifications; separated antennas
Funkmechanik, East Germany	PEM-2	Microwave	70	30	3.0	Like MRA-2 Tellurometer but with 9.0 GHz carrier wave
Laser Systems & Electronics, United States	Rangemaster	Laser	60	5	1.0	Pulse counting in phase measurement (Built for K & E)
	Ranger III	Laser	12	5	2.0	Similar to Rangemaster
Metrimpex, Hungary	GET-BI	Microwave	50	50	3.0	Similar to MRA-1 Tellurometer.
	GET-AI	Microwave	15	30	3.0	Similar to MRA-3 Tellurometer
Spectra Physics United States	Geodolite 3G	Laser	80	1	1.0	Heavy and bulky, accurate
Tellurometer, Pty., Ltd., South Africa	MRA-1	Microwave	100	50	4.0	Circular CRT readout
	MRA-2	Microwave	50	50	4.0	Master and remote interchangeable
	MRA-3	Microwave	80	15	3.0	All-transistorized; dial or digital readout

Country/Manufacturer	Model	Type				Remarks
	MRA-301	Microwave	80	15	3.0	Military specifications
	MRA-101	Microwave	50	15	3.0	1 kHz reference oscillator
	MRA-4	Microwave	60	3	3.0	35 GHz carrier frequency
	MRA-5	Microwave	50	10	3.0	Military specifications; antennas separated
	CA-1000	Microwave	30	17	5.0	Compact, lightweight
U.S.S.R.	Quartz	Laser	40	20	2.0	Heavy and bulky
	SVV-1	Mercury vapor	12	20	1.0	Kerr-modulation with continuous frequency change
	EOD 1	Mercury vapor	35	15	1.5	Similar to NASM-2 Geodimeter
	RDG	Microwave	30	50	3.0	Similar to MRA-1 Tellurometer
	LUS	Microwave	40	30	3.0	Similar to MRA-3 Tellurometer
Wild Heerbrugg, Switzerland	Distomat DI-50	Microwave	150	20	5.0	Pulse counting in phase measurement; separated antenna
	Distomat DI-60	Microwave	150	10	3.0	Use of high modulation frequency $f_s = 150$ MHz to eliminate ground swing
Siemens-Albiswerke, Switzerland	SIAL MD 60C	Microwave	150	10	3.0	Same as Distomat DI-60

TABLE 20.1 Continued

Manufacturer	Model	Emission Source	Range (km)	Accuracy (±mm ± ppm)		Remarks
	SIAL MD 60S	Microwave	150	10	3.0	Same as Disromat DI-60; Separated antenna
Zeiss, East Germany	EOS	Filament lamp	25	5	2.0	Kerr cell replaced by a quartz crystal

laser and combines a theodolite with horizontal and vertical angle measurement capabilities, a direct readout distance measuring component that
corrects the slope distance into horizontal distance to 500 m (the total
measurable distance being 5000 m), and a built-in computer. The total
instrumentation consists of a measuring unit on top of a tripod, a control
and display unit that extends out of the pack on a telescopic rod, and an
optional data-registering unit *Geodat 700* (Fig. 20.13).

Figure 20.13 Geodimeter M700 with Geodat 700 (courtesy AGA Co.).

The digital display includes two sets of seven-digit readouts, and depending on the requirements, the operator may select three different modes with pairs of readings that appear simultaneously on the display panel. The theodolite mode displays the horizontal and vertical angle readings, the set out mode displays the horizontal angle and the horizontal distance readings, and the slope distance mode displays the vertical angle and slope distance readings. The instrument is available either in a metric or an English system and in 400^g or $360°$ angular graduation.

The optics at the measuring unit (Fig. 20.14) resembles the Geodimeter M 6BL with its coaxial optical system and laser attachment. Because relatively short and variable distances are measured, a beam expander is included to vary the divergence of the laser beam from 0.2 to 10 mrad. The beam splitter allows only about 5% of the returned signal to proceed to the eyepiece, while 95% of it is projected to the photomultiplier. Thus about 8% of the light at the laser wavelength of 632.8 nm strikes the eye. The receiving electronics resembles that of the M 8 Geodimeter, but there are two resolvers and no null meter.

The M 700 uses three modulation frequencies, which with $v_0 = 299,792.5$ km/sec and $n = 1.0003086$ introduce the following values of frequencies and unit lengths.

Modulation Frequencies (kHz)	Unit Lengths (m)	
$f_1 = 299.700$	$U_1 = 500.000$	
$f_2 = 29,970.000$	$U_2 = 5.000$	(20.30)
$f_3 = 30,000.000$	$U_3 = 5.005$	

The measuring principle is basically the same as in previous Geodimeter models. The transmitted modulations are mixed with locally generated signals in the receiver with slightly differing frequencies, and the resulting beat signals are used for phase comparison gauged by two resolvers; in the M 700 this automatically results in readings on the display panel.

If measurements are made at distances less than 500 m, only the measuring frequencies f_1 and f_2 are needed. Frequency f_1 gives the coarse distance to 500 m with an accuracy of ± 0.5 m, since the resolvers have 1000 graduations. Frequency f_2 is then selected to obtain the fine distance reading, which is accurate to within ± 5 mm. When measuring with f_2, the resolver starts from the coarse distance determination that is stored in it, giving unambiguous results out to a distance of 500 m. Beyond this distance the conventional procedure of frequency differences is used (i.e, $f_3 - f_2$) to obtain the rough distance, which indicates how many integer

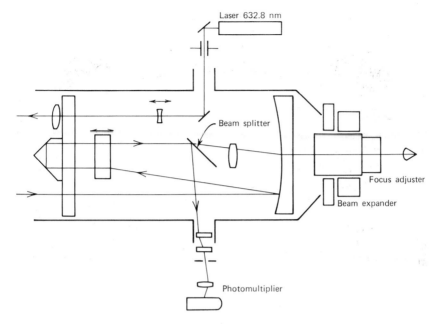

Laser 632.8 nm

Beam splitter

Focus adjuster

Beam expander

Photomultiplier

Figure 20.14 Optical system of Geodimeter M700 (courtesy AGA Co.).

multiples of 500 m are included in it. The expression for it may be written as follows:

$$D' = 1000(L_3 - L_2) \qquad (20.31)$$

where the L_3-reading is obtained the same way as L_2. The final distance equation is then

$$D = n \times 500 + L_2 \text{ m} \qquad (20.32)$$

where n is obtained from Eq. 20.31.

The horizontal and vertical angles are measured by sensing the positions of two pulse or code disks, one for vertical and one for horizontal angles. The electro-optical system consists of optical tracks; these provide pulses that can be counted and then interpolated by the electronics of the measuring unit. Each pulse disk has two tracks of precision marks that are shifted slightly with respect to one another, as in Fig. 20.15. To sense the angle, the pulse tracks of a disk are illuminated by a light-emitting diode (LED), and the image of the pulse tracks is thereby projected through an optical system to the opposite side of the disk. On the detector side of the disk one obtains an image on which the pulse marks from each side of the

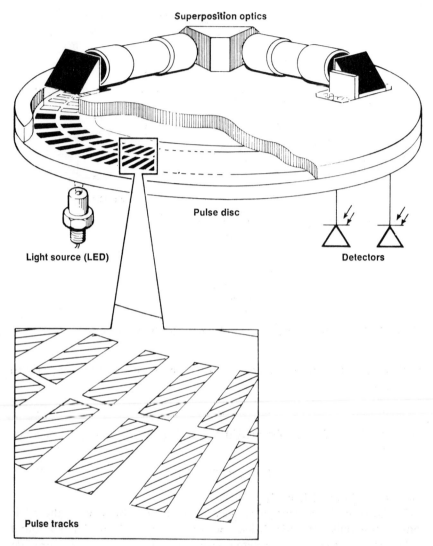

Figure 20.15 Angular measurement in Geodimeter M700 (courtesy AGA Co.).

disk are superimposed. The superimposed tracks move in opposite direction when the pulse disk is rotated. The resulting composite image at the detector side is a moiré effect, which closes off the light completely or lets half the light from the LED pass through whenever the two marks composing the image become precisely coincident. The light reaching the two detectors thus varies in accordance with the two wave trains as seen by the detector semiconductors. Pulse-counting electronic equipment stores the number of periods occurring in the dectector wave trains, and special interpolating circuits compare the light levels at the detectors to resolve the angles further. The standard error of one direction measurement, as claimed by the manufacturer, is $\pm 2''$ in horizontal and $\pm 3''$ in vertical direction. Tests made by *Hallsten* (1971) showed slightly better accuracy in vertical measurements and slightly poorer in horizontal measurements.

The Geodimeter M 700 has a Kerr cell for a light modulator and a Spectra-Physics type 071–1 1 mW He–Ne laser as the light source. Maximum measurable distance with six AGA prisms is about 5 km, with an accuracy of ± 5 mm $\pm 10^{-6}$. Operating temperature range is -30 to $+50°C$, and measuring time for one measurement is 10 to 15 sec. At present, the Geodimeter M 700, together with the Zeiss Reg Elta-14, are considered to be truly tachymetric type instruments.

For automatic registering, *AGA Geodat 700* is available as an optional unit and is housed on top of the front of the carrying case. The Geodat 700 punches all the measurement data on tape from the Geodimeter 700 either in eight-channel ASC11/ECMA code (for direct communication with a computer or plotter) or in five-channel code for transmission by Telex CC1TT No. 2. The AGA Geodat can be operated as a 10-key calculator. The keyboard has 10 digit keys and 6 control keys. If the PRINT key is pressed, all measurement data in the memory of the Geodimeter are registered on tape. The data include horizontal angle, vertical angle, horizontal distance, and slope distance. The keyboard data punch-out can be used for any necessary information. The total weight of the instrument, including the 12 V chargeable built-in battery, is 6 kg. The punch speed is 15 characters per second.

Geodimeter M 76, manufactured by the AGA Corporation, Secaucus, New Jersey, came on the market in 1972. It uses a 2 mW He–Ne laser as its light source. It is fully automatic, and one push of a button produces the final distance in meters or in feet. Atmospheric correction in parts per million is applied manually but computed automatically. The maximum measurable distance, about 3 km with an accuracy of ± 10 mm $\pm 10^{-6}$, is not affected by beam interruptions. Total instrument weight is only 8.5 kg excluding the carrying case.

20.1.2.2.   *Cubitape DM-60 Distance Meter*

Cubitape DM-60 F for units in feet and DM-60 M for units in meters are manufactured by the Cubic Industrial Corporation, a subsidiary of Cubic

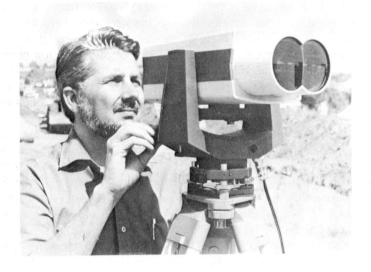

Figure 20.16   Cubitape DM-60 (courtesy Cubic Industrial Corp.).

Corporation, San Diego, California. This lightweight and compact surveyors' instrument (Fig. 20.16) entered the market in 1973. Total instrument weight is about 7 kg, and its dimensions are 12.7 × 17.8 × 35.6 cm. It is fully automatic, and the final distance is obtained by pushing a button. Atmospheric correction in parts per million is computed manually and applied to the final reading during or after the survey.

The Cubitape light source is a Ga-As diode having a wavelength of 0.91 μm. This new type of light emitter became very popular in short-distance measuring instruments in the mid-1960s. It radiates light in the infrared region of 0.875–1.00 μm. For more detailed description of the Ga-As diode, see Chapter 3.

Since the wavelength of the 0.91 μm infrared radiation used in the Cubitape differs essentially from the wavelengths of the light sources described before (tungsten lamp, mercury vapor lamp, and He–Ne laser, e.g.), the built-in refractive index in the Cubitape with reference to $P = 760$ mm Hg and $T = 15°C$ is $n = 1.0002783$. With $c_0 = 299,792.5$ km/sec, the basic or fine modulation frequency is $f_1 = 73.171165$ MHz, representing the unit length $U_1 = 2.048$ m. To extend the unambiguous distance over 2 km, two other modulation frequencies are used: the intermediate and the coarse modulation frequencies, f_2 and f_3, respectively. They are related to f_1 and to each other as follows:

$f_2 = f_1/32$ and $f_3 = f_2/32$. The unit lengths are thus

$$\text{fine } U_1 = 2.048 \text{ m}$$
$$\text{intermediate } U_2 = 65.536 \text{ m} \qquad (20.33)$$
$$\text{coarse } U_3 = 2097.152 \text{ m}$$

and the system starts repeating itself after 2097.152 m.

Each of the three modulation frequencies has a phase-locked reference oscillator. The reference starts the pulse counter, and the returned signal from the reflector stops the counter. The counter interval is thus proportional to the distance from the instrument to the reflector. This measurement is done automatically for all three modulation frequencies and is displayed as the final distance.

Cubitape DM-60 features an automatic self-calibration system. When the instrument makes a phase measurement of the returning signal, it makes a similar phase measurement of a reference signal passing through an internal calibrated path and substracts these phases to obtain a true phase reading. A small torque motor alternately swings the fiber-optics in and out of the transmitted path for each modulation frequency. Its operation may produce audible clicks during the measuring sequence. A variable attenuator in the fiber-optic path allows adjustment of the reference signal to match the received signal. Measurement time is approximately 17 sec without any beam interruption. A beam interruption causes the control logic of DM-60 to halt, and the measurement cycle restarts automatically when the beam path from the instrument to the reflector is clear. A built-in charger operates from 115 V AC and a battery cable is provided for use of other 12 V DC sources. The operational temperature range is -10 to $+50°C$. Maximum measurable distance is about 2500 m, with an accuracy of about ± 5 mm $\pm 10^{-5}$.

20.1.2.3. Hewlett-Packard H-P 3800 and 3805A Distance Meters

The Hewlett-Packard Company, Loveland, Colorado, placed the H-P 3800 Distance Meter on the market in 1971. Models are available giving readouts in feet (H-P 3800A) and in meters (H-P 3000B). The Distance Meter is based on a decimal system in which modulation frequencies are multiples of tens from each other and are used as such instead as their differences. Each decimal and digit is obtained by manually zeroing the phase conditions from right to left five times consecutively, at the end of this procedure the final distance appears as a digital readout in the meter panel.

The light source is the Ga-As diode with a wavelength of 0.91 m. The diode can be directly amplitude modulated to 30 MHz with an alternating current without significant distortion in phase values. Some spurious phase shifts are still introduced in the measurement, however. To eliminate such spurious shifts, the user alternately directs the output of the amplitude modulator through an internal reference path to the detector. The resulting signal serves as a reference for comparison to the external detected signal. Any erroneous phase shift caused either by the modulator or dectector is simultaneously present in both the external and the reference signal and can be eliminated by the phase comparison of these signals.

The "time-sharing" principle of the H-P 3800 is illustrated in Fig. 20.17, which represents the 3800B (*McCullough*, 1972). With the refractive index $n = 1.0002783$ and $c_0 = 299{,}792.5$ km/sec, the basic modulation frequency representing 10 m in distance will be $f_1 = 14.985$ MHz. Three other modulation frequencies are used:

$$f_2 = 1.4985 \text{ MHz}$$
$$f_3 = 149.85 \text{ kHz} \qquad (20.34)$$
$$f_4 = 14.985 \text{ kHz}$$

The frequency that controls the alternating current to drive the Ga-As diode is furnished by a tunable crystal oscillator, where the f_1 crystal is very stable for temperature changes. For long-term drifts the oscillator can be tuned to the correct frequency. The oscillator can also be tuned in such a way that the operator can dial in the refractive index correction, which automatically corrects the modulation frequency. The time sharing between the transmitted signals and the reference signals is made in the chopper beam splitter, which divides the modulated light coming from the Ga-As transmitter diode into two separate beams alternately at a rate of 30 Hz. Consequently, there is an external transmission period of light for 16.67 msec followed by generation of an internal reference signal for another 16.67 msec. The light signal to be transmitted is then focused into a beam that is 5 cm in diameter at the transmitter lens and has a divergence of 1 mrad; this beam is sent to the distant reflector. The internal reference light signal is carried by a fiber-optic light pipe to a variable neutral density filter. This allows the operator to reduce the level of the reference signal until it is equal to that of the received signal at the receiver diode. Both external and internal light signals then pass through an interference filter located just in front of the receiver diode. This filter helps to reject signals of other wavelengths (e.g., visible sunlight) without

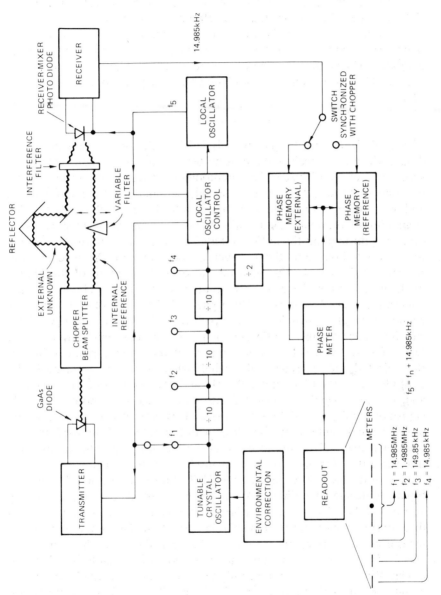

Figure 20.17 Block diagram of H-P 3800B (courtesy Hewlett-Packard Co.).

REFLECTOR

INTERFERENCE FILTER

RECEIVER-MIXER PHOTO DIODE

RECEIVER

14.985 kHz

EXTERNAL UNKNOWN

VARIABLE FILTER

f_5

LOCAL OSCILLATOR

GaAs DIODE

CHOPPER BEAM SPLITTER

INTERNAL REFERENCE

LOCAL OSCILLATOR CONTROL

SWITCH SYNCHRONIZED WITH CHOPPER

TRANSMITTER

f_1

f_2

f_3

f_4

÷ 10

÷ 10

÷ 10

÷ 2

PHASE MEMORY (EXTERNAL)

PHASE MEMORY (REFERENCE)

TUNABLE CRYSTAL OSCILLATOR

ENVIRONMENTAL CORRECTION

PHASE METER

READOUT

METERS

$f_1 = 14.985\,\text{MHz}$
$f_2 = 1.4985\,\text{MHz}$
$f_3 = 149.85\,\text{kHz}$
$f_4 = 14.985\,\text{kHz}$

$f_5 = f_n + 14.985\,\text{kHz}$

291

eliminating the modulation signal from the carrier. None of the beam-splitting, chopping, and filtering processes described will cause phase shifts between the transmitted and reference light signals.

The phase difference between the transmitted and reference light signals is dependent on distance and is measured with the aid of a local oscillator. The oscillator generates a frequency f_s whose value is always 14.985 kHz larger than the modulation frequency f_n, where $n = 1, \ldots, 4$. Thus

$$f_s = f_n + 14.985 \text{ kHz} \tag{20.35}$$

The local signal with frequency f_s is fed into the receiver-mixer photo-diode, which also receives the transmitted and reference light signals in a succession of 16.67 msec. The photodiode converts the received and reference light signals to electric signals, amplifies and combines them, and mixes them with the local f_s signal, introducing 14.985 kHz IF. The combining of these signals actually means that any spurious phase shift occurring throughout the entire system from the Ga-As transmitter diode to the photodiode and the receiver is the same for the received and reference light signals, thus will be canceled. The output of the photo-diode is a combined signal at a frequency of 14.958 kHz with phase alternating every 16.67 msec. This signal is fed to a set of phase memory circuits through a switch that is synchronized with the chopper. This circuitry generates two continuous signals having the same frequency but two different phase values, which represent the original phase values of the reflected and reference light signals when they were received in the photodiode. In the phasemeter the phase difference between the external and reference signals is converted into a direct current whose magnitude is proportional to the differential phase. This current is connected to the input of a null meter amplifier along with the output of the readout. The readout is manually adjusted to bring the null meter pointer to midposition separately for each of the four modulation frequencies.

The entire measuring unit of the H-P 3800 is set on a tribrach with forced-centering capabilities. A power unit is attached to the measuring unit by a cable. The power unit houses the built-in battery charger and a dial-controlled system for atmospheric correction. The dimensions of the measuring instrument are $33 \times 26.2 \times 14.7$ cm, those for the power unit are $17.5 \times 17.5 \times 21.8$ cm, and the corresponding weights are 7.9 and 6.3 kg. With a triple prism assembly the maximum measurable distance is about 3000 m with an accuracy of $\pm 5 \text{ mm} \pm 7 \times 10^{-6}$ at the temperature ranges -10 to $+40°C$ and $\pm 10 \text{ mm} \pm 33 \times 10^{-6}$ at the temperature ranges -20 to $-10°C$ and $+40$ to $+55°C$. All controls and meters are function keyed with graphic symbols and the least count in the decimal drum is 2 mm.

The latest Hewlett-Packard model, *H-P 3805A* (Fig. 20.18) became available in 1974 and also has a Ga-As diode as its light source. It is more compact and weighs less than the H-P 3800. It is fully automatic, and a built-in computer averages the distance measurements without being affected by interruptions of the beam. The digital light-emitting diode display gives the operator the distance and classifies its measuring quality in three ways. A steady numerical display indicates a good, solid measurement within the instrument's specified accuracy; a flashing display means that conditions are such that the measurements are marginal, and a flashing "0" display warns the operator that under present conditions a valid measurement cannot be made. Atmospheric correction is dialed in the measuring unit prior to the measurements, and the distance displayed can be switched into meters or feet. No power unit is used, and the total

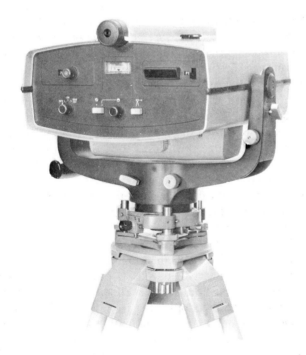

Figure 20.18 H-P 3805A Distance meter (courtesy Hewlett-Packard Co.).

weight of the measuring instrument is 10.1 kg, with dimensions of $33 \times 26.2 \times 14.7$ cm. The least count of the digital readout is 1 mm, and the accuracy of the instrument at its maximum measurable distance of 1600 m is ± 7 mm $\pm 10 \times 10^{-6}$ at the temperature range of -10 to $+40°C$ and ± 14 mm $\pm 30 \times 10^{-6}$ at the temperature ranges of -20 to $-10°C$ and $+40$ to $+55°C$.

20.1.2.4. *Kern DM 1000 and DM 500*

Kern and Company AG, in Aarau, Switzerland, began the production of electro-optical distance meters with the *DM 1000* in 1971. This instrument uses a Ga-As diode with a wavelength of 0.9 μm. Miniaturization in electronic elements and integrated printed circuits keeps the total weight of one piece (Fig. 20.19), including base, battery, and battery charger, to 12.5 kg. The overall dimensions are $35 \times 24 \times 26$ cm. The DM 1000 operates with two modulation frequencies, $f_1 = 14.9854$ MHz and $f_2 = 149.854$ kHz, and the measurements are made separately and automatically with these frequencies, representing the unit lengths of $U_1 = 10$ m and $U_2 = 1000$ m. The phase measurement is based on a pulse-counting method between an internal reference signal and the reflected signal in such a way that any erroneous phase shifts between the signals are automatically eliminated by a built-in calibration line. This instrument is not used in connection with a theodolite but can be separately force centered over a marker point by an optical plummet. The distance reading appears on a six-digit display to millimeters 15 sec after the push of a start key. With a three-prism assembly, the maximum measurable distance is 2.5 km with an accuracy of ± 4 mm $\pm 4 \times 10^{-6}$. The operational temperature range is -20 to $+50°C$.

The newest electro-optical distance meter now available is the Kern *DM 500*, which entered the market in late 1974. It is the most compact and lightweight instrument yet built to be used in connection with a theodolite. The distance measuring unit will be "wrapped-around" the telescope of the Kern DKM 2-AE theodolite (see Fig. 20.20). The power supply case containing four Ni-Cd cells is attached to the legs of the theodolite. Power transmission is arranged by a cable from the power supply to the base of the theodolite, and from there it is fed through the internal theodolite illuminating system to the distance measuring unit. The power consumption is 5 W. The measuring unit weighs 1.6 kg and has the dimensions $15.0 \times 6.9 \times 17.5$ cm. The distance measuring principle is identical to that of the DM 1000, but there is an additional automatic

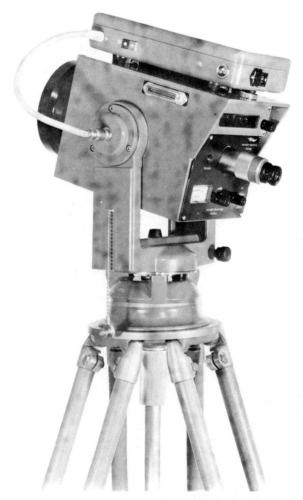

Figure 20.19   Kern DM 1000 (courtesy Kern Aarau Ltd.).

intensity control for short-distance light signals. The maximum measurable distance using three reflectors is 500 m, with an accuracy of about ±0.6 cm.

20.1.2.5.   *Kern Mekometer ME 3000*

The basic research in the development of the present Mekometer was carried out by Dr. *K. D. Froome* and *R. H. Bradsell* at the National

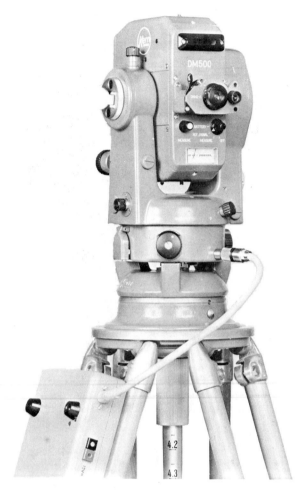

Figure 20.20   Kern DM 500 (courtesy Kern Aarau Ltd.).

Physical Laboratory (NPL), in Teddington, England. The Mekometer ME 3000 (Fig. 20.21), now being manufactured by Kern of Switzerland in cooperation with the British firm COM-RAD, is to some extent the result of the earlier NPL experimental model Mekometers I and II. The ME 3000 entered the market in 1971 and presently is considered one of the most accurate available instruments in the field of electronic distance measurement. It differs from all other instruments described in the earlier sections mainly with respect to the light source and its modulation. A maximum measurable distance of 3000 m with an accuracy of ±0.2 mm± 10^{-6} makes the ME 3000 a valuable tool for construction engineers and researchers in investigating the minute changes in structures caused by strains, and for geophysicists in measuring lateral crustal movements over faults.

The light source is a xenon flash tube with a wavelength of 0.48 μm. Unlike other electro-optical distance measuring instruments, there are no frequency standards in the Mekometer, instead the basic length standards are the UHF cavity resonators filled with dry air and equalized to the outside temperature and pressure. This means that the basic unit length or yardstick is the dimension of the cavity resonator (*Froome* and *Bradsell*, 1965). The measuring principle can be presented in simplified form

Figure 20.21   Mekometer ME 3000 (courtesy Kern Aarau Ltd.).

as follows (*Connell*, 1965; *Burnside*, 1971). The xenon tube is flashed at the rate of 100 Hz, with one-microsecond flash duration. The required modulation frequencies are derived from a standard quartz cavity resonator, and the light modulation is the type of elliptical polarization produced by a potassium dihydrogen phosphate (KDP) Pockels crystal in the modulation cavity. The KDP crystal has several thousand volts applied at the modulation frequency for periods of 40 μsec, commencing just before each flash. These periods are long enough to allow the flash return from the target at the maximum measurable distance of the instrument. Before the light beam from the xenon tube X (Fig. 20.22) reaches the KDP crystal in the modulation cavity, it passes through a polarizer P_1, where it is plane polarized. The KDP crystal causes the light beam to be elliptically polarized at the modulation frequency. Thus when the light leaves the transmitting optics O_1 it is polarization modulated instead of intensity or amplitude modulated, as is the case in all other electro-optical distance measuring instruments. After returning from the distant reflector R, the light beam is led through a movable prism M to the receiving optics O_2. The functioning of this movable prism, thus the controllable length of the total beam path, is explained later.

For phase discrimination the returning light signal is fed to a KDP crystal that is identical to the first one and oscillates in synchronization with it. If

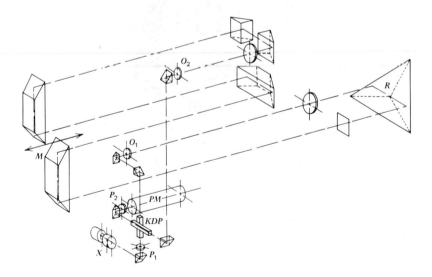

Figure 20.22 Block diagram of optics of Mekometer ME 3000 (courtesy Kern Aarau Ltd.).

the total traversed distance includes an integer amount of the modulation half-wavelengths, the transmitted and returned signals are in phase and the received beam is plane polarized parallel to P_1. The polarization plane of the second polarizer or analyzer P_2 is at right angles to that of the polarizer P_1, and after leaving the second KDP crystal, the plane-polarized received signal cannot pass through the analyzer P_2 and the output of the photomultiplier tube *PM* is zero (or at a minimum). Such an in-phase condition very seldom occurs, and the photomultiplier tube output is minimized by moving the prism *M* within one-half the modulation wavelength, thus altering the total traversed distance. A mechanical calculator coupled to this variable light path converts the displacement into a decimal distance value of the last fractional half-wavelength in the metric system. The displacement itself is measured with high precision optical-mechanical reading devices.

The unit length U_1 of the fine measurement is defined by the dimension of a cavity resonator, which at 20°C and 760 mm Hg is $U_1 = \lambda/2 = 0.30$ m, representing the modulation frequency of $f_1 = 499.5104$ MHz. With this modulation frequency, the fine measurement gives centimeters, millimeters, and tenths of a millimeter in the calculator. Four other nearby frequencies give the unambiguous distance to 3000 m by using the frequency difference method and specific relationships between the frequencies in such a way that the coarse unit lengths are $U_2 = 3$ m, $U_3 = 30$ m, $U_4 = 300$ m, and $U_5 = 3000$ m. The calculator then adds all readings to a single distance display. Since in the Mekometer the dimension of the cavity resonator defines the unit length, the operation of the instrument is almost independent of atmospheric conditions. The pressure and temperature in the cavity resonator are equalized to those at the measuring site. If the temperature, for example, increases, the cavity and consequently the unit length will become longer. According to the old surveyor's rule, too long a tape gives too short a distance. However, the increase in the temperature T (Eq. 6.17) decreases the value of the refractive index n. This, in turn, increases the ambient propagation velocity of the electromagnetic radiation, resulting in too long a distance, and the two phenomena cancel each other. In the Mekometer this compensation is so arranged that a correction unit controls the modulation frequency according to the temperature and pressure conditions at the survey site, and in spite of the variation of the index of refraction, the modulation wavelength, thus the unit length, remains constant. If the average atmospheric conditions along the line to be measured differ essentially from those at the survey site, correction must be computed in high precision surveys.

One distance measurement takes about 2 min. An external 12 V Ni-Cd battery is used, and the power consumption is 18 W. The instrument is housed as an one unit on the yoke and weighs 14.5 kg, with dimensions $46 \times 16 \times 22$ cm.

20.1.2.6. *MicroRanger*

Like the *Rangemaster* and the *Ranger III* described in Section 20.1.1.3, the *MicroRanger* (Fig. 20.23) is manufactured by Laser Systems and Electronics, Inc., for the Keuffel & Esser Company. The newest member of the Ranger series went on the market in 1972. Whereas the other Rangers have a He-Ne laser light source, the MicroRanger uses a Ga-As infrared diode with a 0.915 μm wavelength. This very compact and lightweight instrument is divided into two main units. A head unit that

Figure 20.23 MicroRanger (courtesy Keuffel & Esser Co.).

houses the transmitting and receiving electro-optics weighs about 1.5 kg and can be easily mounted on theodolites, transits, or levels. A control unit houses the display, controls, and computer processing circuitry. The two units are interconnected by a cable. The entire system operates from a power unit attached to the control unit containing a rechargeable Ni-Cd battery. The control unit with the power pack weighs 4.5 kg. The fine reading and coarse reading modulation frequencies are automatically changed and the number of observations is averaged. Phase comparison is accomplished by using the pulse-counting technique. An audible signal indicates where the invisible light beam is on target. The eight-digit distance display can be switched from feet into meters. Four 10-position thumb-wheel rotary switches are used to set instrument and reflector constants into the instrument, and two similar rotary switches set barometer and temperature corrections in parts per million into the Micro Ranger. The operating temperature range is -20 to $+54°C$. The maximum measurable distance is about 1600 m, with accuracies ± 5 mm $\pm 2 \times 10^{-6}$ at the temperature range of -7 to $+43°C$ and ± 10 mm $\pm 4 \times 10^{-6}$ at the temperature ranges of -7 to $-20°C$ and $+43$ to $+54°C$.

Ranger II and *Ranger I* are identical to *Ranger III* (Section 20.1.1.3) except that the displays are seven-digit numbers. The maximum measurable distances are 6 km and 4 km, respectively.

20.1.2.7. *Tellurometer MA-100 and CD 6*

During 1964 a program was initiated at the National Institute for Telecommunications Research of the South African Council for Scientific and Industrial Research to study the feasibility of a practical system characterized by a 1 mm resolution and a maximum measurable distance of 1 km. In 1968 a modified version with slightly longer distance capability and better resolution and accuracy was produced by Plessey S.A. and marketed by Tellurometer Divisions of the Plessey Group of Companies. Basically the *MA-100* (Fig. 20.24), operates on the known Tellurometer principle, with modulation frequencies selected to ensure that their differences from the basic modulation frequency are in a decimal system with multiples of tens. Whereas in the microwave Tellurometers a reference system of modulation frequencies was established in the master units and the remote units, the MA-100 has an internal referencing system of the same basic modulation, and the phase comparison is made between the internal signals and the external signals. The actual phase measurement is performed digitally using the pulse-counting system and a clock oscillator. Ambiguity, or number of full phase rotations, is resolved again by the

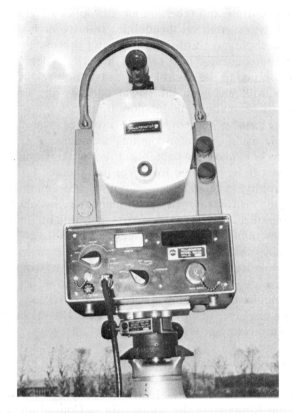

Figure 20.24 Tellurometer MA-100 (courtesy Tellurometer–USA).

general Tellurometer system described in Section 20.1.1.4 (Eqs. 20.20 and 20.21). Readouts appear in a four-digit Nixie-tube in millimeters, millimeters times 10, millimeters times 100, meters, and meters times 10; the units can be selected by means of a switch, and the information is also available in binary-coded-decimal (BCD) form at a socket on the front panel for direct input into tape punch or other recording device. The basic modulation frequency is about 75 MHz, and the actual phase comparison requires a heterodyned beat signal of 10 kHz. The light source of the MA-100 is a Ga-As diode having a wavelength of 0.92 μm. As is common with Ga-As diode instruments, the beam divergence is 0.°25.

Let us now examine the system operation in more detail. To prevent the appearance of small changes in the arbitrary delay (due to phase shift in the electronic circuitry) as changes in the measured distance, an internal reference system is adopted. The reference phase is obtained by a shutter that moves in front of the transmitting diode, blocking off the radiation to the reflector but directing a small fraction of it along a fixed path length into the receiving diode. Since the same circuitry is shared between the internal and the external measurements, any electronic phase error is common to both measurements, thus will be canceled. This, in effect, means that each phase measurement between the transmitted and received light signals is referenced to the stable internal path. With $c_0 = 299,792.5$ km/sec and $n = 1.000274$ representing atmospheric conditions of $T = 15°C$ and $P = 760$ mm Hg, the basic modulation frequency will be $f_1 = 74.9275$ MHz. This yields the half-wavelength of the fine modulation frequency: exactly 2 m. Much as the A and $A-$ patterns were introduced in the microwave Tellurometers, a reference pattern in the MA-100 is produced by changing the heterodyne signal from 10 kHz below the fine pattern frequency to 10 kHz above it. This pattern now gives a reversed sense of phase change with distance, and the difference between this pattern and the fine pattern results in a pattern frequency equal to twice the fine pattern frequency. This difference reading eliminates small beat frequency delay errors and also introduces the needed one meter unit. The other modulation frequencies must be related to the basic modulation frequency as

$$f_2 = 0.9f_1$$
$$f_3 = 0.99f_1 \quad\quad\quad (20.36)$$
$$f_4 = 0.999f_1$$

if their differences from f_1 are to be in a decimal system. The beat frequency itself is kept constant by means of a phase-lock system that controls the reference oscillator whose output is used to adjust the beat frequency. A phase reversal of one of the beat frequencies is made possible to introduce the so-called reverse and forward phase conditions. By taking the mean of the differences obtained from the reverse and forward readings, small phase contamination errors caused by unwanted coupling between the beat frequency channels are eliminated. The phase measurement is performed digitally. The respective zero crossings of the beat frequency signals define a timing period during which pulses from a closk oscillator are counted electronically. The number of pulses counted is a measure of the phase difference between the two signals. An accurate mean value of phase is obtained by averaging over a period of about 3 sec. The phase reading is then displayed on a four-digit Nixie-tube readout. In the coarse pattern frequencies the phase values obtained are automatically subtracted from the stored fine reading, and only two digits for each pattern are recorded and booked. Four digits in the fine readings are recorded and booked, thus giving the last digit in units of 0.1 mm. Unlike most of the instruments described in this section, the MA-100 is not fully automatic. Five separate readings must be taken, including the mean value of the forward and reverse millimeter readings, and the total time needed to complete one measurement is about 2 min instead of a few seconds. However the accuracy of the MA-100 at present is the best of the electro-optical short-range distance meters (excluding the Kern Mekometer ME 3000).

The MA-100 is supplied with an adaptor that fits either in a Wild tribrach optical plummet or in a Kern tripod with provision for mechanical centering. The instrument consists of a single tripod-mounted package containing the optics, associated electronics, and readout; it weighs 17.3 kg. The operational temperature range is −20 to +55°C. The maximum measurable distance with four prisms is about 2000 m, with an accuracy of about ± 1.0 to $1.5 \text{ mm} \pm 2 \times 10^{-6}$. The MA-100 is especially suitable for structural strain measurement and also short-range fault measurements in geophysics.

The latest model of compact surveyors' instruments introduced by Tellurometer in 1973 is the *CD 6*. This instrument has Ga-As diode as the light source with a wavelength of 0.93 μm. It is fully automatic and has a six-digit solid state display. It has continuous ranging facilities and can be hand held or mounted on a theodolite yoke or directly on a tripod. The total weight is 2.5 kg with dimensions of $237 \times 142 \times 74$ mm. The maximum measurable distance with six prisms is 2000 m, with an accuracy of $\pm 5 \text{ mm} \pm 5 \times 10^{-6}$.

20.1.2.8. Distomat DI-10 and DI-3

A short-distance series of Distomat instruments is manufactured by Wild Heerbrugg, in Switzerland. The *DI-10*, first sold in 1968, is a lightweight instrument whose aiming unit is separated from the control unit and can be placed on top of a theodolite telescope (Fig. 20.25) or, with a tribrach, on top of any standard tripod. Its Ga-As diode light source forms a beam with a divergence of 4 mrad at the carrier wavelength of 0.875 μm. The basic modulation frequency (which later in this section is called the terminal frequency) is $f_E = 14.98540$ MHz. By allowing $c_0 = 299,792.5$ km/sec and $n = 1.000282$ which represents the average atmospheric conditions $T = +12°C$ and $P = 760$ mm Hg, the modulation wavelength is exactly $\lambda_E = 20$ m and the unit length $U_E = 10$ m. The fully automatic

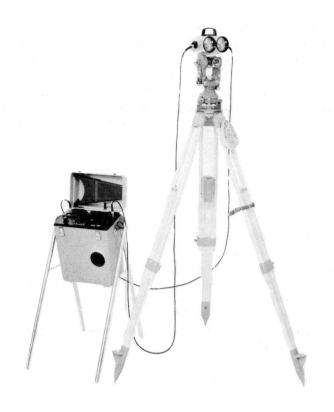

Figure 20.25 Distomat DI-10 (courtesy Wild Heerbrugg Ltd.).

readout is digital/analog, and phase measurement of the fine reading (terminal) modulation frequency is accomplished as follows. The transmitted and received beat signals are compared in the resolver, where the signals are fed into the stator windings. If there is a phase difference between these signals, an alternating current is induced into the rotor winding and is detected in the phase detector. If the output from the phase detecting circuit is fed to a DC motor, the motor can be used to rotate the rotor of the resolver until a balance is achieved and no current runs to the DC motor, which then stops. In this case the transmitted and received signals are in phase; for manual operation this condition represents the zeroing of the null meter.

Finding the multiples of the unit length of 10 m, thus an unambiguous distance to 1000 m, is done by a rather unique method featuring automatically, continuously, and smoothly changing modulation frequency or sweep frequency. This procedure was developed by the *Société d'Études, Recherches et Constructions Electroniques* (SERCEL) in Nantes, France, in cooperation with Wild Heerbrugg. A similar design was reported by *Borodulin* (1961) in the U.S.S.R. for the connection of the Soviet DST-2 distancer. The initial modulation frequency in the sweep cycle is $f_I = 13,48686$ MHz, which is 10% smaller than f_E. We can assume that by using f_I the total distance includes n_I whole wavelengths and the fraction $(L_I/100)\lambda_I$, where L_I is the fractional reading. When the frequency increases and the wavelength shortens to a value of λ_2 at a certain instant during the sweep, the total distance includes exactly $n_I + 1$ of these new waves. In this case the phase difference between the transmitted and received light signals is zero the first time. When the sweep continues, the next zero phase difference is reached by using wavelength λ_3, which is included $n_I + 2$ times in the total distance. When the terminal frequency f_E finally is used, the zero phase has been repeated E integer times plus the residual of the last zero-phase interval $\Delta E/10$. In addition, the total distance to be measured still includes the fraction of the last wavelength λ_E, which will be measured by the resolver as $\Delta\lambda_E = (L_E/100)\lambda_E$.

With reference to *Tikka* (1973), we can write a set of equations for the double distance $2D$ as follows:

$$2D = \left(n_I + \frac{L_I}{100}\right)\lambda_I = \left[\left(n_I + \frac{L_I}{100}\right) + 1\right]\lambda_2 = \left[\left(n_I + \frac{L_I}{100}\right) + 2\right]\lambda_3 = \cdots$$

$$= \left[\left(n_I + \frac{L_I}{100}\right) + \left(E + \frac{\Delta E}{10}\right)\right]\lambda_E + \frac{L_E}{100}\lambda_E \qquad (20.37)$$

From the second and last expressions in Eq. 20.37, we can solve $n_I + L_I/100$:

$$n_I + \frac{L_I}{100} = \left(E + \frac{\Delta E}{10} + \frac{L_E}{100}\right)\frac{\lambda_E}{\lambda_I - \lambda_E} \qquad (20.38)$$

When substituted into the second expression in Eq. 20.37, Eq. 20.38 yields the total distance to be measured:

$$D = \frac{1}{2}\left(E + \frac{\Delta E}{10} + \frac{L_E}{100}\right)\frac{\lambda_I\lambda_E}{\lambda_I - \lambda_E} \qquad (20.39)$$

When the λ_E and λ_I values derived from the given c_0, n, f_E, and f_I values are substituted into Eq. 20.39, a simple distance equation is obtained as follows:

$$D = 100E + 10\Delta E + L_E \text{ m} \tag{20.40}$$

The digital analog DC readout is constructed in such a way that the resolver axis, which is rotated by the DC motor, carries a glass circle with three-digit numbers running in 2 cm steps from 0.00 to 9.98 m (meter, decimeter, and centimeter). A full turn of the fine circle corresponds to 10 m at the terminal frequency of f_E. To register the change in phase during the frequency sweep-up, a second glass circle (the coarse circle) is coupled to the fine circle so that it makes one-tenth of a revolution for a full turn of the fine circle. The coarse circle carries 100 two-digit numbers from 00 to 99. Consequently, each of these numbers corresponds to a multiple of 10 m. The readings of the two circles are projected to a ground glass screen giving a five figure display.

To illustrate the system, we can analyze the following hypothetical reading by using Eq. 20.40 as follows:

$$D = 100 \times 2 + 10 \times 3 + 6.75 = 236.75 \text{ m} \tag{20.41}$$

Another approach also can be used. At the initial frequency f_I the half-wavelength value is 11.111 m and it is included 21.308 times in the given distance. At the terminal frequency f_E the half-wavelength value is 10.000 m and it is included 23.675 times in the given distance. The total number of revolutions made by the rotor of the resolver is then the difference $23.675 - 21.308 = 2.367$, where 23 is the coarse reading obtained from 230.00; the fine reading (6.75) is obtained directly from the cycle value 23.675, yielding the total distance of 236.75 m.

The sweeping up of the transmission frequency is generated by a variable high frequency oscillator that is automatically synchronized with the 9.5 and 11.0 MHz frequencies of two quartz oscillators and is further mixed with one 4.0 MHz quartz oscillator, yielding a sweep of 13.5 to 15.0 MHz. The returning signals with frequencies 13.5 to 15.0 MHz are mixed with the original signals at frequencies of 9.5 to 11.0 MHz, introducing a 4 MHz IF. For phase comparison the 4 MHz IF signal is heterodyned with a signal from a local quartz oscillator of 4.0024 MHz, producing a beat signal of 2.4 kHz to be fed into the resolver. The transmitted 4 MHz signal is also heterodyned, and the resulting beat signal of 2.4 kHz is fed into the resolver for phase comparison. Figure 20.26 illustrates this process. An interruption of the beam during the sweep can cause errors of 10 m in the coarse reading.

The aiming unit contains an internal calibration line. Switching a knob causes the beam of the transmitting diode to be directly reflected to the

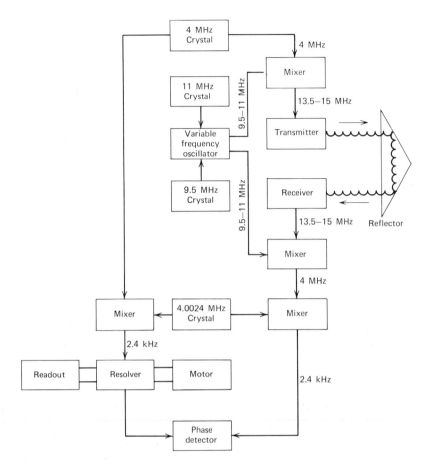

Figure 20.26 Block diagram of DI-10 (courtesy Wild Heerbrugg Ltd.).

receiving diode. With an aid of a drive screw on the control unit, the stator of the resolver is turned until the last three digits of the display are set to zero or to an additive constant due to an internal path within the instrument and within the reflector prisms. Because the transmitting and receiving optics are located side by side, the prisms are displaced laterally by a small amount to take into account the small distance between the optics. The cover at the rear side of a reflector has a spring catch that can be removed for cleaning the prisms. When the nine-prisms assembly (Fig. 20.27) is used, the maximum measurable distance is about 2 km, with an accuracy of ±10 mm. The temperature range is −25 to +50°C. The dimensions of the aiming head are $17 \times 15 \times 10$ cm and those of the control unit are $33 \times 19 \times 36$ cm.

Distomat DI-3 is another joint product of Wild Heerbrugg and SERCEL. It is a typical surveyor's instrument especially suitable for staking purposes because of its capability of displaying horizontal distances with the aid of a built-in computer. Total instrumentation is placed on a tripod together with the theodolite. The control unit is set on the tribrach, the theodolite is placed in a dish of the control unit axis, and the

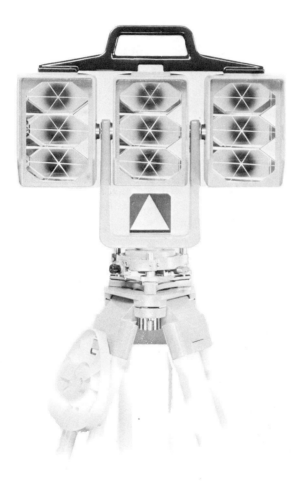

Figure 20.27 Set of nine prisms with target (courtesy Wild Heerbrugg Ltd.)

aiming unit is set on top of the theodolite telescope (Fig. 20.28). Like the DI-10 it has a Ga-As diode light source, but the divergence is about 1.5 mrad. Because of the short distances measured (less than 1 km) only two modulation frequencies are needed, namely, $f_1 = 7.4927$ MHz for the fine reading and $f_2 - 74.927$ kHz for the coarse reading. The corresponding unit lengths are $U_1 = 20$ m and $U_2 = 2000$ m. As in the microwave

Figure 20.28   Distomat DI-3 (courtesy Wild Heerbrugg Ltd.).

instruments DI 50 and DI 60, phase measuring is done by the pulse-counting method.

The slope distance measurement is fully automatic, and the measuring process is activated by pushing a start switch. The instrument first measures the intensity of the return signal; if its level is too high, a servo mechanism introduces a filter. The fine frequency f_1 is then selected, by means of a second servo, an internal calibration line is inserted and measured 100 times. The calibration line is cut out while 1000 external measurements are taken over the distance with the fine frequency f_1. The calibration values are averaged and applied to the external measurements, and the calibrated fine measurement is stored in the built-in computer. After changing to the coarse frequency f_2, the internal calibration line is again measured 100 times before it is switched back to the external path for 500 distance measurements. All measured values are compiled together to give the final result, the measured slope distance, which appears as a six-figure number on an LED display. The whole sequence runs automatically and lasts about 10 sec.

The reduction to horizontal distance is made by the computer in the following way. The operator first transfers the measured slope distance from the display to the computer by setting the *distance selector* switch to either horizontal distance or difference in height, both positions being marked by symbols. The vertical circle reading is now entered in the display by means of five two-way switches representing a 10-button keyboard. After the figures have been entered and checked against the circle reading, the *angle switch* is pressed to transfer the vertical angle from the display to the computer. Depending on the setting of the distance selector switch, either the horizontal distance or the difference in height is now displayed. For extra corrections such as atmospheric correction, correction to sea level, and projection correction, the scale of the display can be varied by an 11-step switch where each step is equivalent to a change of 3×10^{-5}. The instrument operates in metric units, but distances can be changed to feet by the computer. Also, angular units in sexagesimal (360°) and centesimal (400^g) systems can be interchanged.

The temperature range for operation is -25 to $+50$°C. The dimensions for the aiming head are $19 \times 11 \times 7$ cm and for the control unit $20 \times 28 \times 9$ cm. The maximum measurable distance with three prisms is about 600 m with an accuracy of ± 5 mm.

20.1.2.9. *SM-11 and Reg Elta 14*

Carl Zeiss, Oberkochen, West Germany, introduced *SM-11* in 1967 as a combined distance meter having angular measurement capabilities; this was one year before AGA made its first attempt with the M 7T. The distance measuring electronics of the SM-11 are identical to those of the *Reg Elta 14* (Fig. 20.29) which is also manufactured by Carl Zeiss and was put on the market in 1968. The electronics are discussed later in connection with that instrument (*Leitz*, 1969). The angle measuring optic in the SM-11 is similar to the Zeiss Th 4 theodolite with 1^c graduation, the circle readout is obtained by means of a scale-reading microscope. All Zeiss instruments use the centesimal graduation, where the total circle,

Figure 20.29   Reg Elta 14 (courtesy Zeiss Oberkochen).

360°, is divided into 400 grads (400^g) and 1^g in turn is divided into 100^c and 1^c into 100^{cc}.

The Reg Elta has a Ga-As diode as its light source, with a wavelength of 0.920 μm. Since the measurable distance is about 2000 m, there are two modulation frequencies: $f_1 = 14.985$ MHz for fine measurement and $f_2 = 149.85$ kHz for coarse measurement. With a phase resolution of 1:1000, f_2 yields the distance with 1000 m unit lengths within ± 1 m, and f_1 yields the distance with 10 m unit length within ± 1 cm. The changeover between fine and coarse frequencies is made automatically. The built-in program first connects automatically the fine modulation frequency and measures the phase difference between the transmitted and received signals 1000 times, forming the mean value, which is stored in a miniature computer. Then the coarse modulation frequency is used automatically, and phase measurements are made again 1000 times. The mean value is computed, and the consistency between the two sets of measurements is compared and the measured distance is displayed in digital form on six-digit Nixie tubes, three on the left-hand side of the decimal point (indicating hundreds, tens, and meters) and three on the right-hand side (indicating decimeters, centimeters, and millimeters). The modulation frequencies generated by crystal oscillators can be electronically varied in increments of $10^{-5}f$ by means of a stepping switch. This allows the user to adapt the modulation frequency to correspond exactly to the selected unit length $U_1 = 10$ m in ambient meteorological conditions.

The light signal returning from the distant reflector is detected in a photomultiplier tube, and for phase measurement the output voltage of the photomultiplier is compared in phase with the voltage derived from the modulation current. For that purpose the conventional heterodyning principle is applied. Phase measuring is based on the pulse counting principle. The sequence of pulses applied has a frequency exactly 1000 times the frequency of the modulation signals. With the unit lengths of $U_1 = 10$ m and $U_2 = 1000$ m, every pulse thus corresponds to a distance of 1 cm and 1 m, respectively. Because 1000 measurements are averaged, a ± 1 mm readout is justified at the measurable distance of 2.5 km.

The two-step angular measurement procedure is the same for horizontal and vertical angles. For the measurement of integer value of grads, two geared disks with 400 teeth each are meshed with very high accuracy. By way of contact springs, the relative position of the geared disks is scanned by a code disk carrying grad values from 000 to 399 in BCD code. To obtain the fractions of grads, the theodolite (telescope) with another vertical and horizontal axis is pointed at the target from the whole-grad position with the aid of measuring wedges, driven by the precision drives. The interpolation values are contained on the measuring wedges in the form of BCD-coded graduations that are read photoelectrically. The interpolation value is added to the grad value, and together these give the angular value with an accuracy of $\pm 10^{cc}$ (approximately 3 sexagesimal seconds).

Display of the measured data, horizontal angle, vertical angle, and the slope distance can be selected to appear in the digital display by means of a switch. For recording of the results on punched tape, the punching program control automatically calls down the data bits one by one. An auxiliary electronic computer can also be added to the Reg Elta to reduce automatically the slope distance to the horizontal distance and to serve in other special applications. This combination makes the Reg Elta 14 a truly tachymetric instrument. In the special application of Reg Elta 14 presented in Fig. 20.30, the instrument was used to compute and register long throws in the European Track and Field Championship Contest in 1970 in Helsinki. The measuring instrument is set up outside the range of throw. The coordinates of the center of throw (S_0, α_0, and Z_0) are

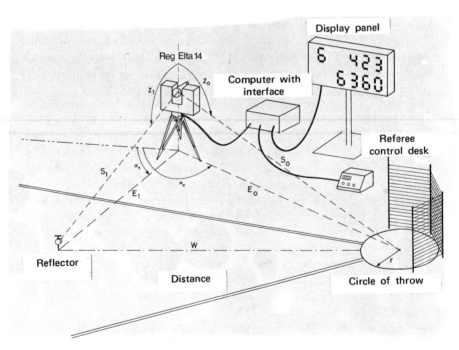

Figure 20.30 Principle of measuring long throws by Reg Elta 14 (courtesy Zeiss Oberkochen).

measured before the competition starts and are stored in the computer, together with the radius r of the circle of throw. After every throw the parameters α_1, Z_1, and S_1 of the point of impact, marked by a reflector, are measured. From these data the computer calculates the distance W of the throw by the cosine law

$$W = [(E_0^2 + E_1^2 - 2E_0E_1 \cos (\alpha_1 - \alpha_0)]^{1/2} - r \qquad (20.42)$$

where $E_0 = S_0 \sin Z_0$ and $E_1 = S_1 \sin Z_1$. At the control desk, the referee declares the throw either valid or invalid and switches the computed value to the display panel. The total process including the punching of the tape takes only a few seconds.

20.1.2.10. Summary of Short-Distance EDM Instruments

Table 20.2 lists EDM instruments capable of measuring distances up to 10 km. It includes most of the makes and brands available at present, together with some earlier experimental instruments that have adapted principles still used today.

Considering the land-to-land electronic and electro-optical distance measuring instrumentation as a whole, the reader can easily recognize the two major initial systems. Among the microwave carrier systems there is no doubt that the Tellurometer system is the basis of most of the instruments developed during the past two decades. Within the series of instruments manufactured by this company, together with several valuable modifications and additions brought in by well-known manufacturers around the world, the following development process is clearly noticeable. The increase of the carrier frequency from about 3 to 35 GHz has made the beamwidth narrower (from 20° to 2°), to reduce the effect of ground reflection. The original CRT readout was changed to dial and digital readouts with the use of the null indicator and finally reached the stage of semiautomation by the use of pulse counting for phase measurements. The maximum measurable distances have been kept between 50 and 150 km, with an accuracy varying between 0.3 and 2.0 cm (excluding the early MRA-1 and MRA-2 Tellurometers).

Although microwave instruments serve primarily for long-distance measurements, the instrumentation using light as the carrier has a much wider scope in range and accuracy. Basically the Geodimeter system, because of its historic development, can be considered to be the initiator in this category, followed by many revolutionary designs by other manufacturers. Instruments in which light is the carrier have been applied fairly equally in long- and short-distance operations, except in the late

TABLE 20.2

Manufacturer	Model	Emission Source	Range (km)	Accuracy (±mm)	Accuracy (±ppm)	Remarks
AGA, Sweden	M 700	Laser	5.0	5	1	Tachymeter type
	M 76	Laser	3.0	5	1	
Askania, West Germany	Adisto S 2000	Infrared	2.0	10	⎤	Same instruments
Askania-Franken, West Germany	Adisto 1000	Infrared	2.0	10	⎦	
Bjerhammar, Institute of Technology, Sweden (1957)	ART	Infrared		10		Tachymeter type
	Terrameter	Tungsten	3.0	25		Experimental; quartz modulator; phase comparison by heterodyne
Cubic Industrial Corp., United States	Cubitape DM-60	Infrared	2.5	5	10	
Hewlett-Packard, United States	H-P 3800B	Infrared	2.5	5	7	
Kern, Switzerland	H-P 3805A	Infrared	1.6	7	10	
	DM 1000	Infrared	2.5	4	4	
	DM 500	Infrared	0.5	10		
	ME 3000	Xenon flash	3.0	0.2	1	Same as NPL Mekometer III; manufactured jointly by KOM-RAD, England

Manufacturer	Model	Source				Remarks
Laser Systems & Electronics, United States	Microranger	Infrared	1.6	5	2	All LSE electro-optical distance measuring instruments manufactured for Keuffel & Esser
	Ranger I	Laser	4.0	5	2	
	Ranger II	Laser	6.0	5	2	Experimental instruments; very high precision; operate independent of the speed of light
National Physical Lab., England	Mekometer I	Xenon flash tube	1.0	1		
	Mekometer II	Xenon flash tube	1.0	1		
	Mekometer III	Xenon flash tube	3.0	0.1	1	
Tellurometer, Pty, Ltd. South Africa	MA-100	Infrared	2.0	1.5	1	
U.S.S.R.	CD-6	Infrared	2.0	5	5	Experimental sweep frequency (1961)
	DST-2	Tungsten	5.0			
Wild Heerbrugg, Switzerland	KDG-3	Infrared	1.2	20	25	
	GD-13	Infrared	6.0	5	5	Sweep frequency for unambiguous distance
	DI-10	Infrared	2.0	10		
	DI-3	Infrared	0.6	5		
Zeiss Jena, East Germany	EOK 2000	Infrared	2.5	10		
Zeiss Oberkochen, West Germany	SM-11	Infrared	2.0	10		Tachymeter type (optical)
	Reg Elta 14	Infrared	2.0	10		Tachymeter type (electronic); distance measuring unit same in both instruments

1960s and early 1970s, when the tendency to produce lightweight, portable surveyors' tools for short distances became predominant. The development steps are mainly concentrated in the use of various light sources (tungsten lamp, mercury vapor lamp, He–Ne laser, Ga–As luminescent diode, and the xenon flash tube). The modulation of the transmitted light has gone through the stages from Kerr cell through Pockels crystals to the direct modulation of the Ga–As diode. Received light signals have been detected and phase measured by a photomultiplier tube and a delay line or, as in many newer instruments, by a photodiode, heterodyned beat signals, and a resolver or a pulse counter. Instruments already available feature fully automatic distance readout displays or automatic displays and recording capabilities for horizontal distances and horizontal and vertical angles. For surveyors and construction engineers, simple and portable instruments with measuring capabilities between 500 and 600 m may be the answer for staking and cadastral survey; geodesists and geophysicists can make their choice among instruments that reach distances to 100 km. The accuracies of all the available modern electro-optical instruments vary between fractions of a millimeter to a few centimeters. The price scales, too, vary significantly.

20.2. LAND-TO-SEA SURVEY

Land-to-sea survey techniques discussed in this section primarily refer to applications in hydrographic surveying, close-to-shore geophysical exploration, and harbor surveying. Maximum measurable distances are usually less than 500 km, and the accuracy obtained is similar to that of geodetic surveying. The boundary line between a hydrographic survey and navigation is arbitrary, and the coverage of electronic navigation in Part III indicates more flexibility in distances and more relaxation in accuracy demands. The land-to-sea survey instruments are grouped according to carrier wave frequency because this variable affects also the distances and geometry involved. For a detailed discussion of error analysis in both circular and hyperbolic configurations, see Chapter 4.

20.2.1 Short Wave and Microwave Techniques

Although some of the systems using long or medium waves were in general use in the mid-1940s, the application of microwave techniques in sea surveying is rather new. The use of microwave instruments in hydrographic surveying was initiated in the late 1950s by Tellurometer, followed by the Cubic Corporation in the mid-1960s. The use of micro-wave instruments and short wave technique in general is limited to distances less than about 100 km because of the line-of-sight condition. The problem of maximum measurable distance was handled in Chapter 8, but as a rule of thumb for instrumentation presented in this section, the following expression is repeated; $D_{max} = 4(\sqrt{H} + \sqrt{K})$, where D_{max} is the maximum measurable distance in kilometers, and H and K are the shipborne and ground antenna elevations in meters.

20.2.1.1. *Autotape DM-40A and DM-43*

Autotape DM-40A entered the market in 1966 and is manufactured by the Cubic Industrial Corporation, San Diego, California. Although it may also be used for mapping purposes connected with helicopter or small plane operations, its primary service is in hydrographic surveying. It is a circular method with two ground stations called responders and a ship-borne measuring system consisting of the interrogator (Fig. 20.31), data converter, plotter, and track indicator. In the distance measuring system,

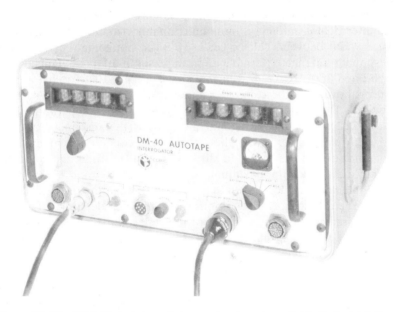

Figure 20.31 DM-40 Autotape interrogator (courtesy Cubic Industrial Corp.).

2.900-3.100 GHz carrier waves are frequency modulated with three ranging frequencies of approximately $f_1 = 1500$ kHz, $f_2 = 150$ kHz, and $f_3 = 15$ kHz to obtain an unambiguous distance of 10 km with the system resolution of 10 cm. The measuring principle is, in effect, the same as that used in the Electrotape DM-20 made by the same manufacturer. Distance is measured by comparing the phases of the transmitted and received modulation waves. In the Autotape this is done automatically for each modulation frequency, and the two range displays appear as five-digit Nixie-tube readouts in the interrogator panel. The information is also transferred to the data converter or computer, which accepts the distance data from the interrogator at a one-second sampling rate. The distances are edited, smoothed, and extrapolated. They are then used to compute position coordinates for the plotter unit and the distance left or right of track or to or from a destination point for the indicator unit. The plotter currently used, a Hewlett-Packard X-Y plotter, provides a real-time track plot of the vessel at 2-second intervals on a previously prepared plot sheet. The indicator consists of two dial displays that indicate the distance deviations from a predetermined line of course and the distance to or from a predetermined point. Autotape can also be coupled with a digital depth sounder and a digital clock, and all data can be recorded on magnetic tape or printed out in tabular form. After the system is in operation, the ground stations may be left unattended.

The interrogator antenna is omnidirectional horizontally, and the vertical beamwidth between half-power is 10°. The horizontal beamwidth of the responder antenna is variable and can be set in 30° increments of 30°, 60°, 90°, and 120°. Its vertical beamwidth is a fixed 10°. To increase the operational distance, the responder antenna should be set at the smallest beamwidth. This is done at the cost of the width of the operation area. Figure 20.32 shows the horizontal patterns of the responder antenna under ideal conditions, where the line-of-sight condition is maintained. Each radial division is equal to 20 nautical miles (nmi). To keep the measurable distance at a maximum, the mission should not require more than a 5° elevation or depression angle as measured at the responder horn antenna. The following two conditions should be considered. (*a*) To avoid the loss of radiated energy because of the reflection from the ground or water between the measuring points, the propagation paths should be maintained at grazing angles. (*b*) To avoid reflected radiation from vertical objects such as large buildings, the position of the responders should be selected to ensure that there are no reflecting surfaces behind the interrogator. These two conditions affect all systems operating with microwave carrier bands.

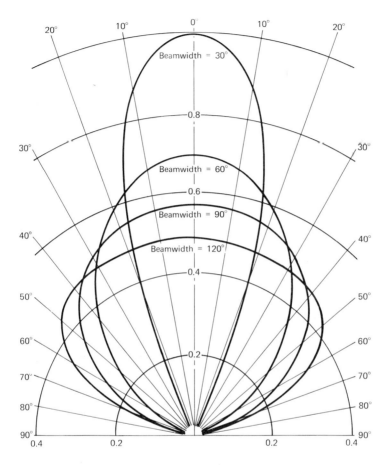

Figure 20.32 Horizontal patterns of responder antenna of Autotape model-40A (courtesy Cubic Industrial Corp.).

The built-in refractive index in Autotape is $n = 1.000320$, and in high precision survey at longer distances, correction should be applied based on observed temperature, pressure, and humidity at the ends of the lines (see Chapters 6 and 7). The power consumption for the interrogator is 8 A at 12 V, for the responder it is 5.5 A at 12 V; the transmitted power is 1 W nominal. There is two-way radio communication between the interrogator and the two responders. The operational temperature range is -10 to $+50°C$. The maximum measurable distance is about 100 km along radio line-of-sight, with an accuracy of $\pm 0.5\,\text{m} \pm 10^{-5}$. The maximum range rate is 290 km/hr, which means that measurements can be made when the interrogator and responder units have relative velocities, with respect to each other, of less than 290 km/hr.

Autotape DM-43 is identical to the DM-40A except that its interrogator unit is capable of measuring distances simultaneously from three responders. This increases the versatility at the system, permitting both hyperbolic and circular configurations to be utilized, with a superfluous observation in the circular case to allow either graphical or mathematical adjustment of the location.

20.2.1.2. Trisponder

The main means of position fixing in exploration, rig positioning, and pipelaying work in the North Sea is the *Hi-Fix* Network, developed and manufactured by Decca Survey Ltd, England and described in the following section. The accuracy of this hyperbolic system cannot be uniform throughout the coverage, and there are places and conditions in which it is insufficient. Thus there arises the need for a system that uses highly portable equipment and has inherently greater accuracy, although with a relatively short maximum range (*Powell*, 1973). Such a system is the *Trisponder* (Fig. 20.33), also manufactured by Decca. The Trisponder is a circular system using basically three stations, one in the survey vessel (the mobile unit) and two (the remotes) on known ground points. The instrument is of the pulsed-type design, where the mobile transmitter sends coded pulses of 0.5 μsec duration to the remote stations at the standard carrier frequency of 9.480 GHz. The transmitters of the remote stations return the specifically coded pulses back to the mobile unit with a standard carrier frequency of 9.325 GHz. The distance measurement in the mobile unit is carried out by the digital counter system so that the leaving pulse starts the counter and the returning pulse stops it. All transmitters utilize a tunable magnetron within frequencies of 9.300 and 9.480 GHz having the pulse power of 1 kW, and all receivers are superheterodynes with frequency range of 9.300–9.500 GHz and with 60 MHz IF.

The mobile station transmits sequentially each second two trains of pulses coded for appropriate remote stations. The remote stations in turn decode their own pulse trains and transmit reply pulses back to the mobile station. At the same time, to minimize possible interference, the mobile station is looking for these particular coded signals and therefore rejects any other signals originating either inside or outside the system. To obtain maximum accuracy, 100 sets of travel times are measured in rapid succession, and the range display presented corresponds to their

Figure 20.33 Decca Trisponder 202A (courtesy Decca Survey Ltd.).

average, with one-meter resolution. This technique is known as the "100 sum mode." Once the Trisponder has warmed up (15–30 min) the range information can be updated automatically (autorange mode) or manually on demand, as required. In the autorange mode the system needs no attention as long as the incoming signals are good. Then the distance measuring unit continues to feed the range data to displays, using standard BCD outputs to such peripheral equipment as a computer and plotter, a digital printer, a tape puncher, or a magnetic tape recorder. Should one of the radio links be interrupted by nonmaintenance of signal line of sight, cancellation of reflected and direct paths, extreme amount of X-band interference, or a heavy rain shower in the radio path, for example, a red warning signal starts to blink, showing the troubled channel.

The standard system can accommodate up to four remotes. The radio pulses from the mobile transmitter are coded to trigger a selected two of these three or four stations, which then send back answering pulses. This versatility of the station layouts will, of course, improve the efficiency and accuracy of the survey. Up to three vessels, each having a mobile unit aboard, can share the system's facilities. When set in the autorange mode, the Trisponder provides a new fix once per second. To do so it gathers within 0.3 sec all the data it needs and displays the ranges until the next fix. It is therefore possible to use two more mobiles in the remaining 0.7 sec.

The mobile station comprises (a) a distance measuring unit (DMU), which contains all the controls and displays for the operation, together with all the electronics for interrogation and identification and for distance measurements; (b) a transmitter/receiver unit (TRU), with omnidirectional antenna, and (c) a 24 V DC power supply. The antennas are horizontally polarized such that the mobile antenna is omnidirectional horizontally and has a vertical beamwidth of 30°, and the remote antenna has a horizontal beamwidth of 87.5° and a vertical beamwidth of 5°. The DMU weighs 11.25 kg, the TRU 6.8 kg, the omnidirectional antenna 0.14 kg, and the directional antenna 0.18 kg. The operational temperature range is −30 to +67°C. The maximum measurable distance is about 80 km, with an accuracy of ±3 m. The maximum range rate is about 180 km/sec, after which a correction for the time lapse between the two ranges displayed must be applied. This instrument may also be used in connection with helicopter or small plane operations.

It sometimes happens that only one site is available for a fixed ground station, such as an isolated offshore platform. This situation rules out the use of the standard circular system. For such a contingency there exists a system known as *Artemis 2*, formerly called *DART* or *Decca Automatic Ranging Theodolite*. This system has a position-fixing accuracy similar to that of the Trisponder, and it also uses microwave transmission to measure range. The position fixing, however, is obtained by the intersection of a range circle with the bearing of the ship measured from the ground station by a split-beam technique to an accuracy of about 2 minutes of arc. The equipment is bulkier and more expensive than that of the Trisponder, but the advantage of using such an instrument in an area with only one known point available is obvious (*Powell*, 1973).

20.2.1.3. *MRB-201 Tellurometer*

The present dynamic Tellurometer system MRB 201 is primarily designed for hydrographic surveying but is also applicable for operations with a hovering helicopter. It is the result of development of several earlier Tellurometer models. Model MRA-2 (p. 267) was modified to serve hydrographic surveying in the following way. The modulation frequencies were changed to the 1.5 MHz region to represent distances in metric units. Three frequencies were used to introduce unit lengths of 10,000, 1000, and 100 m. The master and remote units were not interchangeable, and the control of the remote pattern frequencies were conducted by the master unit. The readout was still the CRT circular time base, with a break caused by the 1 kHz beat signal. The maximum velocity of the vessel was about 36 km/hr (20 knots), to allow the observer to follow the rotating break in the CRT with a plastic cursor device. This modified instrument, *Hydrodist MRB-2*, became available in 1960.

Based on the features of the MRB-2 and the basic electronics of the MRA-3, a new and more advanced model MRB-201 *Dynamic Tellurometer* was developed in 1970. The carrier frequency is 2.8–3.2 GHz with an available alternative range of 3.2–3.6 GHz, the carrier output power being not less than 200 mW. The modulation pattern frequencies are as follows.

Pattern	Master (MHz)	Remote (MHz)
A	1.498470	—
A −	—	1.499470
A +	—	1.497470
B	1.496972	1.495972
C	1.483485	1.482485
D	1.348623	1.347623

With the built-in refractive index $n = 1.000330$, the unit lengths obtained from the frequency difference patterns are 100,000, 10,000, 1000, and 100 m.

The main components of the MRB-201 are the basic unit, the digital range integrator (DRI) unit, and the dial readout (DRO) unit. The basic unit contains all the Tellurometer circuitry except the phase measuring and range display section. It has a thread for mounting on a standard tripod if required. The digital range indicator, which plugs into the basic unit, measures the phase of the incoming signals and displays the range directly on a six-figure cold cathode tube display. This display is unambiguous to a range of 100 km, having a resolution of 0.1 m. The display is initially set up by means of a single, rapid procedure, after which the unit is capable of automatically and continuously displaying the range of the vessel, which has a slant velocity of up to 55 km/hr (30 knots) only. The even-range identification can be achieved at any stage by nulling a meter. The fine pattern reading is updated every millisecond and is virtually continuous. If a temporary loss of signal occurs, the DRI automatically switches in a "rate memory" that updates the display according to the last measured velocity until the signal returns. If no major speed changes or course alterations occur during the period of signal loss, the display will be corrected when the signal returns. As is common with modern position-fixing systems, the readout has been designed with a digital data output that can be coupled to a printer, magnetic tape or punch tape recorder, on-line computer, track plotter, or any other digital recording device. The dial readout unit also plugs into the basic unit and can be used instead of or in addition to the digital range integrator. It is especially suitable for use with the basic unit in the remote stations, which in the Tellurometer MRB-201 system have the advantage of being capable of measuring the base for the circular position fixing. The DRO has no electrical data output. The resolution is 0.1 m, and unambiguous readings to 100 km are obtainable. If the base does not need to be measured, the position fixing can be made with remote instruments without either DRI or DRO attached. If the DRI is not required, an internal battery can be housed in the place provided for the DRI. Figure 20.34 illustrates the plug-in system of the DRI and DRO units into the basic unit. There are different choices of antenna installations available, according to the requirements. In the remote stations on shore, two kinds of fixed antennas can be used—circular, with diameters of 43 or 61 cm, or rectangular, with dimensions of 43×20 cm or 36×25 cm (Fig. 20.35). On shipboard in short-range offshore operations an omnidirectional antenna is suitable because of the all-round directivity which, however, lessens the antenna gain and limits the range. In harbor areas where surrounding structures cause multireflections, the steerable antenna is preferable. The rotation of the antenna is remotely controlled on a control box at the master instrument, and a visual indication of the direction of the antenna is provided for the operator. Steerable antennas are also usable in long-range offshore operations. The following table lists different antennas, their beamwidths, and reachable distances.

Type of Antenna	Approximate Range (km)	Beamwidths (degrees)	
		Horizontal	Vertical
36 × 25 cm rectangular	50	24	20
36 × 25 cm steerable	50	24	20
43 × 20 cm rectangular	50	120	20
43 cm circular	150	20	20
61 cm circular	250	15	15
Omnidirectional	10	360	30

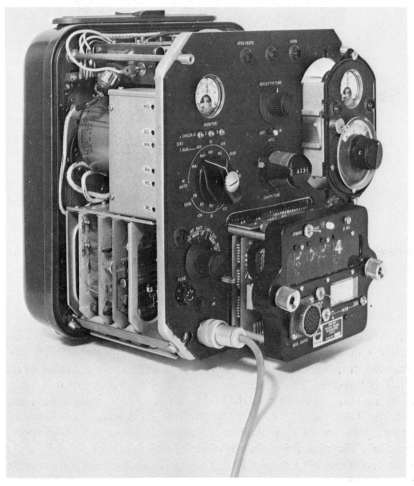

Figure 20.34 MRB-201 Tellurometer (courtesy Tellurometer–USA).

Figure 20.35 Rectangular antenna of MRB-201 (courtesy Tellurometer–USA).

It should be noted that the MRB-201 is a dual system that includes two master units, with the DRI in the survey vessel together with twin antennas. Integral duplex radio voice communication is provided in the MRB-201 system. The power consumption of the basic unit is 2.6 A at 12 V, the basic unit with the DRO takes 2.7 A at 12 V, and the basic unit with the DRI uses 4.0 A at 12 V. The standard measurable distance, assuming the line-of-sight condition, is about 50 km or more with special antennas. The accuracy under dynamic conditions is ±1.5 m and under static conditions ±0.5 m ± 3 × 10⁻⁶. The operational temperature range is −10 to +50°C.

20.2.1.4. *Shoran*

The long obsolete Shoran (*short range navigation*), once used in aerial mapping, has found a new and efficient use in hydrographic surveying and geophysical exploration. The instrumentation has been improved and updated. Shoran was originally designed by the Radio Corporation of America (RCA) for the U.S. Air Force to serve military applications during World War II. Shoran uses an interrogator-responder airborne set, originally symbolized as the AN/APN-3, and two ground beacons designated AN/CPN-2. Shoran is a pulsed-type circular location method. According to the original principle of Shoran, the airborne transmitter generates three trains of pulses of 0.8 μsec duration. Two pulse trains are sent to the two ground stations, known as the High and Low stations (formerly Rate and Drift), on frequencies of 210 and 260 MHz, respectively. The third is used as a *reference marker* in the indicator of the airborne set. Transmissions to each ground station are on separate radio frequency channels. Actually the airborne set has only one transmitter, whose frequency is switched by a commutator alternately from one channel to the other at a rate of 10 Hz; thus each ground station is interrogated every 0.05 sec on this time-sharing basis. Each ground station receives, amplifies, and retransmits the pulsed signals, after a certain delay, to the airborne receiver on a common frequency of 320 MHz (Fig. 20.36). The airborne set then receives the transponded pulses and displays them, along with the reference marker M as deflections on the circular time base of the CRT's type A scan as the indicator. The commutator, simultaneously with the alternation of transmissions to Low and High stations, reverses the polarity of the Low and High pulses L and H so that they appear as inward and outward deflections, respectively, while the reference marker M appears as an outward deflection at the top of the sweep (Fig. 20.37). Two independent measuring circuits are provided, one for each ground station, and the time measurements are obtained by using a so-called set-back method, in which the time of transmission of each pulse is advanced ahead of the reference marker pulse, ensuring that the ground station response pulse is in coincidence with that marker. The time advance is accomplished by using a special phase shifter system so designed that the amount of time advance is linearly related to the angular position of the control potentiometer. Each phase shifter system is mechanically coupled to a digital distance counter indicating directly the range from the aircraft to that particular ground station in statute miles.

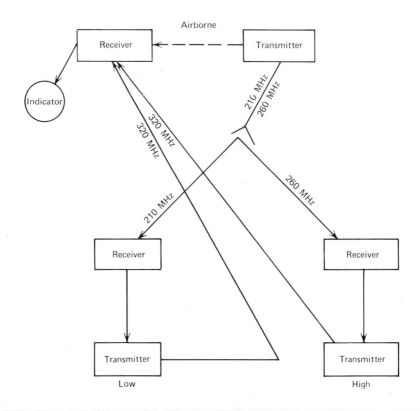

Figure 20.36   Functional diagram of Shoran principle.

The Shoran system is designed on the assumption that the average propagation velocity of electromagnetic energy along the ray path is $c = 186,219.74$ miles/sec. This means that the signal travels about 93 miles and back in $\frac{1}{1000}$ sec, or 100 miles and back in $\frac{1}{930}$ sec. This is considered to be the maximum unit length of Shoran; thus the transmitter of the airborne set generates pulses on the *repetition frequency* of 930 Hz. At this rate there is a complete phase shift in the return signal for every 100 miles of distance between the airborne and ground stations, and the Shoran dials repeat themselves every 100 miles. In the Shoran system three harmonically related *timing frequencies* make the time base revolve once in the time needed for the pulse to travel to the ground station and back at distances of 1, 10, and 100 miles. The timing

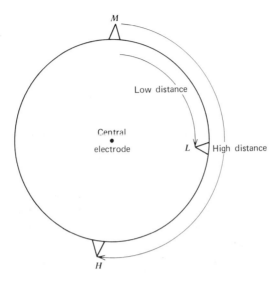

Figure 20.37 Circular time base of Shoran.

frequencies corresponding to those unit distances are:

$$f_1 = 93.109870 \text{ kHz}$$
$$f_2 = 9.310987 \text{ kHz} \qquad (20.43)$$
$$f_3 = 931.0987 \text{ Hz}$$

In about 1949 Shoran as such became obsolete and was replaced by *Hiran* (high accurate Shoran). One of the largest error sources in Shoran was the "rise time" error caused by matching two deflections introduced into the time base by two signals of different signal strength. In Hiran a special *gain riding* system is employed at both ground and airborne stations to reduce the amplitude of the receiver output to a reference level. The signal strength is indicated on an auxiliary oscilloscope, and an operator adjusts it manually. For more detailed information, see *Jacobsen* (1951) and *Laurila* (1960).

The original Shoran and Hiran were bulky and heavy. The airborne set weighed about 100 kg, and the ground installation including antennas and power sources weighed about 500 kg; but the principle was sound and the accuracy was of the order of a few meters. The so-called duct effect at the frequency band Shoran is using makes it possible to reach distances significantly beyond the optical horizon. This phenomenon has encouraged manufacturers to modify Shoran for use in hydrographic offshore surveying. In addition, if ground stations have to be placed inland causing the electromagnetic energy to propagate over land and over water, no velocity change occurs because of the variation of the soil conductivity.

Coastal Survey Ltd., which is a group company of Decca Survey Ltd. in England, has redesigned the original Shoran to meet the requirements of modern electronic technology. Special attention was given to the building of a lightweight modular-constructed transmitter/receiver for use aboard ship. The modified system is called *Digital Shoran* or *Long-Distance Shoran*. Three stations are used, a mobile station onboard a survey vessel and two base stations on land. The mobile receiver offers fully automatic tracking facilities, together with a manual override facility. This mode allows the required positional data to be preset, after which the vessel is conned to acquire the fix. Both analog and digital outputs are provided to allow the use of the widest possible range of peripheral equipment. The frequencies used are 220–300 and 420–450 MHz. The weight of the receiver (Fig. 20.38) is 18.5 kg. The maximum measurable distance is about 320 km (200 miles), with a resolution of 0.001 mile or approximately 1.6 m. The operational temperature range is −10 to +50°C.

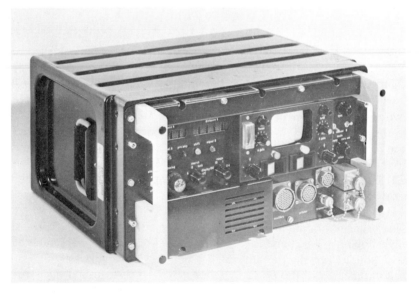

Figure 20.38 Digital Shoran receiver (courtesy Decca Survey Ltd.).

Another modification to the original Shoran was made by the Lorac Service Corporation, a subsidiary of Seismograph Service Corporation, Tulsa, Oklahoma. *Long Range Shoran* has several features similar to the original version. The primary modifications to improve the instrument's suitability for long-distance hydrographic surveying were in the antenna assemblies. High gain antennas are used in such a way that a corner reflector has replaced the open dipole in the mobile unit and a directive yagi antenna has replaced the corner reflector in the base unit. Multiple-vessel operations are possible because of a unique coding feature of the equipment. Since the transmitter in the base station has been modified to serve as a transponder, the received pulses from the mobile station are automatically transmitted as response pulses back to the mobile station. The frequencies used are between 220 and 320 MHz. The output pulse width or duration is 1 μsec, and the pulse repetition frequency is 930 Hz. The maximum measurable distance is about 320 km with an accuracy of about ±4 m, for a 160 km range. The mobile unit is still relatively heavy, weighing about 80 kg.

20.2.2. Instruments Using Medium or Long Waves

Instruments using medium or long waves have certain common features differing from those described in the previous section. All systems employ continuous wave transmission with a phase comparison technique for distance or distance-difference determination. The propagation of electromagnetic energy in the medium or long bands is likely to be affected by changes in velocity caused by the change in conductivity of the ground or water over which the waves propagate (see Chapter 10). Also reflections from the ionosphere affect most of the systems, causing loss of accuracy (see Chapter 9). However, these shortcomings are compensated because waves in those wave bands follow the earth's curvature, and this phenomenon, together with the bigger output powers characterizing these systems, extends significantly the maximum measurable distances.

20.2.2.1. *Survey Decca*

The *Survey Decca*, which is representative of the basic Decca system, is a hyperbolic method employing transmitted signals of very low frequency (around 100 kHz) for phase difference measurements. After successful trials in the early 1940s, the Decca Record Company in London built an operational system that was used extensively by the Royal Navy as a mine-sweeping aid immediately prior to the D-Day landings. In 1945 the Decca Navigator Company was formed to start mass production of the various Decca survey or navigation instruments. At present all Decca survey equipment is manufactured by Decca Survey Ltd., Teddington, England, and in several group companies around the world.

The Survey Decca instrumentation comprises three fixed radio transmitting ground stations, a fixed monitor receiving station, and a mobile receiving station onboard a survey vessel. The transmitting station located at the common focal point of the two hyperbolic patterns is called the master station; the other two, called the slave stations, are labeled *red* and *green* to correspond to the colors of the hyperbolas as drawn on the map. The master station transmits continuous, unmodulated waves at a frequency of f'_M, which is a harmonic of a basic frequency f'_0. Correspondingly, the slave stations R and G radiate continuous waves at frequencies of f'_R and f'_G. All transmitting frequencies are now obtained as follows:

$$f'_M = k''_M f'_0$$
$$f'_R = k''_R f'_0 \tag{20.44}$$
$$f'_G = k''_G f'_0$$

where k''_M, k''_R, and k''_G are numbers of transmitting frequency harmonics in relation to the basic frequency.

The frequency of the master station is kept constant by means of a thermostatically controlled crystal. During the survey the waves radiated by the slave stations are kept locked in phase to signals transmitted by the master station by means of receivers at the slave stations tuned to the frequency of the master station. The Decca Survey receiver has a separate channel for each transmission frequency. If the signals arriving from the pairs of transmitting stations are to be mutually comparable, they must have the same frequency. This is achieved by multiplying the transmission frequencies f' by factors k, in which the coefficients k'' are included as denominators and their lowest common dividends k' as numerators. Consequently, we obtain

$$k_{MR} = \frac{k'_R}{k''_M}, \qquad k_{MG} = \frac{k'_G}{k''_M}$$
$$k_R = \frac{k'_R}{k''_R}, \qquad k_G = \frac{k'_G}{k''_G} \tag{20.45}$$

In the Decca system the coefficients k'' are

$$k''_M = 6, \qquad k''_R = 8, \qquad k''_G = 9$$

whereupon

$$k'_R = 24 \quad \text{and} \quad k'_G = 18 \tag{20.46}$$

and, finally

$$k_{MR} = 4, \quad k_R = 3, \quad k_{MG} = 3, \quad k_G = 2 \tag{20.47}$$

If the basic frequency $f'_0 = 14.7527$ kHz, the transmission frequencies and the so-called *comparison frequencies* will be as follows.

Station	Transmission Frequencies, $f' = k''f'_0$ (kHz)	Comparison Frequencies, $f = kf'$ (kHz)	
Master	88.516		
Red	118.021	354.065	(20.48)
Green	132.774	265.548	

A block diagram of the basic Decca system with these frequencies appears in Fig. 20.39. If the propagation velocity of electromagnetic energy is given the value $c = 299,700$ km/sec, the red and green pattern comparison wavelengths and lanewidths in respect to values (Eq. 20.48) will be

$$\lambda_R = 846.47 \text{ m}, \quad \lambda_G = 1128.64 \text{ m}$$
$$l'_R = 423.23 \text{ m}, \quad l'_G = 564.32 \text{ m} \tag{20.49}$$

The transmitters deliver either 300 or 600 W into the aerials, which can be 21, 30, or 45 m high, according to the needs for portability and range. In this frequency range the proper earth wire system is of great importance. It usually consists of a set of radial wires of lengths equal to that of the antenna mast. The Survey Decca receivers comprise three radio frequency channels, one for each station of a master-red-green chain. The outputs of the master and red slave channels feed multiplier circuits that generate the common harmonic frequency (4 times the master frequency equals 3 times the red frequency). Phase comparison of the two harmonics is effected by a discriminator circuit, and the value of the phase difference is displayed on the red phase indicator or *decometer*. Similarly, the master and green slave signals are compared at a common harmonic frequency (3 times the master frequency equals 2 times the green frequency) to produce the green decometer reading. The receivers operate either with the standard 115 or 220 V AC line, with a 24 V DC battery, or with a generator. The total weight of a Survey receiver with two decometers is approximately 45 kg. The phase indicators or decometers have a clocklike appearance with arms showing the total number of lanes

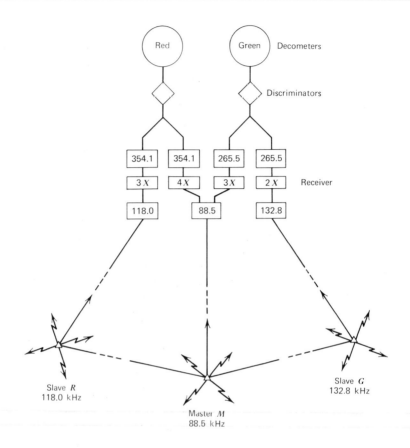

Figure 20.39 Functional diagram of Survey Decca (courtesy Decca Survey Ltd.).

and their fractions. The green decometer counts 18 lanes in a group (Fig. 20.40), and the red decometer 24 lanes in a group according to the k'-values (Eq. 20.46). Each group is identified with a capital letter in a dial window. The hyperbolic charts are printed accordingly. The lane-counting pointers are mechanically geared to the electronically operated lane-fraction pointers and therefore must be manually set to correspond to the proper lane value at any known point. All geometric aspects are presented in Chapter 4. The maximum measurable distance from the ground stations to the survey vessel varies between about 100 and 300 km, according to the type of the transmitter/antenna installation. An analyzed standard error of a setup with base lengths of 50 km and the

Figure 20.40 Survey decometer dial (courtesy Decca Survey Ltd.).

average distance from ground stations to the survey area of 50 km is about ±5 m (*Laurila*, 1956). Since Survey Decca is a multiuser system, an unlimited number of vessels can make the position fixing simultaneously.

20.2.2.2. *Two-Range Decca and Lambda*

Lambda, which entered the market in the late 1950s, was a modified and improved version of the *Two-Range Decca*, an instrument whose service was on a more experimental basis in the early and mid-1950s. Two-Range Decca was designed on the principle that if there is no requirement to use more than one vessel at a time in a survey operation, a Decca chain can be rearranged to introduce circular fixing characteristics. This arrangement has certain valuable features not given by the normal hyperbolic layout. Computation of hyperbolic lattice charts is avoided, and having only two shore stations simplifies the problem of station survey and maintenance in undeveloped areas. Also, it increases the area in which accurate position fixing is possible. Only the red and green slave stations are sited on the coast, and the master station is set up on the survey ship itself, together with the receiver that drives the conventional red and green decometers. The decometer readings are then a function of the *distance* from the ship to the corresponding slave stations.

The basic principle in the two-range layout is the same as in the conventional hyperbolic set. The total number of lanes is given by the known master-slave baseline, which is divided by half the wavelength of the comparison frequency. Assuming that the propagation velocity of electromagnetic radiation is known, the total lane number is therefore a measure of the distance between the master and slave stations. Two-Range Decca consists simply in placing the master transmitter on the survey vessel in the arrangement just described, as well as a receiver. As the master moves, the phase angle at the slave station is automatically held constant by the normal functioning of the slave phase-control circuits; thus given an initial calibration of the slave equipment, it is sufficient to read the appropriate decometer only at the master end of the line. The value of the phase angle between the slave transmission and the master signal received at the slave is preset by manual phase adjustment during initial adjustment of the system. This phase angle is set in empirically by placing the ship at a known point and instructing the slave operator to adjust the phasing control until the desired fractional reading is observed at the ship. This process serves to calibrate the whole system and to compensate for the fixed displacement caused by the receiver's proximity to the master transmitter. The equipment in the Two-Range Decca is identical to that used in a conventional hyperbolic chain except for the shipborne master antenna. The form of this antenna depends on the design of the individual vessel.

The name Lambda is derived from the words *low amb*iguity *D*ecca, and, although its operational principle is very similar to the Two-Range Decca, several essential modifications have been made. Whereas the Two-Range Decca relied solely on mechanical lane counting by geared-down phasemeter pointers to obtain the whole-lane numbers, Lambda uses electronics to measure the "fine" readings or fractions of lanes and "coarse" readings to count whole lanes to a certain value; after this value it employs a lane-identification procedure to obtain an unambiguous distance to about 5 km. Also, harmonics between the master and slave transmission frequencies, thus the comparison frequencies and lanewidths, are not the same as those used in the Two-Range Decca.

Since the radiated and phase-comparison frequencies are harmonically related, it is convenient here to refer to them in a harmonic notation rather than by the actual frequency values (*Powell* and *Woods*, 1963). The master transmitter on the ship radiates a continuous wave signal with a frequency of $12f_0$ where f_0 is the basic frequency of approximately

15 kHz. On shore, the red slave station receives the master transmission and radiates a frequency of $8f_0$ in a manner ensuring that the slave and master signals have a constant phase relationship at a common multiple frequency of $24f_0$. A similar process takes place in the green coordinate ($9f_0$ slave frequency) at a common comparison frequency of $36f_0$. By assigning the values for the basic frequency $f_0 = 14.8$ kHz, and for the average propagation velocity of electromagnetic radiation over water $c - 299,650$ km/sec, the following values for the carrier frequencies, comparison frequencies, and lanewidths as functions of f_0 and c are obtained:

	Carrier Frequencies		Comparison Frequencies		
Functioning	kHz	Harmonic	kHz	Harmonic	Lanewidth (m)
Master	177.6	$12f_0$	—	—	—
Master, lane identification	162.8	$11f_0$	—	—	—
Green	133.2	$9f_0$	532.8	$36f_0$	281.2
Red	118.4	$8f_0$	355.2	$24f_0$	421.8

Clearly all frequencies are harmonically related to the basic frequency f_0, which is not transmitted itself. A decometer can be read with an accuracy better than half a hundredth of a revolution, which, by using the values of the lanewidths tabulated, indicates that the system is sensitive to a change of a meter or two in the ship's distance from the slave station. The lanes are numbered like those of the conventional hyperbolic systems (Fig. 4.1) and are grouped into zones 10 km wide, which is approximately the lanewidth of the basic frequency f_0. For convenience, the decometers are so connected that the readings decrease as the slave stations are approached. The zone, lane, and fraction readings for the two patterns can be converted into desired distance units.

The general measuring principle is the following. The phasemeters on top of the display unit in Fig. 20.41 operate at $24f_0$ and $36f_0$ and determine purely the fractional lane values. Lane integration is carried out by the lower pair of decometers, and these respond to the phase difference between slave and master stations measured at the transmitted slave station frequencies ($8f_0$ for red and $9f_0$ for green). Thus lane integration is performed with respect to a phase pattern three times coarser than the red fine pattern and four times coarser than the green

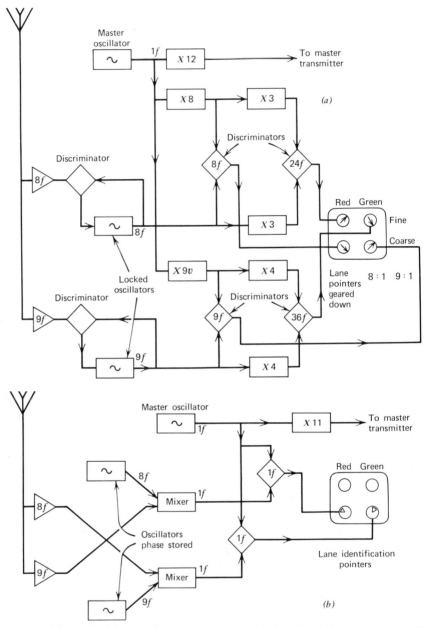

Figure 20.41 Block diagram of Decca Lambda. (*a*) Normal transmission, (*b*) lane identification (from *Powell* and *Woods*, 1963).

fine pattern. This means that the user needs to know only the ship's distance from the slave stations within about ±633 and ±562 m, respectively, to be able to set the lane pointers on the lower meters to the correct numbers when starting the survey. The lane identification process is initiated when the user at the shipborne receiver presses a button; the immediate effect of this is to stop the normal $12f_0$ master transmission from the ship and replace it by a signal of frequency $11f_0$. This signal triggers the lane identification circuitry into action for about 1.3 sec. By using the coarse pattern frequencies, a beat frequency of $(9f_0 - 8f_0) = f_0$ is produced, and the phase comparison with this frequency reduces the ambiguity by factors of 8 and 9 compared with the coarse frequency or slave transmission frequency "lanes"; thus the user has to know the distance between the ship and the shore stations only to about ±5 km.

More detailed descriptions of the measuring principle of these three patterns can be obtained from the block diagram (Fig. 20.41). The master oscillator forms the source of the master transmission and is located on the same chassis as the receiver itself. The signals from the red and green slave stations at frequencies of $8f_0$ and $9f_0$ are received and amplified in separate RF channels containing two-stage crystal filters. Noise-free replicas of the filtered signals are generated by means of locked oscillators employing temperature-controlled ovens. This process encompasses the generation of new sine waves of the slave transmission frequencies, which are then phase locked with the incoming signals. This is done using a discriminator circuit and an auto-resolver. For fine measurement, the slave frequency in the "red" channel is multiplied in the receiver by a factor of 3 and is compared at $24f_0$ ($3 \times 8 \times f_0$), with the twenty-fourth harmonic of the f_0 master oscillation. Similarly the slave frequency in the "green" channel is multiplied by a factor of 4 and is compared at $36f_0$ with the thirty-sixth harmonic of the master oscillation. For the coarse measurement, with reference to the red phase reading, the eighth harmonic of the output of the master oscillator is compared with the output of the $8f_0$ oscillator locked to the slave signal of that frequency. The rotor shaft of the red lane-counting decometer therefore responds to a pattern of lanes whose width is three times greater than that of the fine $24f_0$ lanes. The lane-counting pointer is driven by gearing from the phasemeter rotor in the ratio of 8 to 1; therefore it makes one revolution per zone, which means one revolution for eight red coarse-frequency lanes. Accordingly, the green meter, through the same procedure, has $9:1$ gearing between the rotor and the lane pointer. As mentioned earlier in the general description of the measuring principle, the lane identification system relies on comparison of signals at the basic frequency of f_0. Since it is impossible to transmit a frequency of approximately 14 kHz, which is close to the Omega band, from the stations, a beat note signal is used that is derived directly from the slave transmitting frequencies. Since the master transmitter and receiver are located on the same chassis, the f_0 master signal is obtained directly from the master oscillator. The corresponding f_0 information from the slave stations is provided by interchanging their frequencies, so that red radiates with $9f_0$ and green radiates with $8f_0$ for lane identification. The momentary functioning of the lane identification circuitry, in addition to converting the $12f_0$ to $11f_0$ master transmission, also includes a reversal of the slave frequency harmonics. Given a means of memorizing the phase of the normal slave transmissions while the counterchanged frequencies are being transmitted, an f_0 beat note whose phase represents that of the slave f_0 oscillator can be derived from each slave station. By this method the effect of radiating low frequency signals from the slave stations is simulated without employing any additional frequencies. Similar phase-locking procedures described as taking place in the master receiver are also used in the slave stations.

The phase difference between master and slave at f_0 is displayed by a phasemeter mounted concentrically with the appropriate lane-counting pointer. The phasemeter drives a sector-shaped pointer that on receipt of a lane identification transmission, should move to enclose the lane-counting pointer (Fig. 20.42).

The master transmitting antenna onboard the ship is a base-insulated tubular mast about 14 m high; the receiving antenna system is comprised of two vertical antennas, each 8 m high. The slave stations have receiving antennas similar to those onboard ship, but the transmitting antennas are masts about 30 m high. The power consumption in the slave stations is about 1 kW. The maximum measurable distance varies according to the atmospheric conditions, but a standard distance of about 150 km can be maintained with an accuracy of about ±20 m. Distances up to 500 km have been measured.

20.2.2.3. *Hi-Fix and Sea-Fix*

In the early 1960s the need became evident for a system primarily designed for close-to-shore hydrographic surveying with higher accuracy than that obtained by using the Survey Decca or Lambda. Also, the increasing congestion in the broadcast frequencies required a design of instrumentation operating on one frequency only. These demands led to the development of the Decca *Hi-Fix* system, which operates on the frequency band between 1.700 and 2.000 MHz. The geometric configuration can be either the hyperbolic, with multiuser capability, or the

Figure 20.42   Lambda decometers with lane identification (courtesy Decca Survey Ltd.).

circular, with one-user characteristics. Both take advantage of time-sharing principles to apply the phase comparison technique directly to the transmitted waves. The two slave stations operate independently of each other, and they send the same transmission signals for short periods of time according to the timing sequence controlled by the master station. The sequential signal train has a repetition rate of 1 Hz and lasts 1000 msec: the master sends a trigger signal of 100 msec, followed by the master and the first and second slave signals, 300 msec each. This procedure requires very stable and accurate storage circuits to store the phase of a signal during the periods of no transmission.

Lane identification calls for a second transmission frequency that does not have to be harmonically related to the first frequency. The second frequency is always 10% lower than the first. Thus if the first frequency is 2.000 MHz representing a lanewidth of about 75 m, the second frequency will be 1.800 MHz. The difference of these frequencies is used to produce the beat frequency of 200 kHz with a lanewidth of about 750 m. The standard measurable distance is about 150 km. The instrument error is about ±0.01 lane, which is equivalent to about ±0.8 m on the baseline.

The *Sea-Fix* is developed from Hi-Fix for short-range (up to 70 km) hydrographic operations. The frequencies are the same as in the Hi-Fix, but the sequential triggering rate has been increased to 5 Hz. The fully transistorized Sea-Fix uses miniaturized components. Compact and lightweight, it can be left unattended for long periods ashore or buoy mounted. The accuracy is of the same order as that of the Hi-Fix.

20.2.2.4. *Hi-Fix/6*

The newest version in the Hi-Fix system is the Hi-Fix/6. The name is derived from the fact that one chain can comprise up to six stations, all employing the same single radiated frequency or the same pair of frequencies, when coarse as well as fine position readings are required. These frequencies can be selected within the band of 1.6–5.0 MHz in a manner described later, and the measurements depend on the ground wave mode of propagation. Range and range accuracy are essentially of the same order as for the 2 MHz Hi-Fix system, since the principal limiting factors are those imposed by the propagation phenomena associated with the frequencies concerned. However, since the new Hi-Fix/6 can be used for hyperbolic, circular, or compound position-fixing form, geometric accuracy is improved and the overall range increased. A Hi-Fix/6 chain consists of a "prime" with up to five "secondary" stations. These transmit in a repeated sequence starting with and controlled by the prime, although the prime need not be the master station so far as the patterns and charts are concerned.

The frequency synthesizers used in the transmitting and receiving equipment enables the selection of any value within the 1.6–5.0 MHz band for both the basic or fine frequency f_1 and the coarse frequency f_2 in 100 Hz steps. Under the conditions of frequency congestion now prevailing, it was considered essential to provide virtually instant choice of radiated frequency values and to abandon the use of quartz crystals. The required frequency is selected by setting a thumb-wheel switch to the corresponding heterodyne frequency value, since the receiver and phase control units in the Hi-Fix/6 use superheterodyne circuits. The IF has the round-figure value of 100 kHz; thus to select f_1 and f_2 the user sets in values $f_1 + 100$ kHz and $f_2 + 100$ kHz, respectively. The freedom to change continuously the frequencies f_1 and f_2 in 100 Hz steps exists only within a predetermined 10 kHz band. This is the passband of the RF filter. If, on changing chains, the new frequencies are found to be outside the pass-band of the RF filter circuit board in use, it is necessary to substitute another board. The change takes a matter of seconds and presents no operational problems. Some restriction in the choice of frequency values will exist when two Hi-Fix/6 chains are closely adjacent. To avoid the changes in the mutual interference conditions caused by different shapes of coastlines, for example, a frequency separation not less than 1000 Hz should be assumed for adjacent chains.

As in previously discussed Hi-Fix systems, the lane patterns are pro-duced by comparing the phase of the transmissions at the radiated frequencies. Within the stated frequency band for Hi-Fix/6, the lanewidth varies from 30 m at the highest frequency to 94 m at the lowest, and it is often desirable to resolve the closely spaced ambiguities that these figures imply without depending on the lane-counting action of the displays. Accordingly, the transmission format provides for superimposing a coarse pattern on each fine pattern by radiating from each station a phase-locked signal on a second frequency f_2. In effect, the receiver subtracts f_2 from f_1, thus obtaining from each station a relatively low frequency that is the basis of the coarse lane to which the ambiguity-resolution readings refer. The value of f_2 can in principle lie between 80 and 98% of f_1, resulting, respectively, in coarse lanes 5 and 50 times wider than the fine lanes. In practice, a resolution factor of 50/1 would require extremely stable propagation conditions for acceptable reliability. Also, there would be little advantage to be gained at the other end of the scale from a factor smaller than 5/1. Therefore, values between 10/1 and 20/1 are probably optimal. The choice of suitable frequency values is simplified because the coarse lanewidth need not be an exact multiple of the fine (*Powell*, 1974).

As in the earlier Hi-Fix model, one common fine frequency is transmit-ted, and a time-sharing principle serves to distinguish two slave transmis-sions in the master receiver for phase comparison. As the sequential transmission is organized, the transmission sequence for the six station hyperbolic chain lasts 260 msec and starts with a 20 msec trigger signal radiated from the prime station. The trigger is followed by two 20 msec signals from the prime and from the secondary stations 2 through 6, in turn. Nominally each station sends f_1 frequency in the first of its two 20 msec slots and f_2 frequency, which together with f_1 produces the coarse pattern, in the second 20 msec slot. The transmitted data rate of about 4 Hz can be increased when the chain comprises less than six stations. For phase comparison of the successive signals, the receiver has 12 continuous-running "phase memory" oscillators, one per station for each frequency. The phase of each oscillator is updated every time the corresponding signal is received. Once the secondary stations and their receivers are phase locked with the prime station's main transmission, they do not depend on continuous reception of the trigger signal from the prime station. This is possible because the oscillator that memorizes the prime signal serves as the local time source for controlling the sequence switching. The Hi-Fix/6 receiver (Fig. 20.43) is simple to operate and highly automated, leaving to the user such functions as the suppression of trigger-signal reception in the event of interference and the interpretation of the coarse pattern readout under fringe conditions. The receiver has three in-line numerical displays, each giving the integrated fine lane and fraction reading for a selected station pair, together with the coarse reading.

The Hi-Fix/6 station configurations allow many combinations for simultaneous use. In hyperbolic position fixing, any number of ground stations between three and six can be utilized, and an unlimited number of users can operate simultaneously. The following configurations are available: 3 stations and 3 baselines, 4 stations and 6 baselines (Fig. 20.44a), 5 stations and 10 baselines, and 6 stations and 15 baselines. In circular position fixing, the number of users and the freedom to use both coarse and/or fine readings are limited because one transmitting station is located in each survey vessel. The limitations are as follows: one vessel is able to operate with coarse/fine patterns focused on *any* two of the five

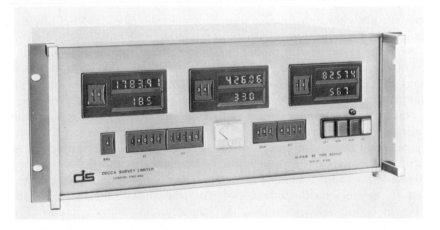

Figure 20.43 Hi-Fix/6 receiver (courtesy Decca Survey Ltd.).

4 STATIONS – 6 BASELINES

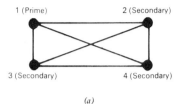

1 (Prime) 2 (Secondary)

3 (Secondary) 4 (Secondary)

(a)

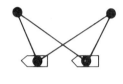

(b)

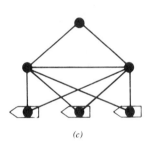

(c)

Figure 20.44 Hi-Fix/6 system configurations (courtesy Decca Survey Ltd.).

fixed stations; two vessels are able to operate with coarse/fine patterns focused on two fixed stations (Fig. 20.44b); three vessels are able to operate with two fixed stations, but only *one* vessel will be able to use the coarse/fine patterns and the other two vessels will not have any coarse/fine pattern facility; four vessels can operate with the fine pattern *only* focused to two fixed stations. In compound operations three fixed ground stations generate hyperbolic patterns with unlimited users. In addition, up to three vessels can operate utilizing circular configuration focused on *any* two of the fixed stations. These combinations produce the following restrictions: with one vessel operating in the circular mode, coarse/fine patterns are available to *all* users; two vessels operating in the circular mode are restricted to the fine pattern only, whereas coarse/fine patterns are available to all hyperbolic users; with three vessels operating in the circular mode *all* users (including hyperbolic users) are restricted to the fine pattern (Fig. 20.44c).

The power consumption of Hi-Fix/6 is about 12 A at 24 V for transmitting stations with the radiated power 75 W maximum and 7 A at 24 V for receivers. The maximum receiver speed or range rate is 4 lanes per second. The typical maximum measurable distance over sea in daylight is about 300 km, with an accuracy of about ±1 m on baseline.

20.2.2.5. Pulse/8

The *Pulse/8* is a new offshore hydrographic Decca Survey system based on the *Loran-C*, which is a combined continuous wave and pulsed system. Pulse matching solves ambiguities in the coarse time measurements, and the fine time differences are measured by comparing the phases of transmission frequency cycles inside the pulses. The operating frequency is 100 kHz, with a spectrum between 90 and 110 kHz. The spectrum is needed to build up pulses with the leading edge as steep as possible and a slowly descending tail; this type of pulse is obtained by adding a number of continuous sine waves in the form of a Fourier series. Eight pulses form groups, and intervals between pulses in one group are 1000 μsec in duration. These groups repeat at pulse repetition intervals from 20 to 100 msec in 100 μsec steps for identification purpose between chains. For a more detailed description of the *Standard Loran* and Loran-C, see Chapters 24 and 25.

As a complementary unit with or as a separate system from the Loran-C, the Pulse/8 is intended for medium ranges (less than 500 nm) either in the hyperbolic or the circular mode. Although a Pulse/8 chain may be configured to operate as a Loran-C chain on one of the established repetition rates, its main application using the *Decca S 501* receiver is to provide temporary coverage over selected exploration areas. The system uses transmission antenna masts 45 m high in low noise areas; mast height can be extended to 100 m. The optimal base lengths vary between 250 and 400 km. The radiated peak power is 150 W using the shorter antenna and 1 kW, using the long antenna. This power, together with the propagation characteristics at the radiated carrier wave band, yields a maximum measurable distance of about 300–500 nm over seawater and 150–300 nm over land. The accuracy within 300 nm is about ±50 m.

A standard system employs from three to a maximum of six stations in a given chain. The transmitters radiate in turn phase-coded pulse groups in precise time registration with each other at a pulse group repetition rate specific to that chain. The Decca Survey receiver type S 501 then determines from any two or three pairs of received signals position lines derived from the time differences between them. Hyperbolic position fixing utilizes the conventional principle described in Chapter 4. Pulse/8 and Loran-C transmitters may also employ the circular configuration to determine a position fix. This mode requires only two transmitters whose precise transmission times are known at the receiving station. A *cesium* atomic clock at the receiver measures the time between the transmission and the arrival of the signals from the two stations, and from this elapsed time the distances of the receiver from the two transmitting stations can be computed.

The S 501 receiver was designated a precision position-fixing instrument for use with both Pulse/8 and Loran-C transmissions. It features the latest development in solid state devices, and printed circuit boards facilitate rapid maintenance. The S 501 is available as a single or double unit with a digital readout. Packaged as a double unit (Fig. 20.45), the Pulse/8 provides complete backup facilities, since each receiver has the ability to track three position lines simultaneously by using one receiver on Pulse/8 and the other on Loran-C transmission. If the atomic standard is employed, one unit will operate in the circular mode using two Pulse/8 or Loran-C transmitters or a combination of both. The Pulse/8 transmitters use the recently developed technique of pulse compression, which has made possible the design of a practical and reliable solid state pulse transmitter with greatly reduced overall dimensions. The complete ground station setup comprises eight compact units, together with the cesium oscillator, transmitting antenna, and AC supply.

Figure 20.45   Decca Pulse/8 receiver (courtesy Decca Survey Ltd.).

At each land-based station a cesium frequency standard is used to provide synchronization to the master station by means of a *timing receiver* included as a backup against failure of the cesium standard. At the slave station this timing receiver receives the master signal and generates a phase-locked signal that provides local timing of the slave functioning. Since a cesium standard is normally used for timing at each station, once the proper time relationships of the signal radiations between master and a given slave have been established, the slave is not synchronized through the timing receiver but rather runs free on its cesium standard. The timing receiver monitors chain performance— specifically, time differences between the received master signal and that of the particular slave. Since this time difference must be constant, any variation is a result of differences between the master and slave cesium standard and/or of the propagation path variations and must be adjusted out at the slave station periodically. The correction is accomplished using incremental timing adjustments in the timing unit of the slave transmitter. The system has several advantages. The whole chain will not go down if the master leaves the air because the slaves will continue to operate and the users can switch over to a station other than the master in the circular mode. The timing receiver can be used as a chain monitor, and it provides redundancy in the master/slave phase-lock operation if the cesium standard fails. Finally, the timing stability of the slave is not affected by local noise or interference.

The antenna normally used with the Pulse/8 receiver is a 2.5 m Citizen's Band whip antenna, which screws directly into the top of the antenna amplifier. A wire antenna 9–15 m long may be used also if mounted not less than 45° to the horizontal plane. Power consumption of the S 501 receiver is 1 A at 115 V; whereas the corresponding figure for the transmitters is 35 A at 115 V. The pulse length is 400 to 450 μsec and its shape is 65 μsec to peak and 500 μsec after peak down 60 dB. The timing accuracy controlled by the cesium frequency standard is about 10^{-12}.

20.2.2.6 *Autocarta*

The *Autocarta* is not actually distance measuring equipment, but it seems to allow a hydrographer such complete automation that it is included in this text. Autocarta was manufactured and introduced in the early 1970s by one of the Decca group companies—Decca Survey Systems, Inc., Houston, Texas. It is available in three different forms or combinations of units. *Autocarta/X* is a complete system designed for independent survey vessels. It provides (*a*) left/right display to helmsman; (*b*) continuous record of ship's track to the scale of the survey, showing fix marks at required intervals; (*c*) printout pattern readings, *X-Y* position, time and depth at each fix; (*d*) computer-drawn chart showing soundings in fathoms and feet or meters and decimeters; (*e*) punched paper tape with position and depth of each sounding. *Autocarta/B* is designed for sound-boats of ships carrying Autocarta/X. It gives facilities *a* and *c* in addition to a magnetic tape containing position and depth data for transfer to the Autocarta/X in the parent vessel for final processing. *Autocarta/P* is designed for vessels returning to the same port each night and having access to a shore computer. It gives facilities *a* and *c* in addition to an IBM-compatible magnetic tape or eight-channel punched paper tape containing the true *X-Y* position and time and depth of each sounding selected to the scale of the survey.

Autocarta/X (Fig. 20.46) converts hyperbolic or circular positioning information into *X-Y* positions on a survey grid, which may be on any transverse Mercator or Lambert conformal projection. Final mapping accuracy is claimed to be of the same order as that obtained by a conventional method, and accuracy is maintained through any alteration of course or speed. The system will work at any scale within the accuracy limitations of the positioning system and at any speed up to 30 knots,

Figure 20.46   Autocarta system (courtesy Decca Survey Ltd.).

provided the speed-scale factor does not exceed the ratio of 1/200. In practice, for example, this means that at 10 knots the allowable scale is 1/2000 and at 30 knots it is 1/6000. The helmsman has continuous indication of distance from any desired track, up to 1000 feet or meters. The sense of the display is reversible; and a fine scale is used when within 100 feet or meters from the track. A continuous track plot on the scale of the survey promotes navigational safety. A permanent record enables any part of any day's work to be replotted at any time in the future. Means are provided to enable the vessel to continue to work when any single peripheral unit is inoperative.

20.2.2.7.   *Raydist*

The basic heterodyne CW principle employed in *Raydist* was first used successfully in 1940 at Langley Field, Virginia, when equipment designed by *Charles Hastings* (1947) was put into operation as a precise means for measuring the true ground speed of aircraft. After World War II the instrument underwent intensive development as means for over-water surveying, primarily for hydrographic and geophysical exploration purposes. The early models were manufactured by the Hastings Instrument Company and sold or leased through the Raydist Navigation Company, Hampton, Virginia. Emphasis in early 1950s was on developing instruments that could use hyperbolic configuration to fix or track survey vehicles, although ranging equipment was also employed for testing the speed of the superliner *SS United States* as early as 1952. *Type E Raydist* was designed for tracking a small continuous wave transmitter placed in a manned or unmanned vehicle; all the controls and measurements were made by a ground-based receiver. In *Type N Raydist* all transmitters associated with survey operations were located at fixed positions on ground, and only the receiving and phase-measuring unit was required on the user vessel. Thus the system was unsaturable, which allowed an unlimited number of users to operate simultaneously. *Type R Raydist* was purely for distance measurements; therefore two independent systems with separate sets of frequencies, or some means of multiple heterodynes and filters, made possible circular location for one user at a time. *Type ER Raydist* combined one range measurement and one hyperbolic measurement to give fixes in a hyperbolic-range grid pattern. In this system only two ground stations were needed, and the measurements were made at the mobile master station. *Type DM Raydist* is a transistorized instrument and operates on a principle similar to that of type ER, but it uses a circular-elliptical grid pattern. The disadvantage of all the models listed is the number of frequencies needed, usually four. The advantage of all Raydist systems is that phase locking between the master and slave stations is unnecessary because of the reference-type transmission principle.

20.2.2.8. Raydist DR-S and DRS-H

The basis for the later Raydist models operating independently of any differentiated systems is the type *DR-S Raydist*. The following very simplified description of this new type (*Hastings* and *Comstock*, 1969) refers to the block diagram in Fig. 20.47. The survey vessel carries a small CW transmitter operating on a typical medium frequency band of about 3 MHz and designated f_m. This carrier is received at each base station ashore, where it is heterodyned by a second CW transmitter located at the base station and operating on a frequency of approximately $f_m/2$. Each base station CW transmitter is tuned a few hundred hertz above or below (depending on the station) $f_m/2$, to develop the desired heterodyne tone. The survey vessel also has the *Raydist Navigator*, where the phase measurements are made. The Navigator receiver receives the two base station carriers as well as the f_m carrier from its adjacent transmitter and

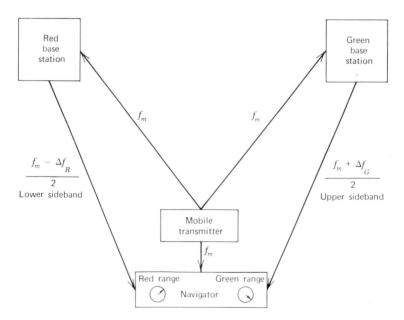

Figure 20.47   Functional diagram of Raydist type DR-S (from *Hastings*, 1971).

develops the same two heterodyne tones developed at the red and green base stations, respectively. The tone at the red station is returned to the Navigator in the form of a single sideband (SSB) on the red station carrier, where it is compared for phase with the same tone developed directly in the Navigator. As the distance between the Navigator and the red base station changes, the relative phases of the two red station tones will change, thus producing an indication of changes in the red range. The same process occurs with respect to the green station to produce the green range. The DR-S is primarily designed for circular location, and as many as four users can operate simultaneously. If a completely nonsaturable system is required (i.e., if an unlimited number of users must be able to operate in the system simultaneously), the CW transmitter operating on frequency f_m is placed ashore as a third station, and each vessel then carries only a Navigator and data handling equipment.

The *Raydist DRS-H* system is manufactured and sold by the Teledyne Hastings-Raydist Company, Hampton, Virginia. This new Raydist is a compact, lightweight system with solid state silicon circuitry. It operates primarily with two range configurations. The long-range system extends to about 250 miles in daytime and 150 miles at night over seawater. The antenna system in this case includes an aluminum tower 30 m high and a whip weighing about 55 kg. In the medium range system distances of about 75 miles can be reached. The antenna consists of a telescopic aluminum whip 10 m tall with two extensions 1.8 m long weighing about 11 kg. In principle and operation it is similar to the DR-S model, but a new and improved *Navigator model ZA-75C* has replaced the one used in the DR-S. The Raydist Navigator, model ZA-75C, is the central component of a Raydist mobile station and receives the various radio frequency signals of the Raydist system in which it is operating. It translates them to audio frequency signals by means of modulation, frequency multiplication, selective mixing, and detection. The audio signals are amplified and processed to provide the driving voltages necessary to operate the associated position indicator. The electronic circuits that provide the position data are contained on 25 circuit cards, assuring rapid maintenance of the Navigator unit. The Navigator is designed to give position information in terms of range coordinates for circular location when a Raydist CW station is in operation aboard the moving vessel. Hyperbolic coordinates are displayed when the CW station is placed in operation ashore, as described in connection with the DR-S. Hyperbolic coordinates are also displayed when the Navigator is operated in a Raydist "T" system, described in the following section. No changes are required in the Navigator when changing position data modes between circular and hyperbolic layouts. For operation in the Raydist "T" system it is necessary to exchange only three color-coded printed circuit cards in the Navigator. Interconnecting cables allow the position indicator to be mounted for optimum use. The ZA-75C Navigator operates with 2 A at 24 V DC. The power consumption of the CW transmitter is 5 A at 24 V; the SSB transmitter takes 2.75 A at 24 V.

The frequency range is 1.5 to 4.0 MHz, and the typical system requires only two frequency allocations for continuous operation of CW and SSB stations. Maximum bandwidth for SSB stations with multiparty operation requires an allocation of less than 0.5 kHz. The lane ambiguities are solved by maintaining continuous operation from a known starting point or by choosing other available location methods. Data are displayed in two digital counters and can be manually recorded or fed to a peripheral system common to most of the modern hydrographic survey systems. The repeatable positioning accuracy is about ±0.02 lane, which with the typical carrier frequency of 3.0 MHz is equivalent to approximately ±1 m on the baseline.

20.2.2.9. Type "T" Raydist

Similar to the DRS-H the type *"T" Raydist* is a new design based on the DR-S principle; but emphasis now is on the use of hyperbolic configuration and four stations ashore. This arrangement allows an unlimited number of users to operate simultaneously, and coverage has been extended to more than 10,000 square miles, provided with almost rectangular grid by using the *Halop* coordinate geometry. As an optional instrument the *Raydist Remote Indicator and Halop Converter model GA-55* or *GA-56* remotely indicates the incremental position information from the Raydist Navigator or converts the incremental information to other coordinates. The incremental signals from the Navigator in the circular mode (DRS-H) can be converted to elliptical-hyperbolic coordinates, and incremental signals in the hyperbolic mode (type "T") can be converted to *Halop*-type coordinates. The formation of the Halop coordinates requires the *sum* and *difference* patterns of the original, or primary hyperbolic coordinates. For example, if the red and green phasemeter readings are subtracted from each other, the output readings represent a family of hyperbolas focused on the red and green slave stations. The summation of the original red and green readings represents a family of roughly elliptical lines running across the difference pattern with angles of cut close to 90°. Another advantage in using the Halop coordinates is that locations close to shore stations can be done with favorable angles of cut. The coordinates in the Halop Converter are displayed on two six-digit

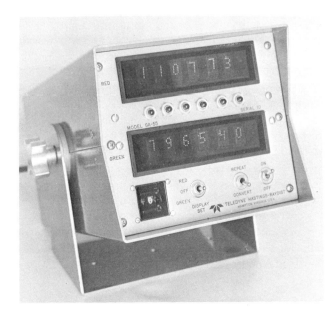

Figure 20.48   Halop converter (courtesy Teledyne Hastings-Raydist).

LED registers (Fig. 20.48), and the selection of the desired coordinate system is accomplished by means of the repeat-convert switch.

The type "T" Raydist resembles the older type N but differs from it in three important ways—namely, the formation of the ground baselines (T-shape), the more conservative use of transmission frequencies, and the different means to transmit the modulation signals. The block diagram in Fig. 20.49 shows the functions of the four-station Raydist "T" system. Here A and C are the CW transmitters from which A transmits continuous waves at the frequency of f_m and C transmits continuous waves at the frequency of $f_m + \Delta f_1$, where Δf_1 is an audio frequency of about 400 Hz. In SSB station B the signal from station A is received and heterodyned with a frequency of $(f_m + \Delta f_3)/2$, where Δf_3 is another audio frequency differing from Δf_1. Station B then transmits the resulting audio frequency heterodyne as a positive single sideband, $(f_m + \Delta f_3)/2 + \Delta f_3$ of the primary measurement frequency generated at SSB station B, to the mobile unit. There, in the Navigator *model ZA-84* (Fig. 20.50), this carrier is received, together with the carrier from the CW station A and the same heterodyne tone developed at SSB station B is developed. Phase comparison is then made by using the two heterodyne tones. A similar

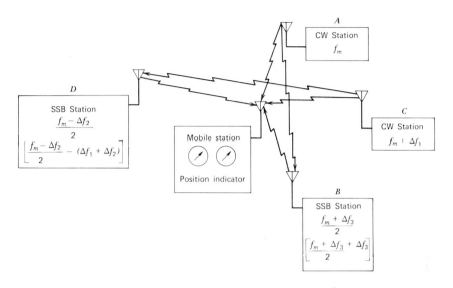

Figure 20.49 Functional diagram of Raydist type "T" (from *Hastings*, 1971).

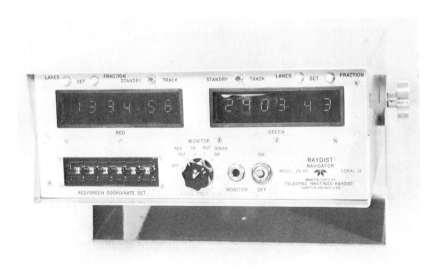

Figure 20.50 Raydist Navigator ZA-84 (courtesy Teledyne Hastings-Raydist).

process is occurring between stations *C*, *D*, and the mobile stations, but with a slightly differing audio frequency Δf_2.

The maximum offshore distances are the same as in the type DRS-H Raydist—250 miles in daytime and 150 miles at night; the accuracy on the baseline is about ±1 m.

20.2.2.10. *Micro Omega*

Micro Omega is one of the applications of the basic Omega system, generally called the differential Omega. It is manufactured by the Teledyne Hastings-Raydist Company and it operates in two modes. The Omega mode is used when operating beyond the 250 mile signal range of a Micro Omega station; the Micro Omega mode is for operations within the range of *one* Micro Omega station. The Raydist Micro Omega Navigator, model ZA-88 (Fig. 20.51) provides switchable operation in either mode, and when the vessel comes within the operating range of a Micro Omega station, the operator throws the mode selector switch to the Micro Omega position. Then the line of position (LOP) indicators continue to read the same two LOPs, but their readings are now automatically corrected almost for all propagational errors such as sky wave error and sudden ionospheric disturbances (SID), as the following system analysis indicates. For a more detailed description of Omega systems, see Chapter 26.

Figure 20.52 shows signal flow within the Micro Omega system (*Hastings*, 1971). Signals from each of the Omega stations are received at both

Figure 20.51 Micro Omega Navigator converter, model ZA-88. (courtesy Teledyne Hastings-Raydist).

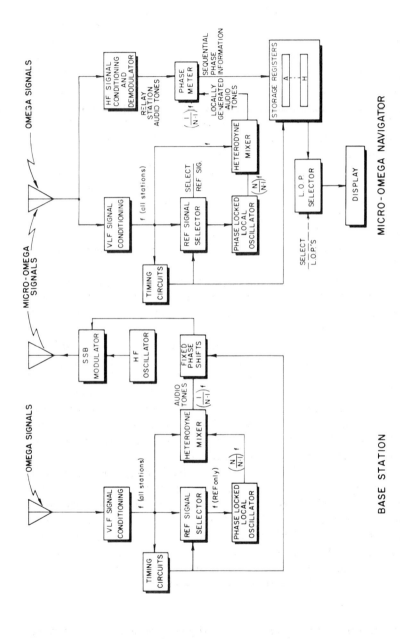

Figure 20.52 Block diagram of Raydist Micro Omega (from *Hastings*, 1971).

the Micro Omega base station and the Navigator. A local oscillator at the base station is phase locked to one of the received signals, usually the strongest, at a frequency $N/(N-1)$ times the Omega frequency f. Each of the Omega signals is heterodyned with the local oscillator to generate a beat frequency of $1/(N-1)$ times the Omega frequency. The value of N is selected to produce an audio beat note of approximately 400 Hz. Thus the sequence of Omega signals generates a corresponding sequence of audio tones that bear phase relationships to each other equal to those among the corresponding Omega signals. The audio tones, in turn, are broadcast throughout the Micro Omega coverage area, employing SSB amplitude modulation on an HF carrier.

The Micro Omega Navigator contains circuitry that duplicates the local oscillator and heterodyning process performed at the base station. Thus audio tones corresponding to each of the Omega signals are also generated within the Navigator. The audio tones with a beat frequency of $(1/N-1)f$ received from the relay or base station are applied to a phasemeter for comparison with the corresponding internally generated tones of the same beat frequency. Since the distance between the vessel and the Micro Omega base station is quite short compared with the distances from the local operating area to the Omega stations, phase errors completely cancel. This is easier to comprehend by visualizing that all propagation errors cumulated over the long distance from the Omega stations are fed almost simultaneously to both the Micro Omega base station and Navigator receivers; all errors will be present in the phase comparison and will cancel out. To obtain the local C-H hyperbolic coordinate (Hawaii–Japan), for example, the C and H phase outputs are applied to a differential and the resulting value is displayed in digital form in the position indicator of the Navigator. Fixed phase shifts are introduced at the local Micro Omega relay station to restore the fixed constants, corresponding to local coordinates.

The shipboard navigational receiver weighs 11.8 kg, the shore station receiver 3.6 kg, and the shore station transmitter 14.5 kg. The power consumption of the shipboard receiver and the shore station transmitter is 75 W drawn from a 24–28 V DC battery; the shore station receiver consumes 50 W from a similar power source. Overall accuracy depends on the formation of the hyperbolic patterns covering the survey area, but a positional accuracy of about ±100 m has been obtained during tests made in Chesapeake Bay and 200 miles seaward.

20.2.2.11. Lorac

Lorac (from long-range accuracy) was originally designed and manufactured by the Seismograph Service Corporation, Tulsa, Oklahoma, in the late 1940s; later models have been manufactured by a subsidiary company, Lorac Service Corporation. The system operates at the medium frequency band of 1.6–2.5 MHz, using a phase-measuring technique based on the heterodyne principle and introducing two families of hyperbolic patterns. All transmitters are located ashore, allowing an unlimited number of users to operate simultaneously. In the older model, the type "*A*" *Lorac*, three ground base stations were needed—the central, the red, and the green stations. The type "A" operated on a signal switching principle, which means that the red and green stations were required to serve double duty in transmitting the reference signals in addition to their own measuring signals. The new type "*B*" *Lorac* requires a fourth ground transmitter to perform the reference functioning, but the signal transmission is continuous and only one reference receiver is required instead of two. Remote control operation is much simpler with the "B" system, which also can be adapted to lane identification. Lane identification enables a mobile vehicle to establish its position within the hyperbolic grid with initial reference only to the vehicle's standard dead-reckoned position. Remote control permits activation and deactivation of all base stations from the reference station, thus requiring only a single operator for the entire system. Within the Lorac series of radiolocation systems, the type "*B*" having lane identification and remote control is the model *UBUE-1000*, the model *UBUB-1000* lacks these features. The type "B" Lorac has become the standard system used by Lorac Service Corporation. The principle of operation of the system (three CW base stations and an AM reference station) is discussed next.

The center station transmits on a frequency of f_1. The green station transmits on a frequency of $f_1 + 135$ Hz, and at the same time the red station transmits a frequency of $f_1 - 315$ Hz. The reference station transmission frequency f_2 is preferably separated from f_1 by at least 20 kHz. At the mobile receiver, the position detector-amplifier is tuned to the frequency f_1 with a bandwidth sufficient to receive the green, center, and red station signals but narrow enough to reject the reference station signal. According to the superheterodyne principle, simultaneous reception of the carriers of the green and center stations produces a difference frequency of $(f_1 + 135) - f_1$ or the beat note of 135 Hz, which is called the green position signal. Likewise, the red position signal is produced as the difference of $f_1 - (f_1 - 315)$, or 315 Hz. After detection, these two position signals are separated and all other tones are rejected. The green position automatic gain control (AGC) filter selects the 135 Hz position tone and feeds it to the green phasemeter. By a similar process the red 315 Hz position tone is fed to the red phasemeter. The AGC action produces phasemeter signals of constant amplitude regardless of the received signal strength.

The position signals remain constant in frequency but are subject to changes in phase depending on the distances between the receiving antenna and each transmitter originating the heterodyne signals. When the receiving antenna moves, the phase of each position signal changes as the distance to the base station changes. For these phase changes to be measured, reference signals are needed. Therefore, at the fixed reference station, a reference receiver is also tuned to frequency f_1 to receive the three CW carriers simultaneously. The 135 and 315 Hz signals are detected and filtered exactly as in the mobile receiver. But these signals remain constant in phase because the reference receiver antenna remains stationary. The reference station then transmits on a frequency of f_2, amplitude modulated with the 135 and 315 Hz tones. At the mobile receiver, the reference detector is tuned to f_2 to receive the reference station signals. The 135 and 315 Hz tones are detected from the reference carrier, separated and filtered in the reference AGC-filter sections, and also fed to the phasemeters. The tones, called the green and red reference signals, remain essentially constant in phase throughout the network coverage area.

The following mathematical analysis of phase propagation is made in respect to the type "B" Lorac. It should be noted that a similar analysis is valid for all hyperbolic systems in which the phase measuring principle yields the same or similar final equation of hyperbolic coordinates, especially in the heterodyne type of instruments. Using the green and center base stations as an example, let the distances, r_1 through r_5 be as presented in Fig. 20.53, and let the green audio signal be designated as Δf_1. If we assume that the phases at the green and center stations at a given time are Φ_{T1} and Φ_{T2}, respectively, the phase of the heterodyned position signal at the mobile receiver, derived from the $T1$ and $T2$ transmissions, is

$$\Phi_P = \Phi_{T1} + \frac{2\pi(f_1 + \Delta f_1)}{c} r_1 - \Phi_{T2} - \frac{2\pi f_1}{c} r_2 \qquad (20.50)$$

Correspondingly, the phase of the heterodyned reference signal at the reference receiver is

$$\Phi'_R = \Phi_{T1} + \frac{2\pi(f_1 + \Delta f_1)}{c} r_4 - \Phi_{T2} - \frac{2\pi f_1}{c} r_3 \qquad (20.51)$$

The phase of the reference signal after it has been sent as an amplitude modulation Δf_1, to the mobile receiver is

$$\Phi_R = \Phi'_R + \frac{2\pi \Delta f_1}{c} r_5$$

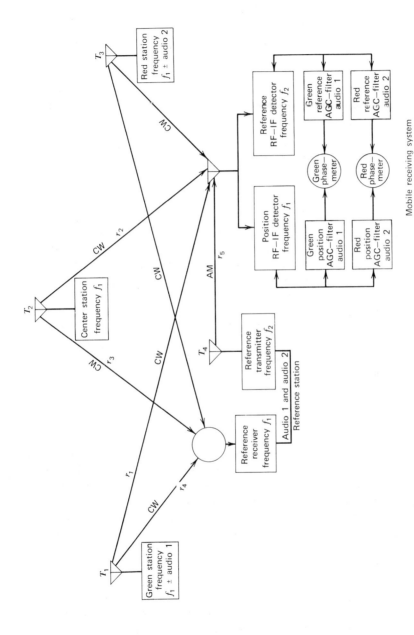

Figure 20.53 Functional diagram of Lorac model "B" (courtesy Lorac Service Corp.).

Thus the phase difference between the position and reference signals becomes

$$\Delta\Phi = \Phi_R - \Phi_P \qquad (20.52)$$

After rearranging the terms and using the simple summation Eq. 20.52 yields

$$\Delta\Phi = \frac{2\pi}{c}(f_1 r_2 - f_1 r_1 + f_1 r_4 - f_1 r_3 + \Delta f_1 r_5 - \Delta f_1 r_1 + \Delta f_1 r_4) \qquad (20.53)$$

Grouping the terms according to their variable and constant characteristics, we have

$$\Delta\Phi = \frac{2\pi f_1}{c}(r_2 - r_1) + \frac{2\pi f_1}{c}(r_4 - r_3) + \frac{2\pi \Delta f_1}{c} r_4$$

$$+ \frac{2\pi \Delta f_1}{c}(r_5 - r_1) \qquad (20.54)$$

where the second and third terms on the right side are constants because f_1, Δf_1, r_4, and r_3 are constants. In the last term $r_5 - r_1$ is a variable quantity, but one must anticipate that Δf_1 is very small compared to the measuring frequency f_1, which is used to compute the hyperbolic coordinates in the green pattern of hyperbolas. Consequently, the reference signal wavelength is very long compared to the dimensions of the survey network, and the last term in Eq. 20.54 may be neglected completely, given an average value, or compensated by a correction carried in the computer programs. By dividing Eq. 20.54 by 2π to obtain the green hyperbolic coordinate (in fractions of a lane) and designating the sum of all constant terms as K, we obtain

$$N_G = \frac{f_1}{c}(r_2 - r_1) + K \qquad (20.55)$$

where c is the average propagation velocity within the network. The foregoing analysis makes it obvious that in the systems employing the heterodyne principle in their phase measurements, the initial phases have no importance; thus no phase locking is necessary.

The phase difference between the reference and position signals is converted into a visual display by a servo-controlled motor resolver combination. The actual display is a rotating drum driven by the phasemeter motor. A decade counter is mechanically coupled to each phasemeter and the counter totals the number of complete turns the phasemeter drum makes. The drum is graduated from 1 to 100, and the complete phasemeter reading in hyperbolic coordinates is therefore the total lane number as read from the counter, plus the fraction of a lane

Figure 20.54   Lorac receiver (courtesy Lorac Service Corp.).

indicated by the drum graduation. The Lorac mobile receiver *UAJB-1000* is illustrated in Fig. 20.54, where the three-digit dial to the left is used for setting in the target point readings when the zero header is used. The solid state circuitry has self-contained power supplies and phasemeter displays in a single container. Lane identification is performed on another identical receiver. Auxiliary equipment such as an analog recorder, zero header, actrac course plotter, and a digital recording system are available for use in conjunction with and driven by the mobile receivers.

In addition to the audio tones (135 Hz for green and 315 Hz for red as stated earlier), values of 155 and 95 Hz or 240 and 600 Hz have been used. The power consumption is 1800 W maximum at the base stations and 2400 W maximum at the reference station. Transmitter antenna power output is about 300 W. The operational temperature range is from 0 to 50°C with relative humidity to 95%. Distances up to 400 km in the daytime and 160 km at night can be measured. The accuracy on the baseline is about ±1 m.

20.2.2.12. *Other Lorac Systems*

The *Lorac Low Power Radiolocation System* is a compact, solid state version of the type "B" system. It is designed for portability, quick installation, and low power consumption. Using 30 W transmitters, each of the three base stations and one reference station operates on either

115 V AC or 28 V DC; power requirements are 300 W at the reference station and 150 W at the base stations. There are three versions of this system. The *UBUJ-1000* is designed for continuous duty with unattended operation on ground-based stations. When the base stations are clock-timer controlled, operation from 28 V storage batteries is possible for as long as a week without recharging the batteries. The *UBUK-1000* adds remote control in cases where continuous operation is not practical and the reference station is manned. Thus the remote station operator is able to control the functioning of the base stations. The *UBUL-1000* adds both lane identification and remote control. Transmitting antennas for the low power stations are telescopic masts 11 m tall; since the masts can be set up by one man, the entire system can be operated by a team of two. The maximum measurable distance is 50–80 km.

The *Lorac Long Range Ranging System URUA-1000* applies circular geometry for position fixing and reduces the number of transmitting stations from four to two. The URUA-1000 system uses a pair of existing Lorac stations without disturbing their normal hyperbolic functions. The system features CW carrier transmission and phase-measuring techniques, and allows several users to operate simultaneously because there are no shipboard transmitters. The URUA-1000 employs the cesium frequency standard for frequency control at the base stations and to generate the reference signals at the mobile receiving station. The position coordinates are results of direct measurements of distances stated in units of lanes. In the cesium frequency standard the frequency shift is constant over a period of time and can be accurately predicted. The calibration or rating the system consists of comparing the drift rates while the mobile receiver remains stationary. The drift rate is usually of the order of 0.05 lane per 24 hours and can be used by the mobile station operator as a daily correction.

The *UCJA-1000 Loran-C Receiving System* is another commercial version of the original Loran-C primarily specified for military use. The UCJA-1000 applies circular geometry, although it is able to use hyperbolic measurements at the same time to increase accuracy and reduce ambiguities. The master transmitting station signal is replaced by the highly stable frequency source, the cesium frequency standard at the receiving station. The instrumentation in the mobile unit thus consists of three basic components: the Loran receiver, the frequency standard, and a minicomputer that converts observed position coordinates to geographic coordinates. Tests of the UCJA-1000 system have produced position fixes with an accuracy of ±80 m at a distance of 1400 km (*French*, 1973).

The *UBJA-1000 Digital Radiopositioning System* is a new digital concept in signal analysis, added to the Lorac series in 1975. It includes an antenna system and a network interface attached to the Hewlett-Packard model 2100 series minicomputer, which has a 16-K memory in one chassis, an analog-to-digital converter, a navigation display, a tele-printer, and an analog recorder. Principally developed for seismic indus-try, this system is claimed to accomplish the following functions: main-taining correct integral lane count; displaying correct position and time, updated at least once each second; providing a graphic record of lanes traversed for reference during postmission analysis; maintaining and providing record number and current time with one-second resolution; providing data smoothing to remove noise transitions (sky wave contami-nation) in the radiopositioning coordinates; continually measuring and providing distance information to a preselected target, showing deviation from a straight line track to the target, and including a fail/safe indicator that controls visual alarms and printout error codes after each record whenever the network signals fall below a predetermined level. Rejection of coherent noise is achieved by a Tchebycheff-type digital filter, and the system gives a printed record of pertinent positioning data at any desired time or position.

20.2.3. The Sonar System

Sonar (for *so*und *na*vigation and *r*anging) lies outside the scope of this book, but because of recent applications that complement the use of modern hydrographic surveying systems, a short introduction is appro-priate. Sonar resembles airborne radar survey operations in most respects; altimetry, position fixing, Doppler navigation, and scan mapping; the main exception is that the medium through which the sonar impulses travel is seawater. Sound or pressure waves propagate through seawater at a speed of approximately $V = 1500$ m/sec; frequencies are usually between 30 and 100 kHz. The stability of sound wave velocity through different depth layers is far poorer than that of electromagnetic energy propagating through the ambient atmosphere, and the power consump-tion is much greater. These shortcomings set certain limitations: for example, such equipment can operate only in shallow water (on the continental shelf) and at distances of a few kilometers, except when using Doppler sonar. The integration of sonar with conventional electronic position-fixing methods is justified because the actual location is made in reference to fixed control points located a few hundred meters below and a few kilometers off the survey vessel, instead of at distances several hundred kilometers inland. In many seismic and oil exploration operations it is more important to locate the vessel accurately within a limited area and to return to a desired prelocated point than to know the absolute (not-so-accurate) position of this area or point. If hydrographic surveying is the main objective, the local area can be tied to the main datum later.

Three main modes of sonar are of interest in civilian use for surveying or geophysical applications—namely, Doppler sonar, range-range circular location, and dynamic positioning. In the *Doppler mode* the velocity of the survey vessel can be determined with respect to the seabed (ground speed) in two perpendicular directions, along the axis of the ship ahead to astern, and along the axis port to starboard. This is accomplished by transmitting acoustic pulses on a frequency f in these four directions at a known angle α with the horizontal. The motion of the ship causes the frequency of the transmitted and reflected signals to vary, and according to the basic Doppler equation

$$V_S = \frac{\Delta f V}{2f \cos \alpha} \qquad (20.56)$$

the velocity of the vessel V_S is obtained by observing the frequency shift Δf. By integrating time in the computations as well, the distance traversed by the vessel along the axes just stated is determined. Because of the "drift" of the ship, the true course and speed "made good" over the seabed can be calculated by applying the ship's heading data into the velocity components. Another notable application of Doppler sonar is in furnishing the ship's speed information as an important input when making satellite fixes. In addition, in satellite navigation Doppler sonar displays continuous position data between satellite passes.

In the circular *range-range mode* at least two transponders are needed; three or more give redundant observations, thus allowing least squares adjustments to be applied for better accuracy. The exact location of the transponders with reference to each other are made during the time of day when any available long-distance electronic surveying system gives the best relative accuracy. Using the sonar principle, the survey vessel measures the slant distance R (Fig. 20.55), taking the time difference elapsed between the transmitted and returned sound signals and the average velocity of the sound waves. The depth H_T of the transponder is known and the horizontal distance S is directly obtained from the triangle *ABC*. By operating with this local circular mode independently or by integration with the shore-based surveying system of any mode, round-the-clock survey activities are ensured. If the transponder is used

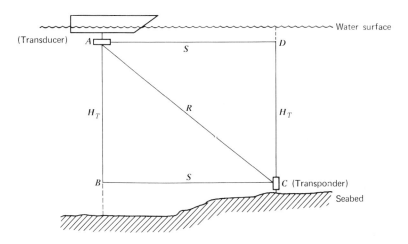

Figure 20.55 Principle of sonar distance determination.

primarily as an acoustic marker to be relocated for dynamic positioning, for example, a direction-finding facility must be established in addition to ranging. Two hydrophones are mounted on opposite sides of the survey vessel. Since the distance d (Fig. 20.56), between the two hydrophones is small compared to the slant distance to the transponder on the seabed, the directions of the transmitted and received sound signals can be assumed to be parallel. The procedure of measuring the time difference of the arriving signals from the transponder at the two hydrophones gives the distance difference AC. Since d is known, the angle Θ between the heading of the ship and the direction of the transponder can be computed, and the ship homes in on the marker by zeroing the angle Θ. *Dynamic positioning* is an important application when a survey vessel must be kept exactly and continuously above a preselected point on the seabed. In this case another pair of hydrophones is placed along a line parallel to the longitudinal axis of the ship and as far from each other as possible, to form a baseline of maximum length. The transponder is placed at the desired location on the seabed, such as close by a drilling hole. If the ship is positioned exactly above the transponder, the distances from each of the hydrophones to the transponder are equal, and the signals from the transponder will arrive at both hydrophones in phase. Any phase shift between the arriving signals indicates drift of the vessel from the predetermined location and will be zeroed by correcting the vessel's position. The sensitivity of such a system depends on the frequency used, thus on the wavelength of the sound propagation. The *echo*

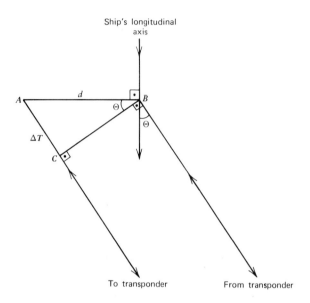

Ship's longitudinal axis

To transponder   From transponder

Figure 20.56 Principle of sonar direction finding.

sounder is a sonar whose hydrophone is directed to ensure that the axis of the directivity is vertical; consequently the depth of the seawater is obtained from the elapsed time difference between the transmitted and received sound signals from the seabed. The *side-scan sonar* is primarily used for ocean bottom exploration rather than for distance or depth measurements. The ocean bottom can be scanned by tilting a transducer on one side of the vessel, forming a fan-shaped beam that is very narrow (1–2°) in the horizontal plane and of the order of 40° in the vertical plane. The vertical plane is perpendicular to the heading of the vessel. A dual side-scan is produced by mounting two transducers in a towed body called the "fish," whose height from the seabed is adjustable, for optimum results.

In a shipborne unit the pressure waves are generated by a vibrating surface operating at a preselected transmission frequency. This surface converts the electrical power into sound power by contact with the seawater medium. The surface element that transmits sound waves through seawater is called the transducer. It has also the reverse capability of converting the sound power into electrical power when made to vibrate by sound pressure waves. When the transducers act as receivers they are called hydrophones. As previously mentioned, the width and shape of the beam play important roles in various sonar applications. As in the case of emission of microwave electromagnetic energy; the beamwidth depends on the dimension of the transducer and the wavelength transmitted. A circular transducer is used to produce a conical beam, and the approximate beamwidth in degrees between half-power limits is obtained from the following expression (*Ingham*, 1974):

$$\alpha^0 = \frac{65\lambda}{d} \tag{20.57}$$

where λ is the wavelength of the transmitted or received frequency and d is the diameter of the radiating surface, both in the same units. This kind of transducer is typical of most Doppler sonars. A narrow, fan-shaped beam is produced by a rectangular transducer with a height of L_H (referring to the vertical spread of the beam) and a width of L_W (referring to the horizontal spread of the beam). The corresponding beam dimensions are then

$$\alpha_H^0 = \frac{50\lambda}{L_H} \tag{20.58}$$

and

$$\alpha_W^0 = \frac{50\lambda}{L_W} \tag{20.59}$$

This kind of transducer is used exclusively in the side-scan sonar. As an example of the application of Eqs. 20.57–20.59, let the average sound velocity be $V = 1500$ m/sec and the frequency used be $f = 60$ kHz. The corresponding wavelength will then be $\lambda = 2.5$ cm. In the case of Eq. 20.57 a beamwidth of $\alpha = 15°$ is desired. Substituting the data into Eq. 20.57, we find that the needed diameter of the radiating surface yields $d = 10.8$ cm. In the case of Eqs. 20.58 and 20.59, beam dimensions of 40° and 2° are desired. Similar analysis shows that the dimensions of the radiating surfaces will be $L_H = 3.1$ cm and $L_W = 62.5$ cm, respectively.

The seabed itself can act as a responding "device" to send the reflected sound signals back to the ship's hydrophones, as in the case in the Doppler, side-scan sonars, and echo sounders. The man-made beacons usually belong to one of two design categories. First is a simple, active beacon called the "pinger," which is a programmed beacon that transmits to the ship's hydrophones short and uncoded pulses at accurately predetermined times at a rate of about 1 Hz. These beacons or markers are often used for dynamic positioning. The more sophisticated and versatile beacons, called transponders, remain passive until interrogated by coded pulses from the ship's transducers. These beacons are used for high precision surveys of various geometric modes.

The key to system accuracy is the propagation velocity of sound waves. The velocity is a function of water temperature, salinity (salt content), and pressure or depth below the surface. Several formulas are derived for the general expression of the velocity in terms of the parameters named, but they differ mainly in the units used. In the following expression (*Horton*, 1957)

$$V = 4422 + 11.25T - 0.0450T^2 + 0.0182D + 4.3(S - 34) \quad (20.60)$$

V is the velocity of sound in seawater stated in feet per second, T is the temperature in degrees Fahrenheit, D is the depth below the surface in feet, and S is the salinity in parts per thousand. If the surface temperature of water is 60°F (15.5°C) and the salinity is 34‰, the velocity of sound through the surface layer according to Eq. 20.60 would be $V = 4935$ ft/sec (1504 m/sec). All the parameters vary diurnally and seasonally and are unstable even during a day's survey because of the many external factors (rain, tidal streams, local currents, etc.). As we found when studying electromagnetic radiation, the average velocity along the sound path is needed. If we assume that the general propagation velocity of sound waves, as a function of depth, is a smooth polynomial of second order, we can use a method similar to that described in Chapter 7. This approach requires the adjustment of several sample observations at known depths, together with Eq. 20.60, for the computation of the average velocity. As an alternative, three observation depths—surface, middle depth, and the seabed—may be sufficient for applying the Simpson rule. It must be kept in mind, however, that most of the integrated radiopositioning and sonar systems used by hydrologists and explorers assume shallow water of a few hundred meters and surface operations limited to within a few kilometers.

There are great variety of sonars available today for navigation and exploration purposes. The following system is selected as a sample because of its seemingly advantageous design for limited-area hydrographic surveying. The *Decca Survey Aqua-Fix* features an interesting dual functioning that extends the measurable range with little loss of relative accuracy. For dynamic positioning or marking a point of interest, a standard transducer-to-transponder double acoustic signal path is employed. For more precise hydrographic surveying only one acoustic signal path from the transducer to the transponder is applied, and the signal is returned to the ship's receiver by way of a surface radio buoy. Each radio buoy is connected to the corresponding transponder by a cable and has its own coded tone on a common radio frequency; all work on a common acoustic frequency. The buoys are designed to transmit on radio upon receipt of an acoustic pulse and acoustically upon receipt of a radio pulse. The ship has a single acoustic transmitter with a separate modulator for each radio buoy. The advantages of this arrangement are the greater depth capability of the transponders (because they are functioning like hydrophones), the greater range (due to the ability to transmit more acoustic power from the ship), and the greater accuracy (due to shorter water path for a given range). Since no acoustic signals are received by the ship, the performance is unaffected by ship's noise. When the transponder is used as an acoustic marker, the Aqua-Fix employs direction-finding facilities such as those described earlier in this section. The standard transponder provides operation in the "listening" mode for about one year after deployment. When continuously interrogated, its maximum operating life is 3 weeks. The maximum operation depth for the transponder is about 750 m. The control and display unit controls both the data fed to the hydrophones and the signals picked up from the transponders; it also presents the slant range in a digital display. The total positioning accuracy within 1000 m is about ±2 m.

20.2.4.   Summary of Land-to-Sea Survey Systems

Because of the great variety of systems and designs now available in land-to-sea survey operations, not all could be included in detail in the previous sections. To complement the analysis, however, short descriptions are given of some instruments characterized by interesting design history or present day practical use.

   E.P.I., standing for electronic position indicator, was designed and built in the Radiosonic Laboratory of the U.S. Coast and Geodetic Survey in the mid-1940s. It was used extensively in establishing a connection between islands off the Bering Sea and the mainland in the early 1950s. The system uses circular geometry to combine the medium frequency pulse techniques of Loran with the measuring features of Shoran. The E.P.I. has only one frequency for all transmissions, $f = 1850$ kHz, which was found to give the most desirable range characteristics. To distinguish pulsed signals from different ground stations, a special delay system is employed in which delay times are always significantly greater than the pulse's travel time to ground stations and back. The pulses arriving at the ground stations could not be used as such to trigger the transmitter in action, as in case of Shoran, because of the long rise time of the leading edge of the pulse due to the long wavelength. This difficulty was solved by securing precise synchronization between the ship's and the ground station's pulse-generating frequencies by comparing the two pulses in an A scan of the ground station CRT. This method yielded correct pulse rates and phase relationships. Distances to 500 km are measurable with an accuracy of about ±50 m.

MAP (for marine autotraverse positioning) is a close-to-shore positioning system using the radar principle and radar frequencies. The location of a survey vessel is obtained from two passive reflectors placed on known points ashore. The azimuth of the radar beam is manually selective, picking only the desired shore targets. In areas where the target is indistinguishable, a noise-maker is attached to the target. This device is a low-powered transmitter, and its purpose is to distinguish the target from background or other strong targets. The display may be presented visually on a radar screen, digitally by an automatic readout, or graphically by automatic plotter. The maximum measurable distance is the line of sight, or with proper antenna installations, about 60 km with an accuracy of about ±15 m.

RANA is a phase comparison system that employs hyperbolic geometry. It was developed by *SERCEL* Radio Navigation Division, Nantes, France, for the French Hydrographic Service and was used in the early 1960s in North African surveying work. Its four transmission frequencies are within the 1600–2000 kHz band, and difference frequencies are formed to produce fine, medium, and coarse lane patterns. Frequencies are selected to ensure that the corresponding lanewidths will be 200 m and 4 and 80 km. If the two coarse patterns are used, a lane identification can be established. The heterodyne principle is operative here, and phase comparison is made by audio tones. Maximum operational range is about 160 km with an average accuracy of about ±15 m.

The *Toran System*, also manufactured by SERCEL, is similar to the model "B" Lorac in that the geometric mode is hyperbolic and the phase differences are measured using heterodyned audio signals. Because the base stations are free-running and phase locking with the master station is unnecessary, a reference ground station sends the reference beat note to the survey vessel as an amplitude modulation of a carrier, differing from those transmitted from the base stations. The frequencies used are between 1.6 and 3.0 MHz. The claimed operational distance is about 700 km, with an accuracy of about ±1 m on the baseline.

The *Toran "O"* is a new version of the basic Toran system. It uses the circular range-range mode; an unlimited number of users can operate simultaneously because there is a single basic frequency f. This frequency is made possible by using the atomic frequency standards in both the base stations' transmitters and the survey vessel's receiver. Base stations A

and B transmit with frequencies $f + \Delta f_A$ and $f + \Delta f_B$, respectively, where Δf_A and Δf_B are audio frequencies and less than 300 Hz, and f is the basic frequency, kept stable with the frequency standards and the synthesizers. In the receiver of the survey vessel, the frequency standard and the synthesizer have dual functions. They generate the local signals with frequencies close to those received from the base stations to be heterodyned, and they generate the local reference beat signals to be used in phase comparison. The maximum measurable distance is claimed to be 1000–1200 km with a measurement resolution of ±1 m. The frequency band is the same as in the basic Toran system.

20.3.   LAND–TO–AIR SURVEY

After World War II the main civilian application of radio aids was in land-to-air navigation systems, although at the same time sea navigation equipment was being tested for use in hydrographic surveying. During the war air navigation systems were divided into two categories. One, in which the location of an aircraft was made by measuring the distances from an aircraft to two ground stations *in the aircraft* was termed the "*H-system.*" A reversal of that system was a British design called "*Oboe,*" which is an acronym of "*o*bserved *b*ombing *o*f the *e*nemy." In this case, the responder beacon is carried in the aircraft, the two measured ranges being controlled from two fixed ground stations. Oboe was successfully used in bombing the Ruhr area in 1942 and 1943 with an accuracy of about ±100 m at distances of about 300 km. After the war the accuracy of Oboe was significantly increased with some modifications, primarily because of the use of C-band frequency in a pulsed type of electronics. It was even used to determine the velocity of electromagnetic radiation at different altitudes by *F. E. Jones* and *E. C. Cornfod* in 1946 (*Hart,* 1948). Another British design in the H-system was the *Gee-H,* where the "Gee" is obtained from the phonetic expression of the first letter of the word "grid." The first successful attempts to use this system in connection with photogrammetric mapping were initiated by the late Brigadier *M. Hotine* in 1943–1945 to produce tactical maps of Southeast Asia (*Hart,* 1946, 1948). After the war the Gee-H system was widely used for photogrammetic mapping of the former British colonial territories in Africa. The accuracy of the pulsed-type Gee-H system was about ±45 m; this rather poor performance was mainly caused by the long wavelength (about 10 m) used in transmissions. The well-known H-system design manufactured in the United States was the Shoran and its improved version Hiran, which were briefly described in Section 20.2.1.4. These became standard instruments for the U.S. Air Force in the extensive survey operations to link the continents for navigational and mapping purposes between the late 1940s and the mid-1960s. Shoran and Hiran have also proved their efficiency and accuracy in the formation of a vast trilateration network in Canada (*Ross*, 1955), and they have been successfully used in the determination of the velocity of light (*Aslakson*, 1951). Except for Hiran, which is used in hydrographic surveying, all these systems are obsolete today and have been replaced by new and more up-to-date systems.

Aeris (*a*irborne *e*lectronic *r*anging *i*nstrumentation *s*ystem), manufactured by *Cubic Corporation* (1962), is the forerunner of the *Shiran* system described in the next section. Aeris is a simultaneous, multidistance measuring system employing circular geometry and a phase comparison technique for frequency-modulation waves on carriers in the UHF band. It consists of four ground responders, an interrogator in the survey aircraft, and a portable data reduction station on ground. The airborne interrogator measures simultaneously and continuously distances to all ground responders and repeats this distance information to the data reduction station. The interrogator unit in the aircraft sends the data signals continuously and the keying signals in sequence to the ground responder stations. Only the responder receiving its own particular keying tone will respond and retransmit the phase information back to the interrogator, where the data are placed on telemetry channels and retransmitted on subcarriers to the data reduction station. In the control point extension procedure, Aeris gives the location of a point in the three-dimensional system. To establish the fix, one must have two distances and the altitude of the aircraft, or three distances. The extra measurements available provide redundant data for the adjustment. The ambiguity of the total distance is solved by using three modulation frequencies simultaneously. The maximum measurable distance and the instrument resolution capability are claimed to be about 250 km and ±1 m respectively.

20.3.1. Shiran

Hiran was used extensively for almost 15 years to form trilateration networks or chains to aid in establishing intercontinental ties. The extensive worldwide surveying programs planned for the future, however, required a more efficient and accurate electronic surveying system, and in

1962 *Cubic Corporation* (1963) made a contract with the Air Force to build the *Microwave Geodetic Survey System*, which was named *Shiran* (*S*-band *Hiran*).

Shiran consists of an airborne subsystem that transmits a continuous wave, phase-modulated signal to each of four ground stations, which retransmit the signal to the airborne subsystem. The airborne subsystem automatically measures the phase shift of the modulation signal and interprets the phase shift as range. Data processing equipment record each range on magnetic tape in a format suitable for computer reduction after each flight.

The ground stations are interrogated sequentially but, in effect, the range measurements are made simultaneously. A special design feature of the equipment that achieves dynamic smoothing of the samples to produce continuous range data in real time is responsible for this simultaneity. The sequencing is performed with keying signals. A unique keying signal is transmitted, and it turns on a particular ground transmitter. When the ground station receives the keying signal, it demodulates the ranging data modulation signal and retransmits this signal at an offset-frequency carrier back to the airborne subsystem, and the station is turned off. A period of guard time, with no modulation, prevents overlapping between the return signal from that station and a return signal from the next station. A different keying tone is then transmitted to turn on the second station of the sequence. A complete sequence of four-station interrogation is repeated 10 times per second. Since the keying tones are generated in the frequency synthesizer in the airborne unit, the set of stations to be interrogated can be selected in the aircraft and changed at will.

Four modulation frequencies are required to yield both accuracy and nonambiguity in distance. The airborne transmitter is modulated by these four signals at coherently related frequencies given later in this section. After the signals are retransmitted from the ground stations, they are received and demodulated in the airborne station. The phase of each modulation signal is compared with its phase before transmission, and the phase differences are directly interpreted by an electronic phasemeter as a binary value of slant range. The electronic phasemeter is a combination of units, including a range memory, a digitizer, and a frequency synthesizer. The frequency synthesizer develops the keying and modulation frequencies, controls the station sequencing, translates the received modulation frequencies for processing in the range memory, and generates the reference phase for the very fine data. The master oscillator is an extremely stable crystal oscillator contained in a proportional temperature-controlled oven. The basic oscillator frequency is $f_0 = 1.3260822$ MHz. In the Shiran system the international nautical mile is the unit of length, and the resolutions are given in feet. The internationally accepted value for the speed of light, $c_0 = 299,792.5$ km/sec or $c_0 = 161,875.0$ nmi/sec, is the built-in value in the instrument, and the modulation frequencies are chosen in a 16^n relationship to finally yield the maximum nonambiguous distance of 500 nmi.

The pattern or modulation frequencies in this system are

$$\text{VF (very fine)} = f_0/2$$
$$\text{F (fine)} = \text{VF}/16 = f_0/2 \times 16$$
$$\text{C (coarse)} = \text{F}/16 = f_0/2 \times 16^2 \tag{20.61}$$
$$\text{VC (very coarse)} = \text{C}/16 = f_0/2 \times 16^3$$

Assuming the values of f_0 and c_0 in Eq. 20.61, the modulation frequencies, the modulation half-wavelengths, and their resolutions are given in the following tabulation.

Modulation Pattern	Pattern Frequency (kHz)	Half-Wavelength (nmi)	Resolution (ft)	
Very fine (VF)	663.041	0.12207	0.7248	(20.62)
Fine (F)	41.440	1.95312	11.5969	
Coarse (C)	2.590	31.2500	185.550	
Very coarse (VC)	0.1619	500.000	2968.85	

The coarse and very coarse frequencies are folded about the fine frequency before transmission, to avoid filtering problems in the receiver. Thus the coarse frequency is digitally mixed with the fine frequency to give an $F + C$ frequency of 44.030 kHz before it is sent to the data adder. The very coarse frequency is digitally mixed with the coarse frequency and then with the fine frequency to form the $F - C - VC$ frequency of 38.688 kHz. This frequency is passed through a bandpass filter and sent to the data adder.

The carrier frequency of the airborne transmitter is derived from the frequency generated by a stable crystal oscillator. The crystal has a frequency of 61.3333 MHz and a stability of 3 ppm. Stability is ensured by enclosing the crystal and the oscillator assembly in a temperature-controlled oven. The actual carrier frequency is obtained by multiplying the crystal frequency by 54, which yields $f_c = 3.312$ GHz. The ground

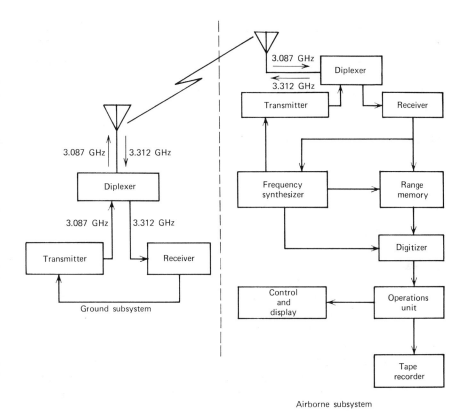

Figure 20.57 Simplified block diagram of Shiran (courtesy Cubic Corp.).

stations retransmit a frequency of $f_c = 3.087$ GHz. A simplified block diagram of the Shiran system appears in Fig. 20.57. The airborne antenna has 360° coverage in azimuth and approximately 8° coverage in elevation. The antenna is either a discone, which is equivalent of a horn having 360° azimuth coverage, or a stacked linear array. The polarization is vertical and the antenna gain is 10 dB. The ground antenna is either a horn having a coverage of 10° in elevation and 80° in azimuth or a stacked linear array in front of a reflector. The polarization is vertical and the antenna gain is 15 dB. The ground antenna and mast can be rotated to the approximate direction of anticipated aircraft maneuvers. The maximum measurable distance depends mainly on the flying height of the aircraft; for example, with an aircraft altitude of about 40,000 ft (12,200 m) the distance in control point extension would be about 450 km, but in the line-crossing operation it would be double, or 900 km.

The power consumption in the ground stations is 11 A at 28 V DC (300 W); in the airborne subsystem it is 20 A at 28 V DC (550 W). The transmitter output power both in the airborne system and on ground stations is 20 W. The airborne subsystem weighs about 440 kg; the weight of the ground instruments without antenna installation and power source is about 40 kg. The standard error of one line measurement 400 km long is about ±3 m, and the corresponding standard error in control point extension is about ±4 m; both figures satisfy Air Force specifications.

20.3.2. Cubic CR-100

The *CR-100 Precision Ranging System* is another product of the Cubic Corporation. It is designed to provide accurate range and range-rate measurements, with compactness and versatility as the primary objects. The CR-100 system consists of a small interrogator and lightweight transponder. These units are the basis of an air-to-ground coherent-carrier tracking loop. To determine range, the system uses CW multiple-tone modulation on the 1–3 GHz carrier band and phase comparison techniques. In position-fixing applications the circular mode is employed, and the system uses a digital code that serves as the transponder "select call." Up to 256 transponders can be addressed without interference. In the range-rate mode the radial velocity of the aircraft is measured by determining the round-trip Doppler frequency shift of the carrier signal transmitted between the interrogator and transponder. The system also features a two-way voice data link and an in-flight calibration capability.

The range and range-rate subsystem is designed to operate as part of a total installation that may include an inertial navigation component and minicomputer. The interrogator is built to operate with most general purpose computers using either serial or parallel interfaces. The power consumption for the interrogator is 40 W standby and 120 W on transmit; the transponder takes 6 W standby and 30 W on transmit. The transmitted power is 4 W in both units. The weights of both units are 11.3 and 19.0 kg for the interrogation and the transponder, respectively. Maximum measurable distance is about 320 km with a range accuracy of about ±1.5 m and a range-rate accuracy of about ±3 cm/sec. The allowable radial speed of the aircraft is 1500 m/sec.

20.3.3. Tellurometer MRB-301

The *MRB-301*, which is manufactured by Tellurometer (Pty) Ltd. is a high-speed positioning system almost identical to the MRB-201 used for hydrographic operations. Whereas the MRB-201 is capable of operating at a vessel speed of little more than 50 km/hr, the MRB-301 can be used in an aircraft flying 900 km/hr. Therefore, the main difference between these two systems is the readout design. In the MRB-301 the readout has been referred to as the analog-to-digital converter (ADC) because it effectively converts analog phase information into digital form. Range measurements can be carried on simultaneously up to three ground remote stations, but typically there are two ground remote stations and two airborne master units. Each airborne unit feeds its own steerable directional antenna with two separate carrier frequencies in the 2.8–3.2 GHz band. The antennas can be mounted on the belly or wings of the aircraft and can be individually steered by the operator. Four modulation frequencies from A to D ($A-$ and $A+$ in the remote) are needed; they are exactly equal to those used in the MRB-201 (Section 20.2.1.3) and are not repeated here. By working with the general Tellurometer principle of frequency differences, the distances are obtained in metric units and are unambiguous to 100 km.

In any high-speed operation it is essential to have virtually continuous and unambiguous information, and for this reason all modulation frequencies ideally should be continuously transmitted (*Marshall*, 1971). Since this is extremely costly and difficult to achieve in practice these frequencies are sequentially transmitted within a total period of 2 sec. The fine $A+$ pattern is interspaced between each coarse measurement to obtain the maximum resolution and is available every alternate 250 msec period. This sequential switching is controlled by the programmer interface unit (PIU). The primary function of the ADC is to convert the pattern phase information to binary coded decimal form, and this is accomplished with a crystal-controlled clock and other circuits that take into account the various dynamic factors involved. In addition, the clock provides sequential readings on each pattern, integrated over a 200 msec period. There is a dead period of 50 msec between patterns. The patterns are switched in the sequence A, $A-$, A, D, A, C, A, and B, which gives a total measuring time of 2 sec; then the next sequence begins automatically. A facility called the "A pattern override" is also incorporated to enable the user to obtain continuous A pattern readings. This provides greater accuracy for a short period of time. A further function of the clock is to time the pattern selection program by way of the PIU.

Since the MRB-301 system is designed to operate at high aircraft speeds and all pattern information is not simultaneously available, the coarse pattern information has to be adjusted according to the slant range velocity before being compared with the fine A pattern information. The

ADC is designed to operate with an on-line computer that can be programmed to give the slant range velocity from the rate of change of the A pattern; a facility can be built into the computer program to enable the prediction of ranges before they actually occur. The ADC has no readout of its own but can be operated directly with an on-line printer or magnetic tape recorder. In this way the phase information can later be processed into full range readings. All information provided by the ADC is available either in parallel or two-digit serial form.

The basic unit of the master instrument is identical to that of the MRB-201. The ADC plugs into the front panel of the basic unit using the space occupied by a battery box in the MRA-3 and by the DRI in the MRB-201. The PIU plugs into the basic unit compartment, which houses the DRO in the MRA-3 and MRB-201. To allow measurement of up to three ranges simultaneously, the master units must be interconnected to synchronize pattern switching and data output. The serial data from all master units are supplied by way of the ADC of the control master, from which they can be demanded at a rate determined by the processing or recording device used. The antennas are similar to those used in hydrographic surveying, and relatively wide beamwidths may cause severe ground reflections. In high-speed dynamic positioning, however, the range changes rapidly; thus the difference between the direct and reflected ray path varies. This produces an error similar to the sinusoidal ground swing, which is present in static microwave distance measuring instruments and can be averaged out over a reasonable period. The maximum measurable distances depend on the type of antennas used, and they vary between 60 and 150 km. The carrier output power is 200 mW, and the current consumption is about 2.5–3.0 A at 12 V for the basic unit and 2.0 A at 12 V for the ADC. The operating temperature range is -20 to $+50°C$. The pattern frequency selections can take one of the following forms: (a) manual selection at master and remote stations, (b) manual selection at master and automatic selection at remote station, (c) automatic selection both at the master and remote stations in the sequence stated before, or (d) automatic selection of an A pattern only by means of an override control available to the operator or computer. The automatic pattern selection at the remote station is achieved by discriminating tones transmitted by the master unit in the 39–70 kHz range.

The accuracy is claimed to be ± 1 m $\pm 3 \times 10^{-6}$. Several trials carried out in Australia (*Marshall*, 1971) have involved line-crossing techniques between points whose coordinates had been measured in previous years using a combination of astronomical observations, Tellurometer traverses, and triangulation. One single crossing over the 30,458.1 m long line between the stations of Kookaburra and Loca yielded the line length +1.2 m longer than the predetermined value. Another line with the known distance of 104,486.2 m between the stations of Boomahnoomoona and Kookaburra was crossed seven times. After adjustment the standard error of one crossing was ±1.8 m and the standard error of the mean (the line measurement) was ±0.7 m.

20.4. LAND–TO–SPACE SURVEY

The concepts of conventional geodetic surveying and even more sophisticated electronic applications changed drastically when the Soviets orbited the first man-made earth satellite *Sputnik* 1 on October 4, 1957. The foundation of real three-dimensional geodesy was set and the first tracking of an artificial satellite was made by American scientists by optical means, using direction determinations instead of distance measurements. A special satellite tracking device, the *Baker-Nunn* camera, was designed and built to the specifications set forth by scientists of the Smithsonian Astrophysical Observatory. The camera is named for James G. Baker, who designed the optical system, and Joseph Nunn, who designed the mounting and mechanical systems. The first camera unit was completed in time to photograph Sputnik 1 less than two weeks after its launch. The Baker-Nunn still provides the most sensitive optical means of observing artificial earth satellites. For example, the first Vanguard satellite, a 6 in. sphere, has been photographed at a range of 5600 km, and the Orbiting Geophysical Observatory has been shot from 37,000 km. The Baker-Nunn camera follows the satellite on its orbit and keeps the image in the center of the photo frame. Two more lightweight cameras were designed in the early 1960, one for the U.S. Air Force (PC-1000) and one for the U.S. Coast and Geodetic Survey (DC-4). Both were stationary and were so oriented during the pass that the set of images of the passing satellite appeared in the middle of the photo frame.

In the early 1960s distance measurements to artificial earth satellites became more prominent. The *U.S. Navy Transit Satellite System*, often also called the *Satellite Doppler Navigation System*, was first presented at the ·Advisory Group for Aerospace Research and Development (AGARD) meeting held under NATO auspices in Istanbul in 1960; it became fully operational 1964. It is described in detail in Chapter 28. At about the same time *Secor*, another system that utilizes radio wave transmission, was initiated. The instruments using radio waves were followed by satellite laser tracking and finally by lunar laser ranging systems.

20.4.1. Secor

Secor is an acronym for *se*quential *co*llation of *r*ange, which describes its method of operation. The system is an electronic distance measuring device in which four ground stations sequentially and on a time-sharing basis interrogate a satellite-borne transponder. The basis of the ranging technique is the determination of the phase shift of modulation waves during propagation to and from the transponder. Each ground station, designated DME (distance measuring equipment), records its range data on magnetic tape simultaneously with signals from an electronic clock and other data. The development of Secor began in 1960 when the Army Map Service (AMS) procured four prototype stations from the Cubic Corporation (*Cordova*, 1965). System tests were begun during the spring of 1961 and were continued until the fall, when results being obtained raised questions about the validity of the basic concepts of the system. The entire program was then halted for reevaluation by direction of the Office of the Chief of Engineers, U.S. Army.

In early 1962 the Army Corps of Engineers, Geodetic, Intelligence and Mapping Research and Development Agency (GIMRADA) was directed to appoint an evaluation team composed of its own scientists and experts, as well as members from the Ballistic Research Laboratories and the National Bureau of Standards. This team made a complete analysis of the system and found the basic concepts valid. It therefore recommended various equipment modifications to ensure reliable operation and to meet the desired system accuracy. In the accuracy specifications set, the total of internal errors was not to exceed a standard deviation of ±5 m at a range of 1600 km or ±20 m at a range of 6500 km. The tracking of the first Secor satellite began on January 12, 1964, and continued until April of that year, when this experimental phase ended. After the testing periods, the ground stations were moved to their first operational positions in Japan and nearby islands and the next phase of Secor began in the fall of 1964.

In the ground stations (one of which is always designated the master), the transmitters operate at a frequency of 420.9375 MHz, providing up to 2 kW power output. The carrier is phase modulated for ranging data, and transmissions are pulse modulated at the rate of 20 Hz. This means that each ground station measures the distance to the satellite in sequential order 20 times per second, and the total cycle of interrogations requires 50 msec, where one cycle is equal to one complete sequence of interrogations from all stations in a Secor network. The 50 msec interrogation cycle is divided into four time zones of 10 msec each, with a 2.5 msec space between zones. An overlap of interrogations from two or more DME stations causes data interference proportional to the amount of overlap; therefore interrogations from each station must arrive at the satellite at different times. To accomplish this, a transmitter key delay control is provided at each station to alter the time of transmission with respect to the timing signal given by the DME station acting as master. Taking into consideration that all signals travel at the same speed but the range to satellite is constantly changing, the system automatically determines the time of transmission for each station; thus signals arrive in proper sequence at the transponder.

The following example illustrates the interrogation cycle timing for a hypothetical case, where distances have been selected to correspond to round figures of times. Let us assume the following distances from the ground stations and approximate transit times to the satellite.

Station	Distance (km)	Transit Time (msec)	
Master	3747	12.50	(20.63)
DME 1	7494	25.00	
DME 2	5621	18.75	
DME 3	3747	12.50	

The master station establishes the 0.0 second time reference, which means that the signal from master will reach the satellite at +12.5 msec. This will be the beginning of the interrogation cycle in the satellite, and the first interrogation must be completed within 12.5 msec of arrival, which is +25.0 msec. The signal from DME 1 requires 25.0 msec to reach the satellite and must be there at +25.0 msec for the second slot on the time scale. To accomplish this the signal has to be transmitted at 0.0 sec, which means that it is transmitted simultaneously with the master signal. The signal from DME 2 requires 18.5 msec to reach the satellite and must be there at +37.5 msec. Consequently, it must be transmitted at +18.75 msec. Finally, the signal from DME 3 requires 12.5 msec to reach the satellite, and to complete the interrogation cycle at +62.5 msec it must arrive at +50.0 msec. Therefore, its transmission time must be at +37.5 msec. From this example it becomes evident that any combination of ground station transmission times could occur during the satellite pass over the Secor network, the only important factor being the signal arrival times at the satellite.

Although the tracking takes place sequentially to a fast-moving satellite, all range readings ultimately correspond to the same time at the

satellite. This is made possible by the following synchronization procedure (*Prescott*, 1965). During the interrogation period the range on each
DME channel reaches its analog servo, which is electronically damped to
ensure that it keeps a smooth and continuous record of phase angle. Each
analog servo is monitored by a digital servo that keeps a binary representation of the range in terms of meters. To achieve simultaneous range
measurements from the four ground stations, the master station emits a
special pulse every 50 msec with its ranging signal. This pulse, retransmitted by the satellite, is received at all stations including the master. On
reaching the stations, the pulse causes the contents of each digital servo to
be immediately recorded on the magnetic tape.

In each ground station the 420.9375 MHz carrier is phase modulated
much as it is in the Shiran system. A highly stabilized crystal-controlled
oscillator generates a basic modulation frequency of $f_0 = 1,171.065$ kHz.
Four modulation or pattern frequencies are derived from the basic
frequency in the following way:

$$
\begin{aligned}
\text{Very fine} \qquad & f_1 = f_0/2 \\
\text{Fine} \qquad & f_2 = f_0/2^5 \\
\text{Coarse} \qquad & f_3 = f_0/2^9 \\
\text{Very coarse} \qquad & f_4 = f_0/2^{12}
\end{aligned}
\qquad (20.64)
$$

The resolution in each pattern is based on an uncertainty in the phase
measurement of about ±0.2 electrical degrees in the very fine pattern and
about ±0.7 electrical degrees in the other patterns. The Secor system
operates in metric units, and by assuming the standard value of the
velocity of light, we obtain the following pertinent data.

Modulation Pattern	Pattern Frequency (kHz)	Half-Wavelength (m)	Resolution (m)
Very fine	585.533	256	0.25
Fine	36.596	4,096	16
Coarse	2.288	65,536	256
Very coarse	0.286	524,288	2,048

In a procedure resembling that for Shiran, modulation frequencies are
mixed into a form in which the spread of the frequency spectrum is not

too wide. Thus we have the following combinations:

$$f_1 = 585.533 \text{ kHz}$$
$$f_2' = f_1 - f_2 = 548.937 \text{ kHz}$$
$$f_3' = f_1 - f_3 = 583.245 \text{ kHz} \qquad (20.65)$$
$$f_4' = f_2' + f_4 = 549.223 \text{ kHz}$$

The extra 20 Hz pulse signal is equal to a simulated half-wavelength of 7500 km with a resolution of approximately 41.83 km. The distance equation(s) now can be written in the general form employed throughout this text:

$$D = (n_i + L_i) \frac{\lambda_i}{2} \qquad (20.66)$$

When the numerical constants are introduced, such expressions yield

$$D_4 = (n_4 + L_4) 524{,}288 \text{ m}$$
$$D_3 = (n_3 + L_3) \ \ 65{,}536 \text{ m}$$
$$D_2 = (n_2 + L_2) \ \ \ \ 4{,}096 \text{ m} \qquad (20.67)$$
$$D_1 = (n_1 + L_1) \ \ \ \ \ \ \ 256 \text{ m}$$
$$D_5 = L_5 \times 7{,}500{,}000 \text{ m}$$

In Eq. 20.67 L-values are the fractional values of the modulation half-wavelengths after the integers n have been extracted. All D-values refer to the same distance, which is determined in succession to finally yield the total, fine distance; the subscripts serve only to clarify the following example of the measuring principle (which in reality, of course, is accomplished fully automatically). Let us assume that the following fractional readings exist:

$$L_1 = 0.277$$
$$L_2 = 0.142$$
$$L_3 = 0.884$$
$$L_4 = 0.985$$
$$L_5 = 0.206$$

First, the 20 Hz pulse gives the extended coarse reading distance as

$$D_5 = 1{,}545{,}000 \text{ m}$$

from which the n_4 is obtained by dividing D_5 by 524,288, yielding $n_4 = 2$.

The substitution of n_4 and L_4 to Eq. 20.67 gives the second approximation D_4. We continue now, dividing D_4 by 65,536 to obtain n_3, until finally D_1 is reached. The corresponding values will be as follows:

$$n_4 = 2 \qquad D_4 = 1,565,000 \text{ m}$$
$$n_3 = 23 \qquad D_3 = 1,565,262 \text{ m}$$
$$n_2 = 382 \qquad D_2 = 1,565,254 \text{ m}$$
$$n_1 = 6114 \qquad D_1 = 1,565,255 \text{ m}$$

For checking purposes all L-values from L_1 to L_4 are now obtained by dividing the D_1-value by the corresponding half-wavelength values and taking into consideration the fractional part only.

The airborne unit of the Secor system is basically a remote-controlled radio relay device. It is designed either to be launched as an integral part of a satellite that is used for other purposes or as an attached device. In the latter case, the transponder is housed in a separate shell that is attached to the main satellite and is launched by explosive connections when the main vehicle is orbited. The transponder receives discrete command signals and data on a carrier of 420.9375 MHz and retransmits the data on two separate carriers of 449 and 224.5 MHz to provide information for ionospheric correction. At all times the transponder is either on standby or operational. Since the time the transponder can remain in operation is limited by the life of its batteries, a selective-call circuit is incorporated to conserve the power. In the standby state only the circuitry related to the reception of the selective-call signal is in operation, and minimum power is drained from batteries. When a selective-call signal from the ground station is received and detected, the selective-call circuitry commands the operative state. This energizes the transmitting and data handling circuits.

In the ground stations two receivers are used to receive response signals from the transponder at frequencies of 449 and 224.5 MHz. The antenna assembly consists of high frequency and low frequency elements mounted in 3 m parabolas located adjacent to each other. The antenna assembly can be rotated through 720° horizontally and −5° to 110° vertically. Simultaneous transmission and reception are possible using diplexing circuits located in the antenna pedestal, together with the servo system required to position the antenna assembly. The remote antenna control panel contains the controls to power the pedestal, position the antenna assembly, and indicate the position of the assembly. The data handling equipment is composed of the following main units. The extended range timing and sequencing system initiates the master timing signal and synchronizes the slaves' timing signals with it. Also, it controls the station transmission timing and produces an oscillograph picture of the interrogation cycles. The frequency synthesizer forms the modulation and reference frequencies for phase comparisons and translates the transponder frequencies received to servo reference frequencies. The electronic servo detects phase shifts between the received response signal and the reference signal and displays the phase shift as a binary digital value. The processor forms control pulses for the functions related to data collection and identification and sends data to the tape recorder in a predetermined sequence. The VLF timing unit provides the station with accurate timing synchronized to a standard broadcasting signal, such as that broadcast by WWV*; this is important especially in using the orbital mode.

20.4.1.1. *Corrections and Accuracy*

Several corrections must be applied to the edited and smoothed Secor data to determine the final satellite position. After the corrections have been applied, certain random residual errors must be analyzed to obtain the total system accuracy (*Prescott*, 1965; *Cordova*, 1965). Four main sources of corrections and errors are as follows:

1. System calibration.
2. Atmospheric correction.
3. Ionospheric correction.
4. Doppler effect.

System calibration consists of three different phases: (*a*) a combined station and test transponder calibration to establish the reference system, (*b*) a combined station and satellite transponder calibration, and (*c*) satellite transponder-antenna calibration. Since the last two calibrations cannot be repeated after the satellite has been launched, the calibration system is based on an assumption that there is no significant phase drift in either transponder. The accuracy of the Secor system depends solely on the stability of the very fine modulation pattern, and therefore the calibration needs to be done over the VF unit length range of 256 m only. To achieve the ultimate accuracy in calibration, the test transponder has been attached to a precision-designed discone antenna by a cable whose electrical length is known precisely. The distance between the discone antenna and the station antenna system is accurately measured by some

---

* WWV is a radio transmitting station broadcasting standard radio frequencies of high accuracy. It is maintained by the National Bureau of Standards at Ft. Collins, Colorado.

independent means; thus the total calibration distance is the cable length plus the antenna distance.

The station and test transponder calibration is carried out by the following procedure. The station is placed in the calibration mode and all the digital servos are set to zero by the switches in the servo unit. The station is then placed in the operate mode and ranged on the test transponder's antenna. The calibration constant N is obtained by subtracting the range reading D_S at the servos from the measured line length D_L:

$$N = D_L - D_S \qquad (20.68)$$

Since the actual measurements are made to the orbiting transponder, which may have phase delay slightly different from the test transponder, it will also be calibrated together with the ground station to find the constant difference between the phase shifts of these two transponders. If the resulting calibration constant is designated N' and the observed distance D'_S, the calibration constant will be

$$N' = D_L - D'_S \qquad (20.69)$$

and the correction to be applied to the observed calibration constant N_0 in actual measurements is the difference of Eqs. 20.68 and 20.69:

$$\Delta N = D_S - D'_S$$

Furthermore, the total ground station and satellite transponder calibration constant will be

$$N'_0 = N_0 + \Delta N \qquad (20.70)$$

The satellite transponder antenna calibration also relies on the discone antenna in the following way. The test transponder connected to the discone antenna is ranged over a certain distance, and then the same transponder is ranged over the same distance, using the satellite antenna system and the same cable length. Since the discone antenna is assumed to have zero phase delay, the difference in range measurements is due to the phase delay in the satellite antenna. The calibration system shows considerable variation in phase delay with respect to orientation of the satellite antenna. It is anticipated that the satellite's motion around its stabilized axis will tend to cancel out errors if the mean value of the phase delay from different antenna positions is used. The calibrations listed previously are made separately for the high and low transmission frequencies. The satellite transponder and the transponder antenna calibrations are made once in each ground station before launching, and the test transponder ground station calibrations are made in each station before and after each satellite ranging. The total residual standard error of the calibration system is about ±2.3 m.

Atmospheric correction can be taken into consideration by either of the methods analyzed in Section 7.4.2 with a residual standard error of about ±0.25 m at the zero zenith distance. Correction and standard error here refer to the range change caused by the retardation of the velocity of radio waves in the atmosphere. The effect of the curvature of the ray path to the range, even at zenith distance of $\psi = 50°$, is less than 1.0 cm and can be neglected.

Ionospheric correction is more complex and unstable than the correction caused by the atmosphere. Radio waves at the frequency band used by Secor are short enough to be able to penetrate the ionosphere but will be affected significantly by the ionospheric refraction, which retards the propagation velocity and bends the ray path, causing an error to the measured range. The refractive index of the ionosphere is a function of many factors, such as the electron density of the ionosphere, the elevation angle, the length of the line, and the frequency of the carrier wave. The electron density itself and the depth of the F_2 layer of the ionosphere vary day by day, with the time of day, with the latitude, and with the amount of sunspot activity. The density is greatest around midday and can cause an error in the measured distance of about 80–120 m. The density is at its minimum at local midnight, varying very little within 6 hours; thus this period, when the range error is about 5–20 m, should be the best for measurements. The error caused by ionospheric refraction is greatest around the equator and decreases as latitude increases. The dependency of the ionospheric refraction on the radio frequency is used in the Secor system to make the ionospheric correction. This approach is based on the fact that the ionospheric refraction of electromagnetic radiation decreases when the frequency of the carrier signal increases approximately in such a way that the ionospheric refraction is inversely proportional to the frequency transmitted. Advantage is taken of this phenomenon by transmitting at two different frequencies of 449.0 and 225.5 MHz from the transponder. A difference of distance measurements indicates the refractive index by establishing a ratio of refraction effects on each of the dissimilar transmissions. The two frequencies establish a K factor or constant for the distance error calculations in the following expression:

$$R_T = R_H - K(R_L - R_H) \tag{20.71}$$

where R_T is the true range, R_H is the range obtained from high frequency f_H measurement, R_L is the range obtained from low frequency f_L measurement, and K is the frequency ratio f_L/f_H in the transponder.

The following simplified example considers only the transmissions from the transponder and neglects the transmissions from the ground stations. Let $R_H = 1,560,235$ m, $R_L = 1,560,250$ m, and in case of Secor, $K = 0.5$. Then from Eq. 20.71, $R_T = 1,560,227.5$ m. The residual standard error at the zero zenith distance is of the order of $+0.25$ m at night and about ± 1.5 m during daytime.

The *Doppler effect* causes correction in some electronic distance measuring devices because of the difference between the emitted wavelength and the received wavelength. This correction is not necessary in any of the operational modes of Secor because the modulated signals are translated down with a maximum Doppler of approximately 20 Hz, whereupon they enter the servo loops. In the servo loops the phase information is converted into digital form before comparing it to the reference data. Once the phase information is in digital form, the comparison is easily made and is not a function of Doppler.

Since the speed of light is known with an accuracy of about $\pm 3.6 \times 10^{-9}$ (see Section 6.1), there are no big systematic errors left in the Secor system. If the random electronic noise of about ± 1 m of the system is added, the total cumulative standard error at zero zenith distant at night will be about ± 2.5 m and during daytime about ± 2.9 m.

In a typical simultaneous mode to determine the position of a new point U (Fig. 20.58), simultaneous distance measurements $M1$, $M2$, $M3$, and $M4$ are made to the satellite from all four ground stations $K1$, $K2$, $K3$, and U. The space position $SP1$ can then be computed by the

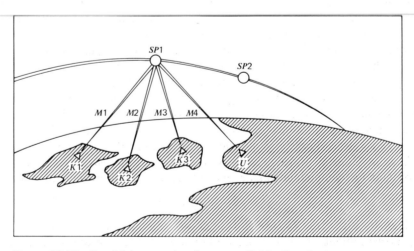

Figure 20.58   Spatial intersection (courtesy Cubic Corp.).

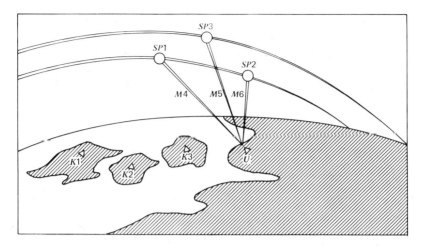

Figure 20.59   Spatial resection (courtesy Cubic Corp.).

intersection of the three spheres of radii $M1$, $M2$, and $M3$ whose centers are known positions. Space position $SP2$ is determined in the same manner, and a measurement is also made from the unknown position U. During a different pass of the satellite, space position $SP3$ is determined in the same manner, including a measurement from U (Fig. 20.59). Finally, the coordinates of U are determined by the intersection of the three spheres of radii $M4$, $M5$, and $M6$ whose centers $SP1$, $SP2$, and $SP3$ are the previously computed spatial positions. Although ranging from the ground stations is done sequentially, dynamic smoothing data techniques permit the fixing of distances from all four stations at a particular instant of time.

20.4.2.   Lasers

The trilateration chains and networks established in the 1950s and early 1960s by electronic long-line measurements were superior to the conventional triangulation measurements with respect to speed of operation, but they still lacked the ability for worldwide coverage. To initiate the world triangulation net and to strengthen the basic conventional triangulation within the United States, in conjunction with the high precision transcontinental traverse for scaling, the Satellite Triangulation Program was started by the Coast and Geodetic Survey in 1963. It became operational in August when observations began at three sites: Aberdeen Proving

Ground, Maryland, Chandler Air Force Station, Minnesota, and Greenville Air Force Base, Mississippi. The stations form a triangle with approximately 1500 km on a side. The data obtained by observing the Echo I satellite at an altitude of 1300 to 1900 km indicated an accuracy of about one part to 750,000 for the determination of the horizontal positions (*North American Datum*, 1971). During the early years of the program, tracking of the satellites was done photographically by such cameras as the Baker-Nunn, the PC-1000, and the DC-4.

Since the atmospheric refraction affects less distance measurements than angle measurements, the potential use of lasers in highly accurate satellite triangulation became evident, and laser ranging on satellite-borne reflectors was initiated by the French in 1965. Since then, the Air Force Cambridge Research Laboratories (AFCRL), NASA, and the Smithsonian Astrophysical Observatory (SAO) have started operating with the laser or with the laser and camera together. Figure 20.60 represents the world satellite triangulation network that is being completed with international cooperation. By using seven stations from the world net (1, 2, 3, 4, 8, 9, and 38) together with 28 stations within the area between Alaska, Greenland, Equador, and Surinam, a first-order geodetic control will be established for the whole of North America. This network will also greatly strengthen the geodetic ties among Alaska, Canada, Mexico, and South America. Outside this continental stretch, Hawaii, American Samoa, and other Pacific islands are tied to the North American geodetic control system through the worldwide geometric satellite network.

The type of lasers SAO uses for satellite tracking are based on the prototype located at Smithsonian's Mount Hopkins Observatory in Arizona. The system has a static-pointing mount, which permits the laser system to operate when the satellite is in the earth's shadow, and it may permit operating during daylight hours. The static-pointing pedestal positions the photoreceiver and the laser to an accuracy of better than ± 0.15 mrad. Because the mount is manually operated, the pulse repetition rate is slow, 1 pulse per minute. The light source is the ruby laser operating with the Pockels Q-switch. The pulse length is 15 nsec, containing about 7.5 J of energy at the wavelength of 6943 Å. The ruby rods and flash lamps are water cooled, and the cavities are purged with nitrogen during operation of the laser. The beam is collimated with a Galilean telescope consisting of a small concave lens and a 15 cm diameter convex lens. The concave lens can be adjusted through a flexible cable to vary output beam divergence from 0.58 to 5.8 mrad. The telescope photoreceiver focuses the returning laser beam onto a photomultiplier tube at

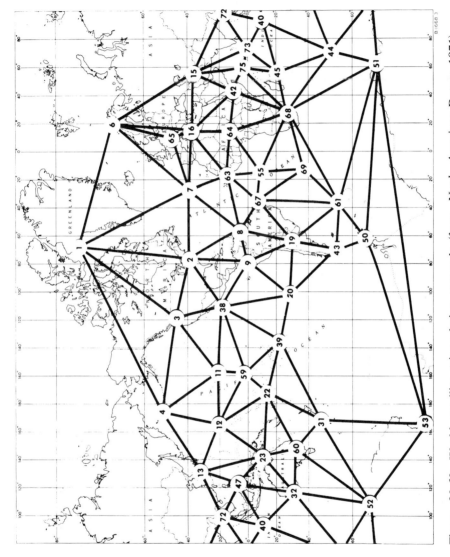

Figure 20.60 World satellite triangulation network (from *North American Datum*, 1971).

the image plane. When the laser beam leaves the Cassegrian focus, it is collimated, passed through a 0.8 mm interference filter, and reduced in cross section to cover the face of the photomultiplier tube. The range counter measures with 1 nsec resolution the time interval between the emission of the laser pulse and the detection of a return signal. The start pulse, corresponding to the laser pulse emission, is produced by a photodiode monitor located near the laser optics, and the stop pulse is the photoreceiver signal. The range counter derives its counting accuracy from a 1 MHz frequency reference supplied by the station clock. The resolution of the system is about ±0.5–0.8 m and the accuracy about ±1.0 m (*Lehr et al.*, 1969).

The AFCRL Geodetic Laser System, designed and developed for measuring distances and directions to a satellite, is based on the use of two ruby lasers and a camera (*U.S. Department of Defense National Geodetic Satellite Program*, 1975). It obtains ranges from station to satellite with a Q-switched laser and determines directions to the satellite by photographing the satellite against a stellar background with a PC-1000 camera, using pulses from a high energy, normal mode laser for illumination. The illuminating ruby laser has a pulse length of 2 msec containing about 250 J of energy at a wavelength of 6936 Å. The pulse is collimated to the width of 2 mrad between half-power limits. The two flash lamps used for pumping are water cooled and the ruby (300 mm long × 19 mm diameter) is cooled with liquid nitrogen to an operating temperature of about 150°K to avoid excessive atmospheric absorption. The maximum firing rate is once in every 3 sec because of the time needed to recharge the capacitors. Ranges are determined by using pulses from a Q-switched ruby laser capable of emitting bursts of 10 pulses per pumping period (400–500 μsec). The pulse length is 30 nsec, containing 0.5 J of energy at a wavelength of 6943 Å. The collimated beamwidth is 1.45 mrad, and the aiming method for both lasers is programmed. The pulses reflected from the satellite are collected by a 22 cm aperture, $f/2$ telescope, and are focused to the photomultiplier. The field of view is 1.7 mrad for operations at night and 1.1 mrad for operations in the daytime. Time is kept by a cesium-controlled clock whose output is compared continuously against Loran-C signals, with an accuracy of about 1 μsec. The travel time counter uses a separate crystal oscillator that is calibrated against a cesium clock, the drift being kept below ±10⁻⁹, which corresponds to an error of ±3 mm at the range of 3000 km. The PC-1000 camera with its field of view of 10° is used to photograph reflections from the satellite. Four or five images from the satellite are obtained with the camera oriented in one direction. The camera is then swung to another position to record another set of images on the same pass; ranges are measured during the entire pass. By averaging the 10 pulse bursts, the standard deviation of the range has been found to be about $m = \pm 0.45$ m; the system accuracy of this multipulse approach is estimated to be about ± 1 m.

A giant step was taken for lunar and earthbound geodetic and geophysical research when the Apollo 11 astronauts Armstrong and Aldrin placed the *lunar ranging retroreflector* (LRRR, or LR^3) on the lunar surface on July 21, 1969 (*Bender et al.*, 1973). Immediate range measurements were hampered for awhile by an initial uncertainty of the exact landing site and by weather problems. On August 1, 1969, the Lick Observatory of the University of California succeeded in obtaining strong return signals. Soon afterward the McDonald Observatory of the University of Texas also obtained signals from the Apollo 11 reflector; then observations were reported by AFCRL in Arizona and by observatories in France and Japan. The Apollo 11 LR^3 contains 100 solid fused silica corner reflectors mounted in a square aluminum panel with a side of 46 cm. Each reflector is 3.8 cm in diameter and is recessed by 1.9 cm into circular holes in the panel for thermal control. The second reflector placed on the lunar surface was built in France and was carried to the moon by the Soviet Luna 17 spacecraft in November 1970. It was mounted on the lunar exploration vehicle Lunakhod 1 and after a few months observations made at the Crimean Astrophysical Observatory in the U.S.S.R. and at the Pic du Midi Observatory in France, nothing further has been reported. Two more Apollo reflector arrays have been placed on the lunar surface, namely, the Apollo 14 array, deployed on February 5, 1971, and the Apollo 15 array, deployed on July 31, 1971. The Apollo 14 reflector assembly is similar to that of Apollo 11 with some minor modifications. The Apollo 15 array contains 300 corner reflectors packed in a compact form; its total dimension is only 104×61 cm. The McDonald Observatory noted signals from the Apollo 14 reflector the same day it was deployed and from Apollo 15 a few days after deployment. The last reflector placed on the moon was a French-built package carried by Luna 21, which landed on January 15, 1973. The reflector was mounted on Lunakhod 2, and observations were also made in the McDonald Observatory, in addition to those made in French and Soviet observatories. The three Apollo reflectors still in use are tilted so that the normal to the panel points roughly toward the earth. They form a triangle with sides 1250, 1100, and 970 km, and are separated enough from one another to provide possibilities in determining accurately the angular motion of the moon about its center of mass, separated from other parts of the problem (*Mulholland* and *Silverberg*, 1972).

The *Q*-switched ruby laser used in the McDonald Observatory has a pulse length of 4 nsec, containing about 3 J of energy at a wavelength of 6943 Å. The repetition rate is one pulse per 3 sec, and the beamwidth is about 1.2 mrad. The transit time measurement consist of two steps. Clock pulses are generated at 50 nsec intervals, and first the integral number of the 50 nsec intervals included in the total transit time must be determined. Second, highly accurate measurements must be made of the time interval between the start pulse generated by the outgoing laser pulse and the first subsequent clock pulse, and between a stop pulse generated by the photomultiplier and the following clock pulse (*Bender et al.*, 1973; *Faller et al.*, 1971). The system accuracy until late 1971 was about ±30 cm, but since that time a new calibration method has increased the accuracy to be about ±8 to 15 cm (*Bender* and *Silverberg*, 1974).

A second lunar laser ranging observatory has been constructed at the 3050 m summit of Mount Haleakala, Maui, Hawaii, by the University of Hawaii Institute for Astronomy, which opened late in 1974. The design, construction, and operation have been closely coordinated with the *Lunar Ranging Experiment* (LURE) team. Figure 20.61 is an artist's cutaway view of the east elevation of the University of Hawaii LURE Observatory. At the north end of the observatory a dome with a diameter of 9 m, made by Ash Dome Inc., houses the receiving telescope, the *Lurescope*. The Lurescope dome is connected to the main building by an enclosed hallway. The main building, containing the control and timing electronics, the laser and transmitter system, and the support facilities, is a 9×19 m prefabricated steel structure. The transmitter system is located near the south end of the observatory and is constructed on two levels (*Carter*, 1973). The lower level is a laboratory environment that houses the laser and interfacing optics, the lunastat drive and pointing electronics, and the pointing optomechanical instrumentation. The upper level enclosed within a 7 m diameter dome contains the primary optics of the refractor feed telescope and the *Lunastat* (Fig. 20.62). The isolated instrument pier extends from beneath the laboratory level to the upper level. All the optomechanical components of the transmitter system, including the laser, interfacing optics, pointing instrumentation, feed telescope, and Lunastat, are mounted on this common pier.

The instrumentation proper consists of the following main components: (*a*) the laser, (*b*) the Lunastat, (*c*) the Lurescope, and (*d*) the event timer system.

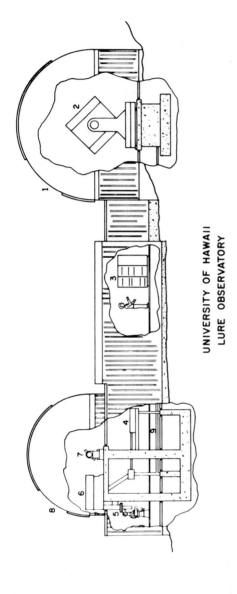

UNIVERSITY OF HAWAII
LURE OBSERVATORY

1. 9-Meter Diameter Dome
2. Lurescope
3. Control & Timing Electronics

4. Laser
5. Pointing Instrumentation
6. Feed Telescope

7. Lunastat
8. 7-Meter Diameter Dome
9. Calibration Light-Link

Figure 20.61 University of Hawaii LURE Observatory (courtesy Institute for Astronomy, University of Hawaii).

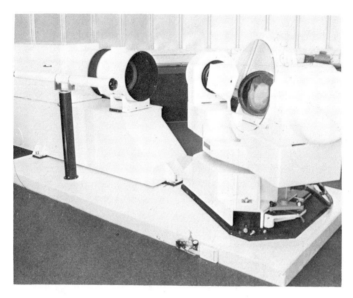

Figure 20.62 Lunastat and pointing instrumentation (courtesy Institute for Astronomy, University of Hawaii).

In the solid type neodymium-yttrium-aluminum-garnet (Nd-YAG) laser developed by GTE Sylvania, Electronic Systems Group, Mountain View, California, the parent material is the yttrium-aluminum-garnet crystal doped with neodymium. The 3.8 mm diameter × 70 mm YAG rod is pumped continuously by two tungsten-halogen incandescent lamps. After several amplification stages, the generated 0.2 nsec pulse contains approximately 0.5 J of energy at a wavelength of 10,640 Å. This infrared radiation is frequency doubled by a cesium-dideuterium-arsenate (CsD*A) frequency doubling crystal to produce the desired 5320 Å light in the green region of the spectrum. Photomultipliers are more sensitive to this light, and the required pumping energy is smaller. The laser operates at the rate of 3 pulses per second, each pulse having the pulse length of 0.25 nsec and containing 0.25 J of energy at a wavelength of 5320 Å.

The collimated laser light is routed to a diverging lens system for beam expansion, and after recollimation it is transmitted toward the Lunastat. The Lunastat was constructed by Goerz-Inland Systems Division, Kollmorgen Corporation, Pittsburgh, Pennsylvania. A 70 cm diameter, ultralightweight (25 kg), fused quartz optical flat is the principal optical component. The mounting is the altitude-azimuth type with stabilized casting and maximum symmetry of construction. The electronic logic, control, and readout units are standard items, and both manual and computer control models are available.

Faller's group at Wesleyan University, Conn. built the Lurescope (*Faller,* 1972). The Lurescope consists of a bundle of 19 cm aperture refractor telescopes that are optically coupled to form a single image. For the laser ranging, a second field stop is placed in the common focal plane. The light passing through this spatial filter is collimated and routed through an optical filter; after beam splitting, the two components are directed to two photomultipliers. The Lurescope has an altitude-azimuth mount, the vertical axis has oil bearings, and the horizontal axis has mechanical bearings.

Currie's group in the University of Maryland developed a clock system that is capable of measuring the epoch of individual events with a resolution of 100 psec and, by differencing, determine the interval between two events occurring within a few seconds of each other to approximately 200 psec. The crystal oscillator, Austron-Sulzer model 1250, provides a 10 MHz driving frequency to the event timer unit. The epoch of the event timer is set to UTC via WWVH* and Loran-C broadcast, and the frequency of the crystal oscillator is monitored and adjusted by way of Loran-C and VLF broadcasts.

The estimated resolution of one range measurement is about ±0.5–0.6 nsec in time, or about ±15 cm in distance. For a set of observations, the uncertainty in determining the mean value can be estimated as ±0.1 nsec or ±3 cm (*Carter,* 1973).

At this writing (late spring, 1975) the latest proposal to expand the lunar laser ranging program is to build transportable lunar ranging stations, which could observe for 3 to 6 weeks at chosen sites. Moving from site to site, they could monitor their geocentric positions relative to the fixed lunar laser stations with accuracies of a few centimeters in all coordinates. It is expected that such stations operating over several years could plot three-dimensional drift vectors at a large number of sites with accuracies better than one centimeter per year (*Bender* and *Silverberg,* 1974). The receiver/transmitter would be 0.8 m in diameter, the average transmitted power 1 W, and the total weight about 2300 kg. At the pulse repetition rate of 3–10 pulses per second, the pulse length would be 0.2–0.6 nsec, containing about 0.1–0.3 J of energy at the wavelength of 5320 Å.

---

* WWVH is a National Bureau of Standards operated transmitting station near Kekaha, Kauai, Hawaii.

The applications of the lunar laser ranging program are many, and in addition to such extraterrestrial investigations as those mentioned earlier, samples of the terrestrial research are given by quoting *Dr. P. L. Bender*, of the National Bureau of Standards and the University of Colorado, about the desirability of the "quasi-earth-fixed" coordinate system and on the motion of the tectonic plates:

The following criteria seem desirable for quasi-earth-fixed coordinate system:
(*a*) The system should be as invariant as possible with respect to changes in the number, distribution, and data acquisition rates of observing stations in different parts of the world. (*b*) The system should facilitate the rapid determination of sudden changes in the position or motion of individual stations, as well as sudden changes in polar motion and the rotation of the earth. (*c*) The system should be approximately fixed with respect to the mantle in some average sense .... The relative motion of the major plates has been discussed widely, and most models agree quite well. Recently, there has been considerable interest in the question of determining motions relative to the mantle .... It seems desirable to keep an open mind and to assume that some of the phenomena will behave differently than we expected based on simple models. It should be remembered that much of the information used in determining the relative motion of the plates is based on the integrated motion over periods of 100,000 years or longer.*

* From: IAU Colloquium, No. 26, Torun, Poland, August 26–31, 1974.

ELECTRONIC NAVIGATION

TWENTY-ONE

INTRODUCTION TO NAVIGATION

Navigation, one of the oldest applications of the mathematical sciences, presents two main problems: namely, determining the position of the ship at any time moment, and ascertaining its correct heading to the next position. As long as solid land is visible, these problems are easy to handle, but when the ship is at sea with only the horizon for reference, navigators are on their own. The beginning of successful navigation takes us to the era of about 1000 years B.C. when the natives of Crete sailed from their island directly to Egypt, a distance of more than 500 kms. They used the sun and the stars, knowing that when the sun was at its highest point at noon, it pointed toward the south. Of the stars, *Kochab* (β *Ursae minoris*) was the guiding star because at that time its position deviated only about 7° from the polar direction. Being unable to detect such a small rotation around the celestial pole, the navigators took this star to be the center of rotation and assumed that it pointed toward the north (*Tuori*, 1965). Consequently, the direction was satisfactorily determined, and the next step was to understand the shape of the earth. A scientifically satisfying explanation of the spherical shape of the earth was given by the Greek philosophers *Pythagoras* (ca. 580–500 B.C.) and *Parmenides* (ca. 504–450 B.C.). The famous astronomer *Hipparkhos* (ca. 150–100 B.C.) developed a method to express the location of a point on the earth with the aid of the geographic coordinates, latitude (ϕ) and longitude (λ). Figure 21.1 represents the case during the Equinox (which itself had little practical value because there were no almanacs giving the declination of the sun), and we see that the elevation of the pole equals

405

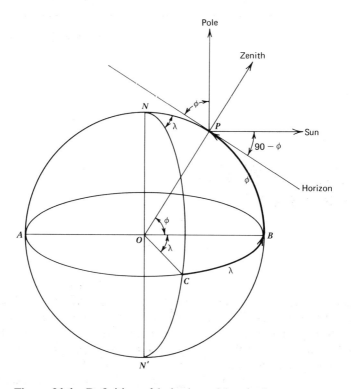

Figure 21.1   Definition of latitude and longitude.

the latitude ϕ. Shipboard measurements of the elevation of Kochab, however, were very inaccurate without proper instruments, and the longitudinal differences could not be found without clocks. *Eratosthenes* (276–195 B.C.), experimenting in Egypt, had used very primitive methods to determine the size of the earth and had succeeded surprisingly well (his value differed only about 16% from present standard length of the meridian). Thus the longitude differences were determined based on the estimated east-west travel distances, and the foundation of present celestial navigation was established.

From the days of these early attempts, a long way was still ahead even for conventional navigation. In 1569 the Dutch inventor *Gerhard Krämer*, whose Latin surname was *Mercator*, published his famous world map drawn in the projection carrying his Latin name, but the mathematical explanation and the proper tables were published by an Englishman, *Edward Wright* in 1590. The Mercator projection, which presents the constant-azimuth course of the ship as a straight line (rhumb line), is still used today in most navigation charts. Another Englishman, *John Topp* published the Seaman's Calendar in 1601; it contained the ephemerides of the moon and the sun at certain intervals and also coordinates of the most important stars. The *sextant* appeared in England in 1731, when *John Hadley,* basing his work on earlier experiments of *Hooke* and *Newton,* developed an instrument that was capable of measuring angles accurately even on a ship's unstable deck. In 1761 an English watch-maker named *Harrison* built a chronometer that was running only 15 sec slow in 156 days when tested in 1764.

At the beginning of the twentieth century, the invention and first commercial use of radio aided navigators as an accurate timing control. Under conditions of international cooperation, several transmitting sta-tions started sending time signals so frequently that chronometers could be calibrated at least once every hour. This period may be considered to be the beginning of an entirely new era in navigation history, the era of *electronic navigation.* This new kind of navigation became "self-sufficient" and could be exercised without observing celestial bodies. In its first form, which still is one of the most widely used, electronic navigation served primarily ships and aircraft from ground-based stations. Distances, direc-tions, and distance differences became the basic parameters to form different configuration modes. In navigational terminology, a *line of position* (LOP) is a locus of points established along a line of constant distances from a reference station, along a constant azimuth from a reference line, or along a line of constant differences from two stations. Thus an LOP can take several forms: (*a*) it can be circular, with the reference station at the center; (*b*) it can be a straight line, with the LOP being established by an angular relationship to a line going through the station to the north, or (*c*) it can be a hyperbola along a line of constant distance difference in relation to two ground stations. Of these three geometric modes, the circular mode is called Rho (P), the angular mode is called Theta (Θ), and the hyperbolic mode is called hyperbolic. To establish a fix, one LOP must intersect another LOP in any desired combination, such as Theta-Theta, Rho-Theta, or Theta-hyperbolic within a given system.

In the next chapters the navigation systems are presented in the following order: (*a*) ground-based systems, (*b*) airborne systems, and (*c*) satellite systems. Among the ground-based systems there is no strict line between hydrographic surveying and navigation, and since fairly extensive treatment of these systems was given in Section 20.2, emphasis is placed on long-distance and worldwide systems in the following chapters.

TWENTY-TWO

CONSOL–CONSOLAN

Consol is a Theta system giving single radial lines of position originating from a ground-based transmitting station. Developed by the Germans during World War II, it was used extensively for long-range submarine navigation under the name *Sonne*. After the war the system was modified by the British, who called it Consol; this was followed by an American adaptation, *Consolan*. Although the system is less flexible and sophisticated than most of the modern navigation methods, Consol is continuously used in Norway, Sweden, Ireland, Great Britain, and Spain, and there are Consolan stations in Nantucket, Miami, and San Francisco. The main advantage in using Consol (or Consolan) is especially important for small boat users: that is, no expensive receiving equipment is needed. The ground station sends "dot" and "dash" signals in sequence, and any standard receiver capable of being tuned to the system frequency range of about 190–516 kHz can be used for navigation.

The ground station has three antennas A, B, and C (Fig. 22.1), oriented in a straight line and usually spaced at equal intervals of three wavelengths (*Day*, 1966). By alternately changing the amplitude and the relative phase angle in the antenna currents among the three antennas, a multilobe radiation pattern is developed. The radiation pattern consists of alternate sectors of instantaneous dot signals and longer dash signals. Each sector is approximately 12° wide and is separated from the adjacent sector by constant tone or equisignal lines (*E*), produced electrically by slowly rotating the pattern in the directions given in Fig. 22.1. One cycle of the rotation lasts one minute; thus, for any given one-minute cycle, the

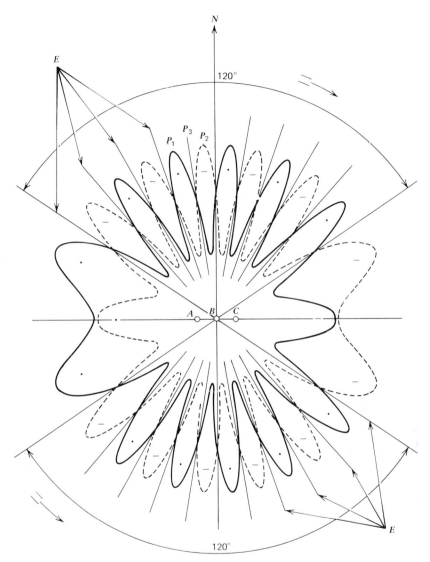

Figure 22.1 Consolan lobe patterns (from *Day*, 1966).

lobe in a given position is replaced by the one adjacent to it. There are 60 dots or dashes in each sector, and since the Consol is an aural system, the operator must count the number of dots and dashes during any one-minute cycle to establish his position to a given radial of 60 within each 12° sector. Thus an observer at P_1 (Fig. 22.1) would notice a series of dots, an equisignal, then a series of dashes. An observer at P_2 would notice a series of dashes, an equisignal, then a series of dots; and an observer at P_3 would note an equisignal, a series of dots, then another equisignal.

In using the Consol system the navigator must follow a few simple steps to avoid ambiguities in counting the signals. The equisignal line will not be sharply defined, and it may appear to be several seconds long when the blending of two sectors is occurring. This may cause loss of signals during the blending period, and the correct number is obtained by the following procedure:

1. Count the number of dashes or dots heard *before* the equisignal tone.
2. Count the corresponding number of signals heard after the equisignal tone.
3. Add the total dots and dashes heard on each side of the equisignal tone and subtract from 60.
4. To find the correct radial in that sector, take half the difference obtained in step 3 and add it to the amount obtained in step 1.

For example, the navigator may have counted 13 dashes, equisignal, and 41 dots. The simple rules can be put into the form of the following expression:

$$Da_c = Da + \frac{60 - (Da + Do)}{2} \tag{22.1}$$

where Da_c is the correct number of dashes, Da is the counted number of dashes, and Do is the counted number of dots. The substitution of the given values into Eq. 22.1 yields $Da_c = 16$, which means that the navigator's position is along the 16-dash radial of the dash sector in which he happens to be (Fig. 22.2). Two conclusions about position fixing in the Consol system can now be made. First, the navigator must know his position closely enough to be able to identify the dash or dot sector he is in. Second, to obtain a fix he must have two sets of radials available from two separate ground transmitters. Otherwise, the Consol serves only as a homing device.

The effective range at which reliable reception can be obtained depends on the receiver noise level caused by noise within the circuitry and atmospheric noise during transmission. With experience, the operator learns to determine carrier wave dots and dashes at distances from which voice communications would be completely unintelligible. The accuracy of the system depends greatly on the geometric considerations of the lobe patterns. Although each sector includes equal amount of dots and dashes,

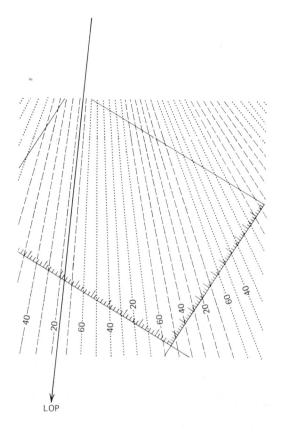

Figure 22.2   Fixing LOP = 16 in Consolan lattice chart (from *Day*, 1966).

the width varies as Fig. 22.1 indicates. The width of a sector is narrowest, about 12°, in the direction perpendicular to the base formed by the antennas, thus giving the best accuracy, but when approaching the direction of the base, the sector width increases rapidly, making the direction determination invalid. Also, close to the transmitting stations the sector width becomes so narrow that great ambiguity exists in finding the correct sector. Within these limitations it can be said that the effective range of either Consol or Consolan is between about 50 and 2500 km. The accuracy of the system on daytime is about ±0.3° and at night about ±0.7°.

TWENTY-THREE

THE DECCA NAVIGATOR SYSTEM

The *Decca Navigator System* manufactured by the Decca Navigator Company, Ltd., London, is very similar to Survey Decca (Section 20.2.2.1) in principle and in operation. The main differences lie in the number of slave stations, the radiated power, and the use of lane identification in the navigation system. The system is hyperbolic-hyperbolic, and the first chain became operational in 1946; subsequently ten chains were added in Europe, four in Canada, and one in the New York area, giving good coverage throughout the waters of West Europe and the eastern Atlantic seaboard. With two chains in the Persian Gulf, one in Bombay, and one in Calcutta, Decca is also operative in the Eastern Hemisphere. The Decca Navigator System, which is available on a rental basis was approved internationally by the second meeting of the International Marine Conference on Radio Aids to Navigation in May 1947.

The conventional Decca chain consists of three slave transmitting stations (red, green, and purple) symmetrically located around the common master station. The baselines joining the master station to the slave stations vary in length according to the requirements, but best results are obtained with baselines between 130 and 200 km long. To get the fix, one selects two stations giving the best intersection angle of hyperbolas, using the third reading as a check and as a redundancy to improve accuracy. At the beginning of the voyage or when entering the coverage area of a Decca chain, the decometers are first set by hand to the correct values; thereafter they count the LOPs traversed by the ship. To indicate the

412

appropriate lane identification scale to be illuminated, telling the user which one to read. To increase the accuracy of the lane identification, a $6f_0$ signal is also extracted successively from the slave transmissions and is compared with the master in a $6f_0$ discriminator. If fed straight to a decometer movement, the $6f_0$ pattern would give rise to six pointer rotations per zone. This rotation is indicated by a six-arm pointer that appears within the fan-shaped sector pointer, indicating the correct lane reading.

Of the transmitting stations, the master station has very high frequency stability of about 10^{-6} and the slave stations' phase-locking circuits have stability of about 5 nsec. The functions of the master station are to transmit normally a frequency-stabilized CW signal, to initiate the necessary changes in transmission and receiver circuit connections just before the lane identification transmission, and to bring up a second transmitter during lane identification. The slave stations normally transmit signals on their respective frequencies and keep these transmissions automatically phase locked to the master signal. Also, on reception of the triggering signals from the master, the slave stations either are turned off or transmit on extra frequencies for lane identification as required. Figure 23.1 shows the normal operation and the regrouping of frequencies at the ends of the baselines during the half-second period of lane identification performed for each color at 20 sec intervals. It is seen that for lane identification two transmitters are required at each station except at the purple station, where three transmitters are necessary. Power supply requirements of the transmitter and essential technical equipment are met by generating sets embodying both AC and DC drive motors. In the event of AC mains power failure, the generating sets are switched automatically to DC drive, deriving power from a storage battery. The Decca Navigator System uses guyed lattice masts more than 100 m tall; an extensive copper wire earth system radiates from the mast and contains a total length of wire of about 10 kms. The radiated power varies according to the needs and can be up to 800–1000 W.

In 1964 the Decca Navigator Company introduced a new receiver, *Mark 12*, which uses the *multipulse system*. This system entails the transmissions of additional composite signals from the Decca ground stations. The increase in radiated information is the basis of the system, and pulses derived from these signals are formed within the receiver. Their phase difference is directly related to the ship's position. The technique ensures that the accuracy of the pulse is little affected by adverse radio conditions; thus the navigator has a more accurate meter display of both lane identification and position line information. With the

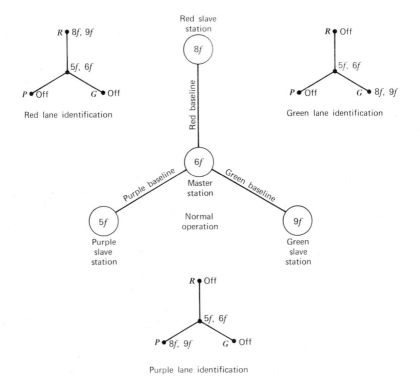

Figure 23.1   Lane identification principle of Decca navigator system (courtesy Decca Navigator Co., Ltd.).

multipulse system, the rate of lane identification signals from Decca transmitting stations is trebled, the full sequence of red, green, and purple signals now occurring three times in a minute. The lane identification meter has the accuracy to give an independent position fix, and the change from one chain to another or from one chain to a position line of another chain can be made by a turn of the chain switch. No referencing is necessary when changing the chains. The range of the Decca Navigator system is about 500–600 km with an accuracy of about ±5 m on the baseline, or an average of ±30–50 m in navigation at the maximum range.

Dectra (for *Dec*ca *tra*cking and *ra*nging: also called *Trunk Route Decca*) is a modified Rho-Theta system developed by the company as a navigational aid especially for North Atlantic air traffic. In its original form, Dectra consisted of a pair of transmitting stations located in Newfoundland, combining a tracking pattern of hyperbolic position lines whose axis ran approximately east and west. Position lines running approximately north and south were generated by one of two stations in Canada in conjunction with a third station in the United Kingdom. The receiving equipment in the aircraft detected these two patterns and derived a position fix from them. The revised Dectra, in operation since 1965, employs a double-ended tracking system, which means that the tracking pattern is also radiated from the United Kingdom. This arrangement substantially increases the accuracy of the overall tracking information in that the range over which either pattern is used is only half the total route distance. The revised system also provides a means of coarse ambiguity resolution, since it is possible to switch the receiver from pattern to pattern and compare results.

The tracking pattern (Fig. 23.2) comprises a family of constant phase-difference hyperbolic position lines focused on the originating master station A and slave station B on the west side of the Atlantic. Station A transmits a stable continuous wave signal at a frequency of 70 kHz with a brief break every 10 sec. During each break in the transmission from A, the slave station B sends a signal of the same frequency and is phase locked to A by a circuit capable of memorizing the phase of the master signal during the slave transmission. Since there is a break in the signal (time-sharing principle), respective signals are readily recognized by the receiver, and there is no reason to use separate frequencies as in the Decca Navigator system. Measurements of the phase difference between the two signals in the airborne receiver serves to locate the aircraft on one of the hyperbolic lines. The tracking also can be done by using the pair of stations C and D on the eastern side of the Atlantic, depending on which pair is closer. The ranging pattern of position lines is generated by the slave station C, located in the United Kingdom, and the westerly master station A. In the ranging process the frequency of the slave transmission differs slightly from that of the master transmission. The difference frequency, about 150 Hz, is used for phase comparison, whereas the actual lanewidth or range interval equals one-half-of the slave transmission frequency. The maximum usable distance is about 3000 km, and based on actual testing (*Kelly*, 1963), the errors in the tracking and

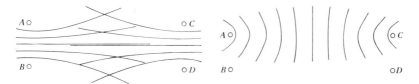

Figure 23.2 Dectra ranging and tracking patterns (Laurila, 1960).

ranging were found to be less than 15 km in the worst location on the test line connecting Shannon, Ireland, to Gander, Newfoundland. The major source of errors is the anomalies in the propagation velocity of electromagnetic radiation. This is not a critical factor from the standpoint of air safety, since all aircraft flying in mutual proximity are affected by this source of errors, and the relative positioning between the aircraft can be determined accurately. When approaching either terminal of the route, the absolute accuracy naturally increases.

TWENTY–FOUR

LORAN-A

Loran is a pulse-type hyperbolic-hyperbolic navigation system developed during World War II. *Loran-A*, still used today, is the predecessor of the more accurate and efficient Loran-C described in the next chapter. During the early 1940s tests were made to find the band of frequencies that gives the best ground and sky wave propagation for use with a system like Loran, which relies on both direct waves and reflected waves from the ionospheric layers. It was found that the lower frequencies produced the most stable sky wave signals at night, whereas the higher frequencies gave more stable sky wave reflections by day. These tests made it obvious that the medium frequencies offered the best compromise solution for long-range navigation performed both at night and in the daytime.

At present Loran-A transmitters employ radio frequencies between 1700 and 2000 kHz; this range of frequencies is known as the *Standard Loran band*. Transmitters radiate pulses 50 μsec long, with a peak power of 100 kW, which gives a ground wave range over seawater of about 1000 km, the corresponding range over land at the earth's surface being only about 200 km. At night the range can be increased by employing the reflected waves from the ionospheric E- and F-layers. Since the F-layer varies greatly in thickness and density of ionization, only the sky waves reflected from the E-layer are reliable enough in their persistence and timing to be used in Standard Loran. When the sky wave technique is used, the distance traveled by the reflected waves is clearly longer than the corresponding ground wave distance; thus the sky wave transmission

419

time is longer than the ground wave transmission time. The single pulse transmitted by the ground station transmitter consequently is received at the surveying vessel as two pulses, the sky wave component coming after the ground wave component by an interval of time called the *sky wave transmission delay*. If the elevation of the E-layer is considered to be constant, the sky wave transmission delay is a function of the "ground wave distance" between the transmitter and the receiver.

The greater the ground wave distance, the less the sky wave distance exceeds it. This phenomenon means that the standard error caused at longer distances by erroneous estimation of the elevation of the E-layer for the determination of the sky wave transmission delay is smaller than at shorter distances. At night the range obtained by using the sky waves over seawater can be as great as about 2500 km. Although it is helpful to be able to measure long distances with a reasonable accuracy using sky waves, the technique has the following disadvantages. The use of sky waves requires corrections to the precomputed ground wave charts and tables. The useful pulse reflected from the E-layer is followed by several useless and anomalous pulses (reflections from F-layer and multiple reflections from both of the layers), which introduce uncertainty into the identification of the desired pulse. Furthermore, the inhomogeneity of the E-layer distorts the shapes of the reflected pulses, making them difficult to match.

Loran-A stations are often installed as a chain consisting of several stations along a coastline. The optimum baseline length when stations are separated by seawater is found to be about 500 km. If stations are separated by land, the maximum baseline length is about 200 km. Since Loran-A is primarily designed for long-range navigation to cover large areas (e.g., the oceans), the network along the coastlines must be extended over several thousand kilometers. Therefore several independent pairs of stations must operate simultaneously, and means for unambiguous pair identification must be provided. Although all Loran-A stations operate in the same band of radio frequencies, between 1700 and 2000 kHz, they do not use the same frequency. Four different frequencies or channels are used, as follows: channel 1, 1950 kHz; channel 2, 1850 kHz; channel 3, 1900 kHz; and channel 4, 1750 kHz. Each channel consists of eight pairs of stations, which are identifiable by a special arrangement of pulse recurrence rates (timing frequencies). Four of the pairs operate at a so-called *low basic rate* (about 25 Hz) and the other four at a *high basic rate*, (about 33.3 Hz). The four pairs operating with the same basic rate are, in turn, separated from each other by the so-called *specific rates*. Under this arrangement one of the four pairs operates exactly at the basic rate and the other three pairs operate with specific rates differing from each other by about 0.25% in the low rate group and about 0.33% in the high rate group.

The shipboard Loran-A operator first selects the channel that gives the best geometric properties. This calls for a special chart made for Loran-A navigation systems and used by tuning the receiver. He next selects the most suitable pairs of stations by synchronizing the timing frequency of the time base of the CRT to correspond to the basic and specific recurrence rates of the pairs in question. The pulse coming in at the correct recurrence rate appears stationary on the time base, while the pulse coming in at a different basic rate can easily be identified and ignored because it makes fast jumps over the time base. Pulses coming in at the same basic rate but at different specific rates drift slowly to the right or to the left along the time base, depending on whether the wrong specific rates are lower or higher than the correct specific rate. Pulses coming in at the same basic and specific rates appear stationary and are moved only by the movement of the vessel.

Since all pairs of stations operate with different specific recurrence rates, it is obvious that special arrangements must be made for the master station, which is a mutual member of two pairs of stations. This is done by making the master station transmit in a so-called *double pulsing* fashion, which means that the master station sends out two independent sequences of pulses, one slightly more closely spaced than the other. The pulses must be triggered by different timers but may be sent out by the same transmitter.

The average position error close to the baselines in daytime when using ground waves is about ±270 m, reaching the value of about ±2 km at a distance of 1200 km. At night, using sky wave propagation at distances between 500 and 2500 km, the average position error ranges between ±2.5 and ±14 km, respectively.

TWENTY-FIVE

LORAN-C

Walter J. Senus

Loran-C is a highly accurate position-determining system that uses the difference in the time of arrival of radio frequency pulses broadcast by three or more transmitting stations. In the 35 years since work began on long-range pulsed navigation systems, major changes have taken place.

In 1940 the U.S. National Defense Research Committee was assigned to develop a long-range precision navigation system to fill the need for all-weather trans-Atlantic convoy navigation (*Hall*, 1947). This initial effort produced the current Loran-A navigation system, which operates in the 1700–2000 kHz frequency band and provides time difference information by means of pulse envelope matching techniques. It was recognized early in the initial Loran program that a low frequency (LF) Loran system would provide greater accuracy and extended navigation coverage. This development fostered the experimental system (LF Loran), which was placed in operation in 1945 (*Jansky* and *Bailey*, 1962). This system used a pulsed signal at 180 kHz to derive navigation fixes. Early experience with this system was good, but testing of the system was terminated at the close of World War II with the cessation of immediate military requirements (*Hefley*, 1972).

A later system, *Cyclan*, appeared shortly thereafter and featured the added sophistication of automatic cycle matching within the pulse envelope. The system provided a two-frequency environment at 160 and 180 kHz. Further development of the Cyclan system stopped when the

Atlantic City Radio Conference (1947) designated the 90–110 kHz band for the development of long-range navigation systems; Cyclan required a total bandwidth of some 40 kHz.

In 1952 work began on the *Cytac* system, which was an automatic long-range bombing system (*Dean et al.*, 1956). The radio navigation portion of the system operated in the 90–110 kHz frequency band and provided time difference information through radio frequency (RF) cycle detection. Tests on Cytac continued until 1955 and demonstrated highly accurate fix information over a very large geographic area.

With the advent of strong military requirements in 1957, the Cytac navigational subsystem was resurrected and renamed Loran-C. This is the system available today, in a much expanded configuration.

25.1. GENERAL DESCRIPTION

The Loran-C navigation system is a low frequency, pulsed signal, hyperbolic position-fixing system; it operates at 100 kHz. The system is composed of a series of chains, each one consisting of a master station and two or more secondary stations separated by distances of up to 2000 km. Currently there are three operational chains in the United States, one on the eastern Atlantic Coast, one in Alaska, and one in Hawaii. Additional chains are located in the northwestern Pacific Ocean, the Mediterranean Sea, the Norwegian Sea, the northern Atlantic Ocean, the central Pacific Ocean, and in Southeast Asia (*Hefley*, 1972). To complement these already established Loran-C chains, the Coast Guard is actively engaged in establishing chains on the west coast of the United States and the Gulfs of Alaska and Mexico *Polhemus*, 1972). These chains will be operational in the late 1970s. Figure 25.1 shows the location of the current and proposed Loran-C chains. Table 25.1 contains salient characteristics for all the existing Loran-C transmitting stations.

A Loran-C line of position is determined by measuring (*a*) the difference in time of arrival of synchronized pulsed signals from a master and secondary transmitting station, and (*b*) the difference in phase of the synchronized 100 kHz carrier within the master and secondary pulses. Figure 25.2 describes a typical Loran-C pulse. The transmission format consists of a group of eight pulses (the master station has nine) transmitted in sequence from all stations in the chain. Figure 25.3 is an example of a typical chain's transmission format. Within the Loran-C receiver, LOP determination is done in two steps. First a coarse determination of position is obtained by establishing a sampling point on the envelope of

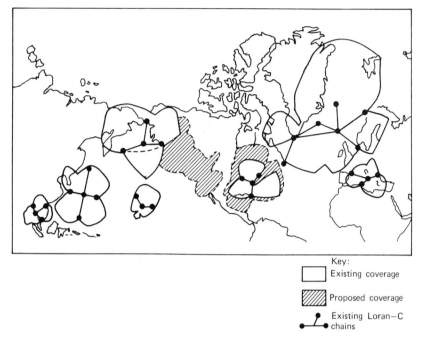

Key:
☐ Existing coverage

▨ Proposed coverage

●—●—● Existing Loran—C chains

Figure 25.1 Current and proposed Loran-C coverage. (courtesy United States Coast Guard).

each pulse and measuring the time difference between sampling points. Second, a fine indication of position is obtained by measuring the difference in phase of the 100 kHz signals at the sampling point. The final time difference reading is formed by adding the fine cycle reading to the coarse envelope reading.

At the transmission frequency of 100 kHz, one cycle of RF energy represents 10 μsec; this cycle ambiguity results in baseline lanewidths of approximately 1.46 km. However, the ambiguity is resolved automatically by the receiver as long as the envelope reading is correct to within ±5 μsec. This technique of determining the proper cycle for the fine measurement is referred to as envelope derivation (*Dean et al.*, 1956). The user does not have to count lanes or know his initial position, as in other phase-measuring systems. Herein lies one of the major advantages of Loran-C over other electronic navigation aids.

Loran-C is primarily a ground wave system, and since it is a pulsed system, the sky wave is easily separated in time from the ground wave. By sampling near the beginning of the received pulse (the third cycle crossing

TABLE 25.1

Station	Coordinates Latitude and Longitude	Station Function	Coding Delay and Baseline Length (μsec)	Radiated Peak Power
U.S. East Coast Chain—Rate 9930 [old rate SS7 (99,300 μsec)]				
Carolina Beach, N.C.	34-03-46.50N 77-54-47.29W	Master		1.0 MW
Jupiter, Fla	27-01-58.85N 80-06-53.59W	W Secondary	11,000 2695.51	400 kW
Cape Race, Newfoundland	46-46-31.88N 53-10-29.16W	X Secondary	28,000 8389.57	2.0 MW
Nantucket, Mass.	41-15-12.29N 69-58-39.10W	Y Secondary	49,000 3541.33	400 kW
Dana, Ind.	39-51-08.30N 87-29-12.75W	Z Secondary	65,000 3560.73	400 kW
Electronics Engineering Center, Wildwood, N.J.	38-56-58.59N 74-52-01.94W	T Secondary	82,000 2026.19	200–400 kW
North Atlantic Chain—Rate 7930 [old rate SL7 (79,300 μsec)]				
Angissoq, Greenland	59-59-17.19N 45-10-27.47W	Master		1.0 MW
Sandur, Iceland	64-54-26.07N 23-55-20.41W	W Secondary	11,000 4068.07	1.5 MW
Ejde, Faroe Islands	62-17-59.64N 07-04-26.55W	X Secondary	21,000 6803.77	400 kW
Cape Race, Newfoundland	46-46-31.88N 53-10-29.16W	Z Secondary	43,000 5212.24	2.0 MW
Mediterranean Sea Chain—Rate 7990 [old rate SL1 (79,900 μsec)]				
Simeri Chichi, Italy	38-52-20.23N 16-43-06.39E	Master		250 kW
Lampedusa, Italy	35-31-20.80N 12-31-29.96E	X Secondary	11,000 1755.98	400 kW
Kargabarun, Turkey	40-58-20.22N 27-52-01.07E	Y Secondary	29,000 3273.23	250 kW
Estartit, Spain	42-03-36.15N 03-12-15.46E	Z Secondary	47,000 3999.76	250 kW

425

TABLE 25.1 Continued

Station	Coordinates Latitude and Longitude	Station Function	Coding Delay and Baseline Length (μsec)	Radiated Peak Power
Norwegian Sea Chain—Rate 7970 [old rate SL3 (79,700 μsec)]				
Ejde, Faroe Islands	62-17-59.64N 07-04-26.55W	Master		400 kW
Bo, Norway	68-38-06.55N 14-27-48.46E	X Secondary	11,000 4048.16	250 kW
Sylt, Germany	54-48-29.24N 08-17-36.82E	W Secondary	26,000 4065.69	400 kW
Sandur, Iceland	64-54-26.07N 23-55-20.41W	Y Secondary	46,000 2944.47	1.5 MW
Jan Mayen, Norway	70-54-51.63N 08-43-56.57W	Z Secondary	60,000 3216.20	250 kW
Southeast Asia Chain—Rate 5970 [old rate SH3 (59,700 μsec)]				
Sattahip, Thailand	12-37-06.91N 100-57-36.58E	Master		400 kW
Lampang, Thailand	18-19-34.19N 99-22-44.31E	X Secondary	11,000 2183.11	400 kW
Con Son, Republic of Vietnam	08-43-20.18N 106-37-57.39E	Y Secondary	27,000 2522.07	400 kW
Tan My, Republic of Vietnam	16-32-43.13N 107-38-35.39E	Z Secondary	41,000 2807.28	400 kW
North Pacific Chain—Rate 5930 [old rate SH7 (59,300 μsec)]				
St. Paul, Pribiloff Is., Alaska	57-09-12.10N 170-15-07.44W	Master		400 kW
Attu, Alaska	52-49-44.40N 173-10-49.40E	X Secondary	11,000 3875.17	400 kW
Port Clarence, Alaska	65-14-40.35N 166-53-12.95W	Y Secondary	28,000 3068.97	1.8 MW
Sitkinak, Alaska	56-32-19.71N 154-07-46.32W	Z Secondary	42,000 3284.83	400 kW
Northwest Pacific Chain—Rate 9970 [old rate SS3 (99,700 μsec)]				
Iwo Jima, Bonin Is.	24-48-04.22N 141-19-29.44E	Master		3.0 MW

TABLE 25.1 Continued

Station	Latitude and Longitude	Station Function	Coding Delay and Baseline Length (μsec)	Radiated Peak Power
Marcus Is.	24-17-07.79N 153-58-53.72E	W Secondary	11,000 4284.11	3.0 MW
Hokkaido, Japan	42-44-37.08N 143-43-10.50E	X Secondary	30,000 6685.12	400 kW
Gesashi, Okinawa,	26-36-24.79N 128-08-55.99E	Y Secondary	55,000 4463.24	400 kW
Yap, Caroline Is.	09-32-45.84N 138-09-55.05E	Z Secondary	75,000 5746.79	3.0 MW
Central Pacific Chain—Rate 4990 [old rate S1 (49,900 μsec)]				
Johnston Is.	16-44-43.85N 169-30-31.63W	Master		300 kW
Upolo Points, Hawaii	20-14-50.24N 155-53-08.78W	X Secondary	11,000 4972.38	300 kW
Kure, Midway Is.	28-23-41.11N 178-17-29.83W	Y Secondary	29,000 5253.08	300 kW

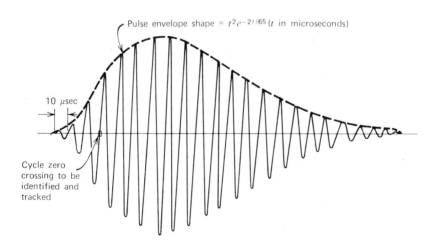

Figure 25.2 The Loran-C pulse. (courtesy United States Coast Guard).

Pulse envelope shape = $t^2 e^{-2t/65}$ (t in microseconds)

10 μsec

Cycle zero crossing to be identified and tracked

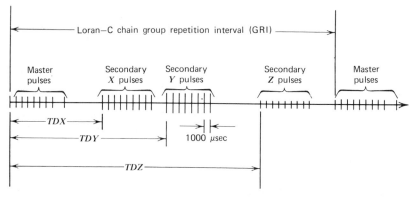

Figure 25.3 Loran-C chain transmission format. (courtesy United States Coast Guard).

point or the 30 μsec point) the Loran-C receiver is able to resolve the delayed sky wave energy from the ground wave energy, allowing navigation solely on the stable uncontaminated ground wave to distances of approximately 2000 km (*Jansky* and *Bailey*, 1962).

To reduce sky wave contamination, within each of the multipulse groups from the master and secondary stations the phase of the RF carrier is changed systematically with respect to the pulse envelope from pulse to pulse. The phase of each pulse in an eight- or nine-pulse group is changed in accordance with a prescribed code and is either in phase 0° or out of phase +180° with respect to the stable transmitter reference oscillator signal. The phase code used at a master station is different from the code used at a secondary.

To minimize simultaneous reception of signals from all transmitters in a Loran-C chain, the groups of similarly shaped RF pulses from the paired stations are time spaced. The spacing is achieved by making the secondary stations wait a prescribed time after the master station transmits its signal. This minimum time, the "coding delay," is predetermined by system propagation times and equipment characteristics. With all the foregoing considerations, the equation for time difference *TD* becomes

$$TDA = \frac{D_A - D_M}{c_p} + CDA + T_c \qquad (25.1)$$

where $c_P = 100$ kHz phase velocity
D_A = distance to transmitter *A* (secondary)
D_M = distance to transmitter *M* (master)
CDA = coding delay for secondary *A*
T_c = propagation prediction error

Particular groups of stations are identified by providing a different group repetition interval or rate for each chain group. Chains are selected by their group repetition intervals (GRIs); for example, one chain transmits a pulse recurrence interval of 9990 μsec and another transmits at 5970 μsec. To select a particular chain, the operator sets a receiver to synchronize on the desired pulse repetition frequency (PRF). This kind of inherent Loran interference is referred to as cross-rate interference.

In addition to the information contained in Table 25.1, other facts are fundamental to understanding and using the Loran-C system. For chart preparation, signals are assumed to propagate with a phase velocity $c_p = 2.997942 \times 10^8$ m/sec and in an index of refraction $n = 1.000338$ for standard atmosphere. The phase of the ground wave is discussed in a later section. Seawater conductivity ($\sigma = 5.0$ ℧/m) is used in calculating baseline lengths. Geodetic distances are calculated with reference to the Fischer spheroid (1960), with an equatorial radius $a = 6,378,166.0$ m, a polar radius $b = 6,356,784.283$ m, and a flattening $f = (a - b)/a = 1/298.3$ (*Bomford*, 1962).

Inquiries pertaining to the Loran-C system can be addressed to:

Commandant (G-WAN/3)
U.S. Coast Guard
400 Seventh Street S.W.
Washington, D.C. 20590

25.2. ACCURACY

Position accuracies in the Loran-C system are functions of three parameters: (1) geometrical considerations, (2) our lack of total understanding of the 100 kHz ground wave propagation phenomena, and (3) instrumentation errors in the transmitting and receiving systems.

The geometrical considerations are identical to those discussed earlier for other baseline radio navigation systems such as Decca. The standard deviation σ in microseconds is primarily determined by the instrumental and the prediction uncertainty factors. It varies from chain to chain and is further influenced by the signal-to-noise ratio in the service area.

In the construction of Loran-C charts, as well as in the automatic, self-contained, computerized Loran-C receiver, calculations are based on station location on the earth's surface and assumptions concerning ground conductivity. A later section deals only with the recent developments in the description and prediction of Loran-C ground waves. We simply state here that worst-case prediction inaccuracies can be up to 3 μsec in magnitude. The ideal propagation conditions exist over all seawater paths, and the higher phase errors as specified earlier are normally experienced over irregular, inhomogeneous terrain propagation paths (*Johler*, 1975).

Instrumentation errors in transmitting stations and user receiving equipment provide a portion of the resultant time difference inaccuracy. In a radio frequency noise-free or low RF noise environment, these errors are 10^{-2} μsec in magnitude, therefore negligible. However, as the signal-to-noise ratio decreases below 0 dB, the standard deviation of cycle time difference will increase at least one order of magnitude. Transmitter stabilities for Loran-C chains are between ± 0.05 and ± 0.2 μsec. This parameter is a function of chain transmitting and timing equipments and the overall operational control of the chain (*Polhemus*, 1972).

25.3   INSTRUMENTATION

All Loran-C transmitting stations consist of a high power transmitter, a timer/synchronizer section, and an antenna subsystem. A simplified functional block diagram of a transmitting station appears in Fig. 25.4. An atomic oscillator (cesium) is the basic timing unit of most stations. The oscillator output is applied to a gated amplifier and then to the power amplifier and transmitting antenna.

The Coast Guard has recently initiated a program directed toward providing all existing and future Loran-C transmitting stations with the new solid state AN/FPN-54 Colac (combined Loran A, C) timer/synchronizer hardware. This will improve system stability at all chains having the equipment. The most common Loran-C transmitter is the AN/FPN-44, which is manufactured by the International Telephone and Telegraph Corporation. This transmitter provides 400 kW of radiated power from a 190 m top-loaded, vertical monopole antenna. Another transmitter, the AN/FPN-45, will provide in excess of 3 MW of radiated power when used in conjunction with a 411 m vertical monopole antenna system.

In future Loran-C stations the antenna system will be of a new design, the sectionalized Loran transmitting (SLT) antenna system. The prototype system is currently at the Carolina Beach master transmitting station. The system is composed of four 190 m towers arranged as support elements for the actual radiating elements. The radiating elements approximate an inverted pyramid whose apex is centered above the transmitter building, located at the center of the square formed with the

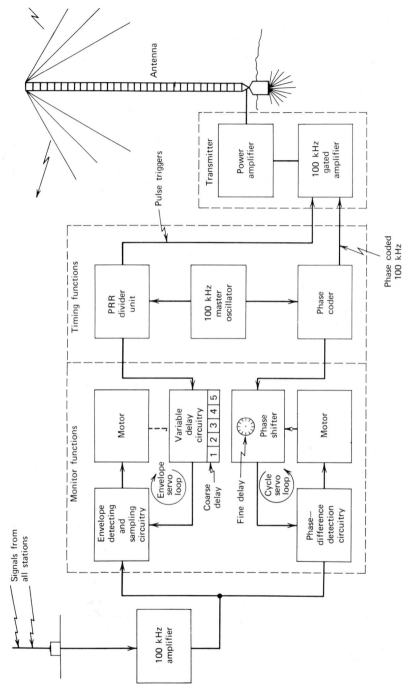

Figure 25.4 Loran-C transmitting station (adapted from "The Loran-C System of Navigation," Atlantic Research Corporation, 1962).

towers in its corners. The AN/FPN-44 will provide more than 1 MW of radiated power when used with the SLT antenna (*Polhemus*, 1972).

Figure 25.5 is a basic block diagram of a hard limited Loran-C receiver. This very popular processing technique in modern Loran-C receiver design was prompted by the decreasing cost of the medium-scale integrated (MSI) circuit read-only memories (ROMs) used for processing control.

The signal received at the antenna is amplified at the antenna preamplifier and sent to the receiver RF section. Here it is processed by a notch filter group that eliminates harmful in- and near-band interference. The notch filter output is applied to a bandpass filter that eliminates RF noise.

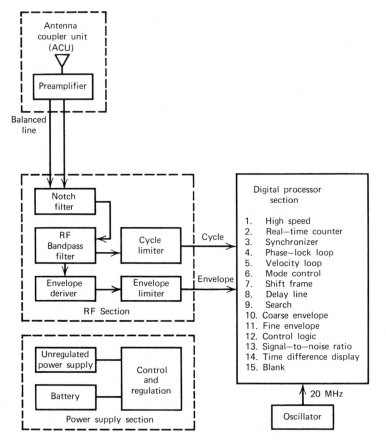

Figure 25.5 Hard limited Loran-C receiver.

At this point the signal is split into two channels, the envelope channel and the cycle channel. The processor takes the output of these two channels and measures the time difference between arrival of the master signal and each of the two or more desired secondaries. If needed, second-order phase-lock loops are provided to correct for vessel dynamics and oscillator drift.

Most Loran-C receivers can process signals that differ in amplitude by as much as 80 dB in a signal-to-noise environment of -10 dB and in the presence of sky waves with amplitudes four times greater than the desired ground wave signal (*Hassard*, 1973). Loran-C receivers have been built in varying degrees of complexity. The highly sophisticated units like the AN/ARN-92 automatic airborne receiver provide automatic pilot and range-range functions, whereas smaller units supply simple time difference digital displays. The user must choose the equipment best suited for his mission.

25.4. PROPAGATION OF LORAN-C ELECTROMAGNETIC ENERGY

The propagation of Loran-C-type signals has been the subject of extensive investigation ever since the initiation of the Cytac program. Most of the theoretical and experimental work in this field has been done by Johler and his associates at the Department of Commerce Laboratories in Boulder, Colorado (*Johler*, 1961, 1963, 1970). A fundamental paper in the field is NBS Circular 573 (*Johler*, 1956). The idealized model is one in which the omnidirectional radiation pattern from the transmitting station is propagated down the spherical wave guide formed by the earth's surface and the lower E- or D-region of the ionosphere. The propagating wave is modeled by the superposition of optical rays emanating from the transmitting source. The ground wave is of immediate interest and is easily separated from the sky waves by this type of solution, allowing detailed examination. At distances of up to 2000 km from the transmitter, the Loran-C wave is all ground wave (*Johler*, 1961). Since this physical reality exists, the resultant wave is sensitive to the geophysical parameters on the earth's surface over which it propagates. Among these are the earth's curvature and ground conductivity. On the other hand, the resultant radiation is rather insensitive to sunrise and sunset effects, which originate in the ionosphere. The ionosphere affects the sky wave, however, and this is why the sky wave is not used for precise navigation.

The computation of a Loran-C time difference grid is based on station locations on the earth and assumptions about ground conductivity and the index of refraction along the path. If stations were located on a perfect sphere having a surface of uniform conductivity, propagation time differences for arrival of signals at a point could be calculated with a high degree of accuracy. These ideal conditions are at best approximated when all propagation paths are entirely over seawater. In reality, propagation paths are either partially or entirely over irregular, inhomogeneous terrain. Computations based on seawater paths can lead to errors between observed and actual positions ranging up to 3 μsec in magnitude (*Johler*, 1975).

Loran-C signal phase prediction is further complicated by changes in index of refraction and ambient weather conditions. These sources of error generate small contributions to the overall error attributed to propagation prediction and are usually ignored in practice. To obtain complete insight into the theory of propagation of 100 kHz Loran-C transmissions, the reader is referred to the references dealing with the subject. Of fundamental importance is NBS Circular 573, and anyone interested in detailed investigations in this area should become acquainted with this paper.

25.5 DIFFERENTIAL LORAN-C AND LORAN-D

The differential Loran-C technique is identical to that described for differential Omega. In differential Loran-C fixed-monitor receivers are located in the geographic areas in which accuracy improvement is desired. These monitor receivers measure any deviation in the time difference from the previously established long-term average time differences at their geodetic positions. Deviations in the readings are caused by variations in the transmission path characteristics and the transmitter timing stability. The amount of variation from the long-term average is used as a differential correction for a nearby area of interest. This differential correction is radio transmitted to all users in the nearby area and applied by the users to their respective receiver time difference readings. Tests conducted on the East Coast Loran-C chain showed fix accuracies of 12.2 m drms (95%) and 4.6 m cep* (50%) (*Goddard*, 1973). These extremely good results were demonstrated over separations of up to 140 km from the nearest monitor site.

---

* Circular error probable.

Loran-D is a lower power, shorter range, modified but compatible version of Loran-C, using transportable solid state transmitting equipment. By using shorter baselines (<1000 km) and groups of 16 versus 8 pulses, the Loran-D system is able to attain accuracies of ± 0.04 μsec rms in its grid stability (*Senus*, 1972). An excellent up-to-date description of the Loran-D system is given in a recent article (*Frank*, 1974). Although primarily a military system, Loran-D offers substantial potential for nonmilitary applications.

TWENTY-SIX

OMEGA

Walter J. Senus

Omega is a worldwide radio navigation system capable of providing hyperbolic position fixing by phase comparison techniques on very low frequency (VLF) continuous wave electromagnetic signals. The environment is a group of eight land-based transmitting stations. The radiation from these stations provides usable information to a variety of land, sea, and airborne users.

Omega was developed to satisfy the need for a worldwide, all-weather continuous navigation system. These requirements have never been satisfied fully, however. Celestial navigation is not all-weather. Other electronic systems do not give global coverage. Inertial systems are expensive and limited in accuracy, and they degrade in time. Current satellite systems do not yield continuous fix information; the sophisticated computers associated with them are expensive.

Omega may be considered to be the oldest VLF radio navigation system, although it has evolved a great deal from its original state. In 1947 J. A. Pierce proposed a hyperbolic system (Raydux) based on phase-difference measurements (*Pierce*, 1947). The original system was conceived to operate in the vicinity of 50 kHz with a sine wave modulation of 200 Hz. Later, around 1955, the Radux-Omega system added the VLF 10.2 kHz transmission to the original Radux signal. Still later the LF radiation was terminated and additional VLF transmissions at 11.33 and 13.6 kHz were added to provide the Omega environment we have today. Omega history thus can be traced almost 30 years.

In the early 1960s Omega entered the development stage, with four transmitters located in Hawaii, New York, Trinidad, and Norway. Of these four original station locations, Norway and Hawaii remain today as full power (10 kW) operational stations. The measurements taken on the original four stations showed the Omega concept to be completely feasible and prompted the operational system implementation (*Swanson*, 1970).

In presenting the Omega navigation system as it exists today, this chapter describes the system's various characteristics and comments on its general usefulness.

26.1. GENERAL DESCRIPTION

In the Omega system, eight VLF transmitting stations are distributed around the globe as in Fig. 26.1. Each station transmits about one second, in turn, at precisely the same radio frequency. Figure 26.2 describes this

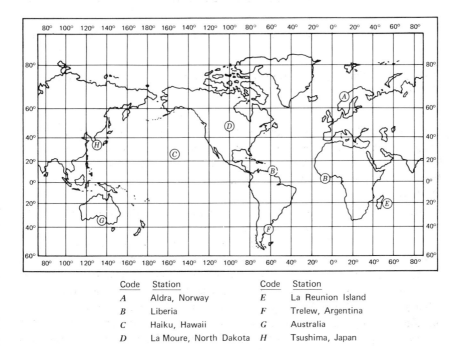

Code	Station	Code	Station
A	Aldra, Norway	E	La Reunion Island
B	Liberia	F	Trelew, Argentina
C	Haiku, Hawaii	G	Australia
D	La Moure, North Dakota	H	Tsushima, Japan

Figure 26.1 Omega transmitter locations. (courtesy United States Coast Guard).

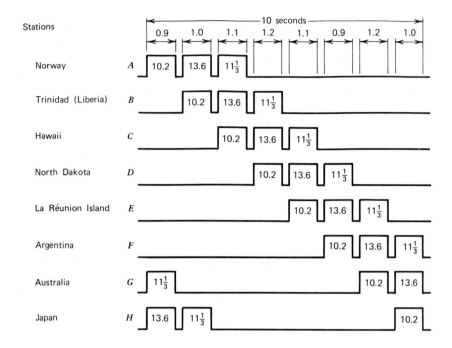

Figure 26.2 Omega system transmission pattern. (courtesy United States Coast Guard).

transmission format. The position of a receiver is established by the intersection of hyperbolic contours defined by the relative radio frequency phase of the transmissions of suitably chosen pairs of stations. The stations in the network have an average separation of approximately 10,000 km.

Figure 26.3 illustrates the navigation principle applied by the Omega user. The navigator can determine a line of position generated by any convenient pair of stations and cross it with one or more lines from another pair or pairs. He must consult charts (*H. O. Publication*, No. 1-N, 1966) and tables (*H. O. Publication*, No. 224) to relate the Omega readings to geographic position. Because there are diurnal and annual changes in the velocity of propagation, the navigator is given a set of predicted propagation corrections to reduce his observations to the proper chart numbers (*Calvo*, 1974). All the information can ultimately be stored in a computer and called on as needed, thanks to more complex hardware.

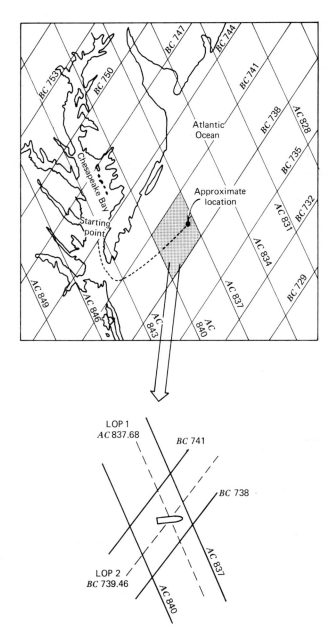

Figure 26.3 Position location using Omega. (courtesy United States Coast Guard).

The Omega of eight VLF radio stations provides a radio signal environment within which the location of an observer can be determined anywhere on the earth. Each station transmits a 10.2, 11.33, and 13.6 kHz continuous wave signal for approximately one second in turn, every 10 seconds, with all transmissions locked to the station cesium standard. The signal field phase is everywhere stationary. The relative phase of a particular set of signals observed at any given point depends on how much farther it is from one of the stations supplying the pair of signals than from the other. Furthermore, the same phase angle will be observed at all points having the same difference from the two transmitters. The loci of such points (hyperbolas) are contours of constant phase (isophase contours) fixed on the surface with respect to the location of the corresponding transmitting stations (Fig. 26.4, see also Chapter 4).

When good phase stability of transmissions was discovered at the very low frequencies (*Pierce*, 1966), the advantage of the Omega technique appeared quickly. The phase stability is in part a result of the dominant

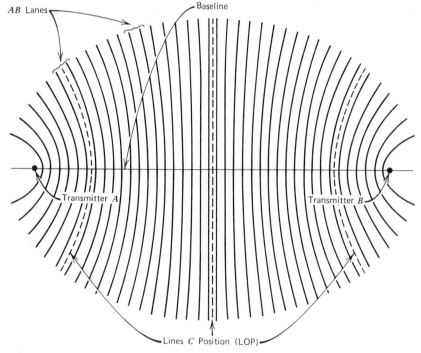

Figure 26.4 Isophase contours for pair of Omega stations (courtesy United States Coast Guard).

mode of transmission that is characteristic of VLF. Fortunately, VLF signal propagation is subject to low attenuation; therefore it is possible to have transmissions that cover the earth, yet are measurable to 1 or 2 μsec. Omega uses these transmissions at 10.2, 11.33, and 13.6 kHz, where only the first transverse magnetic mode (TM_1) is well excited and mode interference effects at sunrise and sunset are much less severe, thus simplifying the production of accurate charts and tables to assist the navigator.

The phase of the radio signal is a periodic quantity. Thus a given phase value repeats itself an integral number of wavelengths in distance. In the Omega system the relative phase angle of the signals of each pair of stations defines not one contour but a family of contours, numbering twice the distance between the stations in wavelengths and spaced one-half wavelength apart on the baseline. Only one contour of the family contains the position of the observer, and a means of resolving the 16 km ambiguity must be provided. Lane identification consists of establishing the initial position of the observer by some external means. Thereafter a count may be kept of the lanes traversed, either through the use of a strip chart or by some more sophisticated electronic means.

Lane ambiguity is also reduced by repeating the transmissions of stations at another frequency integrally related to the base frequency, one-third higher, or 13.6 kHz (*Pierce*, 1968; *Swanson* and *Hepperly*, 1969). The additional signals define another set of zero or in-phase contours about the same stations as foci, with contours spaced more narrowly by the ratio of the wavelengths. The wavelengths are commensurate in a ratio of 4:3. The coincident contours then define a pattern of broader lanes, each broader lane extending over three lanes of the basic 10.2 kHz pattern. The addition of the 11.33 kHz frequency causes a triple coincidence every 9 to 10 to 12 lanes, extending the unambiguous lanewidth to approximately 136.8 km.

Table 26.1 gives the most recent status of the Omega system's development. As can be seen, all stations with the exception of Australia are expected to be on-air by late 1975. In addition to the information contained in Table 26.1, other facts are fundamental to the understanding and using of the Omega system. All charts are prepared with reference to the Fischer ellipsoid (1960; *Bomford*, 1962), which defines an equatorial radius $a = 6,378,166$ m, a polar radius $b = 6,356,784$ m, and a flattening $(a - b)/a = 1/298.3$. The propagation velocities used in chart preparation are: (1) free space (group) velocity $c_0 = 299,793$ km/sec, (2) nominal charted (phase) velocity $c = 300,574$ km/sec, yielding a nominal ratio $c_0/c = 0.9974$.

TABLE 26.1 MOST RECENT STATUS ON OMEGA SYSTEM DEVELOPMENT

Station Letter	Location	Antenna Type	Geodetic Coordinates	Projected on-air Date
A	Aldra, Norway	Valley span	Lat = 66 25'12.39"N Lon = 13 08'12.65"E	Currently on-air at full power
B	Trinidad (Liberia)	Valley span (tower)	Lat = 10 42'06.2"N Lon = 61 38'20.3"W	Intermediate (Trinidad) station to be replaced by Liberia in December 1975
C	Haiku, Hawaii	Valley span	Lat = 21 24'16.9"N Lon = 157 49'52.7"W	Currently on-air at full power
D	La Moure, N. Dak.	Tower	Lat = 46 21'57.20"N Lon = 98 20'08.77"W	Currently on-air at full power
E	La Réunion	Tower	Lat = 20 58'26.47"S Lon = 55 17'24.25"E	Expected on-air date, December 1975
F	Trelew, Argentina	Tower	Lat = 43 03'22.53"S Lon = 65 11'27.69"W	Expected on-air by late 1975
G	Australia	—	—	All status pending the signing of bilateral agreement between United States and Australia
H	Tsushima, Japan	Tower	Lat = 34 36'53.26"N Lon = 129 27'12.49"E	Expected on-air at full power by mid-1975

26.2. INSTRUMENTATION

The largest complex of equipment is found at the transmitting stations. Each station may be considered to comprise three functional subsystems: timing and control, transmitting, and the antenna. A functional block diagram of these subsystems appears in Fig. 26.5. The timing and control subsystem is responsible for signal generation and phase control. The fundamental signal source is a precision cesium beam frequency standard with a stability of 5×10^{-12}. Within this subsystem a format generator furnishes the necessary synthesis instructions to the frequency synthesizer. The synthesizer provides output wave forms at 10.2, 11.33, and 13.6 kHz.

The transmitting subsystem includes the components that amplify the signals generated by the frequency synthesizer. This process is carried out in three stages of amplification: the input, the driver, and the power amplifiers, which boost the signals to their design power levels of 150 kW at the antenna system input.

The antenna subsystem consists of the devices that properly tune the antenna and radiate the signal. Within this subsystem the variometers, a variable inductance device, provide a high-resolution frequency output. The helix, which is a large coil fitted with a number of taps, furnishes a coarse inductive tuning capability. Finally, the antenna itself is a structure to simulate an electrically "short" dipole. The term "short" is used here because the antennas, although quite high (about 370 m), must simulate a half-wavelength to be most efficient. The wavelength at 10.2 kHz is approximately 30 km. Antennas are present in the Omega system in two configurations: (1) the valley span (as is the case in Hawaii and Norway), and (2) the top-loaded, vertical monopole. The valley span structure consists of several cables suspended between two elevated land masses as support and radiating elements. Table 26.1 summarizes the antenna types found at the operational Omega transmitting stations.

With the Omega system nearing completion, a large number of Omega navigation receivers have appeared. Equipment is available from both foreign and domestic sources.

As an example of the diversity and complexity of receiver equipments, a comparison can be made. The Tracor 599 series Omega receiver (Fig. 26.6) consists of a radio frequency portion, a programmer section that provides a format generating capability, a comparator section in which the actual phase measurement is made, and some associated power supplies. This unit is a highly reliable receiver and has been used by the Coast Guard in its worldwide Omega monitoring programs. Unfortunately in itself it gives the navigator little usable information. Phase-difference readings must be taken from the phasemeter, and lane counts

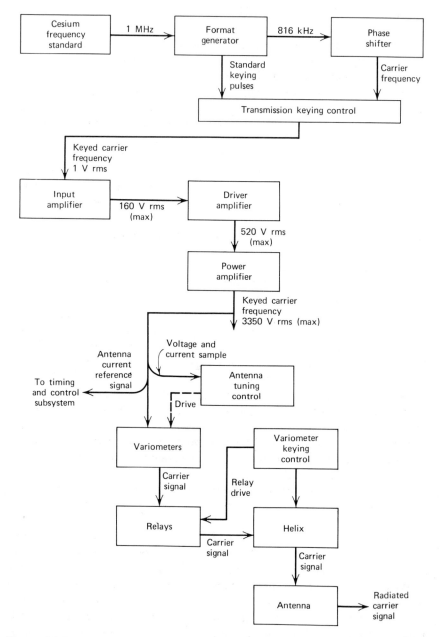

Figure 26.5 Functional block diagram of Omega transmitter (courtesy United States Coast Guard).

must be kept by means of external recording devices. The information in turn must be applied to the appropriate Omega charts and tables to secure a fix in latitude and longitude.

On the the other hand, we have the highly sophisticated CMA-719(c)-AN/ARN-115 Automatic Airborne Omega Navigation system (Fig. 26.7), manufactured by the Canadian Marconi Company. This receiver can be functionally divided into three sections: an RF section (which includes the flush plate antenna shown), a control indicator section, and a computer section. The RF section alone does the entire set of functions described for the Tracor receiver. The computer section provides automatic control of the RF section and computational functions such as coordinate conversion; thus no charts or tables are required. The control indicator acts as interface between the operator and the rest of the receiver. Among other functions this section shows the operator this current position in longitude and latitude on a lighted display. The size of the receiver/computer unit is $19.5 \times 25.4 \times 49.5$ cm and it weighs 99.2 kg; the control indicator is $11.4 \times 14.6 \times 16.5$ cm and weighs 13.2 kg.

The two receivers just described offer the extremes of available Omega receiving equipment. There are receivers at the intermediate levels that have such features as lighted digital displays of LOP and automatic synchronization, thus making the selection of Omega receiving equipment a function of the mission in which it is to be employed.

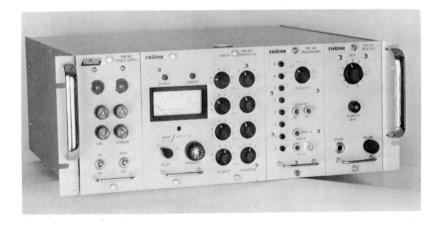

Figure 26.6 Tracor 599 R Omega receiver (courtesy Tracor, Inc.; reprinted from United States Coast Guard's Omega Monitor Station Manual).

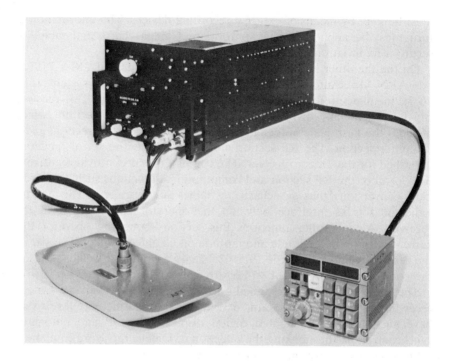

Figure 26.7 CMA-719(c)-AN/ARN-115 Automatic Airborne Omega system. (courtesy Canadian Marconi Co.).

26.3. ACCURACY

Omega navigational errors are generated from numerous sources, including human operator error, equipment-generated errors, uncertainties in the knowledge of radio wave propagation, and propagation prediction correction errors. Of these errors, those which usually limit the system accuracy arise from propagation variation and propagation correction errors (*Swanson*, 1970).

Geometric dilution of position, which is a limiting factor in other electronic navigational aids, does not play a large role in Omega because of the long system baselines; also, the navigator has the option to choose the pairs of stations that offer the best crossing angles (*Swanson*, 1968).

The published accuracies of the Omega system when used in the conventional mode are: less than 1.84 km cep in the daytime, and less than 3.7 km cep at night.

The error limits can be expected to be reduced as prediction techniques are refined and additional monitor data are collected (*Kasper*, 1971).

26.4. DIFFERENTIAL OMEGA

VLF transmissions experience anomalous fluctuations in propagation velocity. These variations in received Omega signals contribute errors to the navigational fix. Investigations into VLF propagation have reported that the anomalies have predictable effects over differential regions, and recent studies have suggested statistical models to describe the distance-dependent correlation relationships (*Kasper*, 1970).

Over a lesser geographic area, the accuracy of the Omega system can be enhanced considerably by a technique referred to as differential Omega. Within the lesser area, the difference between the receiver reading and the charted value can be broadcast as a correction factor to other vessels or aircraft operating in the area.

As a user travels away from the monitor site, there is an increase in the difference between anomalies experienced by his received signals and those received by the monitor; hence the farther a user is from the monitor, the less accurate will be his corrections. Eventually the magnitude of the errors in the basic Omega signals will be the same as that of the errors resulting from using the base station monitor information; therefore the limits of the differential Omega area are set at the distance at which the accuracy offered by the differential Omega correction is no longer better than the accuracy of the unaided Omega system.

A differential area is shown in Fig. 26.8. Omega signals are received from stations *A* and *B* at the monitor site and by a user in the area. These two signals are combined to determine the line of position *AB*. At the monitor site, the received LOP is compared with the computed *AB* pair LOP. The user may be informed of the observed LOP error at the monitor and/or of a fix error by adding another pair LOP for which the monitor will generate corrections.

Error sources affecting the accuracy of differential Omega are: RF noise, instrumentation, tracking lag, transient propagation anomalies, nominal day-to-day ionospheric variations, signal correlation, geometric factors, and differences in the nocturnal and diurnal propagation velocities (*Swanson*, 1974).

Since differential Omega was introduced more than 10 years ago, experiments have been performed by a number of investigators, both American and foreign, including those at Tracor, Beukers Laboratories,

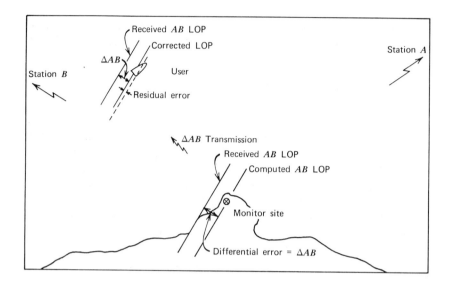

Figure 26.8 Differential Omega concept.

Hastings-Raydist, Navidyne, SERCEL (France), and various Navy research laboratories. Experimental data plus an analytic analysis indicate that peak errors at base station ranges of about 555 km would be approximately 1.1 km in worst-case positions (*Swanson*, 1974). The average error over 1110 km ranges was also reported by Swanson to be 0.74 km. Other investigators have predicted position errors of only 0.26 km (*Kasper*, 1971).

Implementation of the differential Omega technique has varied from a simple voice broadcast of correction factors to automated techniques such as Micro Omega. In the latter, the monitor station transmits a single sideband audio tone whose phase contains the correction factor information. This system claims 90 m accuracy as far as 370 km from the relay station and is fully automated (see Sections 20.2.2.10).

26.5. PROPAGATION OF ELECTROMAGNETIC ENERGY AT VLF

Radio wave propagation at VLF has been studied experimentally and theoretically for many years. Several texts form the nucleus of VLF wave propagation understanding (*Budden*, 1961; *Davies*, 1965; *Galejs*, 1972;

Wait, 1962). Although much is known about propagation at very low frequencies, our present knowledge does not always permit as precise a prediction of fields over complex paths as might be desired. In general, several conceptual models are employed to explain the behavior of VLF radio waves. For distances greater than a wavelength away from the source, we consider two presentations of the propagating energy; (1) a combination of ground wave and multiply reflected optical rays, or (2) a guided wave consisting of a superposition of sustained modes. The second case is analogous to the modes generated in conventional wave guides used at much higher frequencies.

The physical condition encountered is one in which a layer of the ionosphere (the D-region or lower E-region) commencing some 60–90 km above the surface of the earth acts to guide the VLF energy between the surface caused by it and the surface of the earth. Although heterogeneous, the earth's surface is initially considered to be a sharply bounded surface with an effective conductivity that is relatively constant over the VLF band. The ionospheric layer has electrical properties that vary with time, frequency, geographic location, and bearing of the incident radiation. In the first approximation this layer is considered to be a surface having some reference reflecting height and an equivalent reflection coefficient.

Numerous mathematical expressions have been formulated in an attempt to explain fields propagating at VLF. For long paths having only one dominant propagating mode, the wave guide theory with sharply bounded ionosphere yields (*Swanson*, 1972)

$$E = \frac{3 \times 10^5}{h} \left[\frac{P\lambda}{a \sin (d/a)} \right]^{1/2} e^{-\alpha' d} \Lambda \qquad (26.1)$$

where E = received field strength (μV/m)
  h = height of the ionosphere
  P = radiated power (kW)
  λ = wavelength
  a = spherical earth radius
  d = path length
  α' = attenuation coefficient (nepers/Mm); α dB/Mm = 8.68α'
  Λ = excitation factor

Formulas of this type are useful in field strength calculations but yield little information about the overall phase structure so essential to Omega prediction. To extract phase information, a more detailed model analysis is required.

A guided wave mode is considered to exist when the electromagnetic energy is guided between the previously described reflecting boundaries. A simple approximation to the real physical system is found in the wave guide (Fig. 26.9), where the earth is treated as a homogeneous conducting sphere and the ionosphere is replaced by a uniform shell of electrons. The problem was analyzed very carefully (*Wait*, 1962). The total radial component of the field is considered to be a superposition of representative modes:

$$E_r = \sum_{n=0}^{\infty} E_n(r) \qquad (26.2)$$

The result is the following general expression for the space part of the radial component of the electric field:

$$E_r = \frac{A}{h} \left[\frac{d/a}{\sin (d/a)} \right]^{1/2} \sum_{n=0}^{\infty} \delta_n S_n \exp\left(-2\pi i \frac{S_n d}{\lambda} \right) \qquad (26.3)$$

where δ_n = exitation factor
$\quad a$ = earth radius
$\quad d$ = great circle path distance
$\quad \lambda$ = wavelength
$\quad A$ = constant
$\quad S_n$ = special function used by Wait

Equation 26.3 can be reduced to an expression for the dominant first mode of propagation studying an example that served as an approximation to the earth situation (*Cha* and *Morris*, 1974):

$$E_r \cong \frac{A}{h} \left[\frac{d/a}{\sin (d/a)} \right]^{1/2} \exp\left(2\pi u_1 \frac{d}{n} \right) \exp\left(-\frac{i\omega d}{c_{P_1}} \right) \qquad (26.4)$$

where c_{P_1} = phase velocity of and u_1 = attenuation rate for the $n = 1$ mode. The quantity $\omega d/c_{P_1}$ is called the phase and indicates the number of cycles of harmonic space variations that the mode has passed through in a distance d, including any fraction of a cycle. This calculation bore out the idea that only one mode dominates the field after the radiation has passed some 1000 km from the transmitting station. Although sophisticated, this theory does not confer the ability to predict the phase at a receiving point. However it does supply an adequate model from which to base a semiempirical approach to determining the Omega signal phase at a receiving site. In the model, the single dominant transverse magnetic mode is taken to be representative of the Omega field beyond a certain minimum distance from the transmitter.

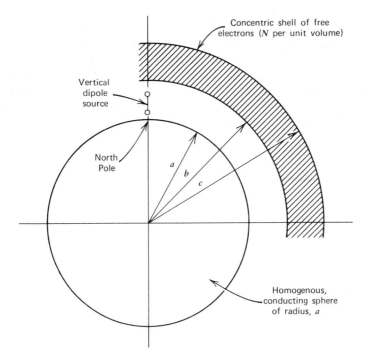

Figure 26.9   The earth-ionosphere wave guide.

In the Omega system distance is measured by determining the number of cycles of phase traversed during a journey. This phase accumulation is then related to distance by using the characteristic phase velocity of propagation. Therefore it is the understanding of the phase character that determines the accuracy of the system. In practice the nominal phase velocity mentioned earlier is used in chart preparation. The charts are then corrected from the predicted propagation correction tables for estimated variations in the nominal phase velocity. These tables account for the real-world geophysically forced variations from the idealized propagation model. The generation of the tables calls for the linear regression of some 14 functional forms to yearly blocks of Omega monitored phase data. Several publications have outlined this calculation in great detail, and no further discussion is made of it here (*Swanson*, 1972).

Two anomalous propagation conditions are common to VLF systems (*Swanson*, 1970). The first, which are short-term conditions, are referred to as sudden ionospheric disturbances. These occur when X-rays in the

2–8 Å band are emitted from the sun during the course of a solar flare. The disturbances are therefore restricted to sunlit paths and their a net effect is that of advancing the received phase on the path. The advance is rather rapid and takes place over several minutes. The return to normal rarely takes longer than half-hour. The second anomaly, polar cap absorption (PCA), causes phase advances both day and night over Arctic paths only. Errors of several miles may persist for several days during PCA events.

TWENTY–SEVEN

AIRCRAFT DOPPLER NAVIGATION

Aircraft Doppler radar navigation is a completely self-contained navigation system based on the well-known classical *dead reckoning* principle, where the position of an aircraft is determined from the track angle and the traversed distance. These quantities are not obtained directly because of such existing factors as wind speed and direction resulting in drift angle. To understand the practical applications, one first must "solve" the *basic navigation triangle* and become familiar with the parameters involved. In Fig. 27.1 the heading of the aircraft is the angle α between the north and the true air speed vector AD; the drift angle δ is an angle between the true air speed vector AD and the ground speed vector AB, caused by the wind speed vector DB. The wind direction from north is ε, and the track angle β is an angle between the north and the ground speed vector AB. Solution of the triangle ABD yields (a) the continuous determination of the speed and direction of wind, thus providing the navigator with valuable information to position the aircraft in a favorable tail wind; (b) the determination of the track angle and the ground speed vector directly; and (c) the reverse computation of the track angle and ground speed vector from the wind speed vector and wind direction.

The heading and the true air speed vector are obtained from the aircraft's conventional instrumentation; the drift angle and the ground speed vector are found from the aircraft's Doppler radar system. Then DB is obtained by using the cosine law of trigonometry applied to AB, AD, and δ. The angle ADB is determined by using the sine law applied to AB, DB, and δ, and the wind direction will be $\varepsilon = \alpha + \gamma$, where α is the

453

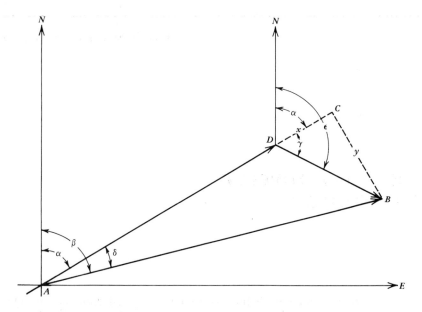

Figure 27.1 Navigation triangle.

heading and γ is the supplement angle of ADB. When the drift angle δ has been observed, the track angle is obviously $\beta = \alpha + \delta$. Finally, if the wind speed and direction are constant over a reasonable period of time, solution of the navigation triangle is also used to permit one form of memory operation in Doppler navigation (*McKay*, 1958). Using the known wind speed vector DB and the known angle γ, x- and y-components of the triangle DCB can be computed; and since AD can be observed the ground speed vector AB and the drift angle δ can be solved from the triangle ABC.

27.1. PRINCIPLES OF DOPPLER RADAR

The general principle of Doppler radar is based on sending electromagnetic waves to the ground by a pair of reflectors mounted so that their beams form a fixed angle Θ (about 45° with each other) and an adjustable angle γ, Fig. 27.2 with the horizontal plane. Because of the Doppler effect, a shift in frequency occurs between the transmitted and received signals, since there is radial motion between the aircraft and the ground. This motion causes a continuous phase shift in the returned signal, and

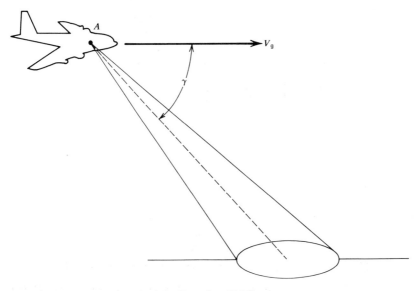

Figure 27.2 Doppler principle (Laurila, 1960).

the difference between the transmitted and received frequencies can be obtained from the following expression by taking into account the need for the signal to travel twice the distance

$$\Delta f = \frac{2V_g}{c} f \cos \gamma \qquad (27.1)$$

where V_g is the ground speed of the aircraft, c is the propagation velocity of the radar signal, and f is the transmitted frequency. By substituting $1/\lambda = f/c$, Eq. 27.1 can be solved for V_g as follows:

$$V_g = \frac{\Delta f \lambda}{2 \cos \gamma} \qquad (27.2)$$

where Δf is the measured frequency shift and λ is the transmitted wavelength.

Since all radar energy reflected from the flat ground at an angle γ undergoes the same frequency shift, a family of constant γ-angle hyperbolas can be drawn (Fig. 27.3). Because of the drift of the aircraft caused by the wind, contacts with the ground of the two beams, separated by an angle Θ, do not lie along a constant frequency shift hyperbola (Fig. 27.4), and frequencies returned from each contact will differ by an amount proportional to drift. By rotating the beam assembly around a vertical

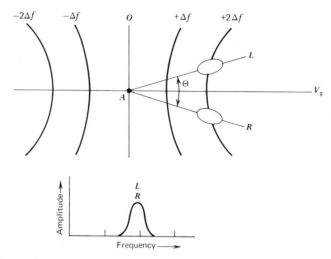

Figure 27.3 Family of constant angle hyperbolas (Laurila, 1960).

axis to ensure the coincidence of the frequency spectra L and R (Fig. 27.3), drift angle δ between the air speed vector V_a and ground speed vector V_g can be recorded.

In 1956 the General Precision Laboratory, Inc., started manufacturing a radar Doppler navigation system called the *Radan*. It used the so-called *Janus* technique to generate microwave pulses simultaneously along the

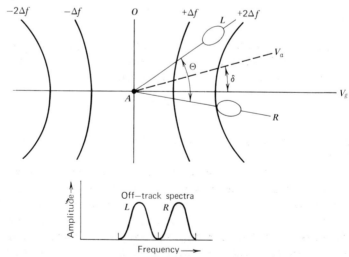

Figure 27.4 Determination of drift angle (Laurila, 1960).

track, both fore and aft relative to the vertical through the aircraft. The energy returned from the forward beam is of frequency $(f + \Delta f)$ and the energy returned from the rearward beam is of frequency $(f - \Delta f)$. The difference frequency, obtained by heterodyning the returned energies, then serves as a measure of ground speed. Then the instrument problem becomes one of measuring a comparatively low audio frequency, which can be done with high precision (*McKay*, 1958; *McMahon*, 1957). For drift measurement, four beams are employed in such a way that two beams are forward looking and two beams are rearward looking. The right front and left rear lobes are transmitted together and constitute the *right beam*, and the left front and the right rear lobes are transmitted together to constitute the *left beam*. If the antenna were off the ground track to the left (as in Fig. 27.4, e.g.) the return from the left beam would indicate a lower frequency than the return from the right beam. The drift computing elements of the system sense this difference in frequency shift and use it as an error signal to servo the antenna back to ground track. Thus the angle through which the antenna has been turned to align it with the ground track equals the drift angle δ. If the γ angles of the beams within each pair are symmetrically oriented, any Doppler shift due to a vertical velocity component will be equal in magnitude and sign for both beams of each Janus pair and will add to zero in the frequency mixing process. Thus the system is insensitive to climb or descent of the aircraft and measures only the ground velocity (*Saltzman* and *Stavis*, 1958).

27.2. NEW DEVELOPMENTS

The *AN/APN-187 Doppler Velocity Altimeter*, manufactured by the Kearfott Division of Singer-General Precision, Inc., Pleasantville, New York, is a solid state, microminiaturized FM-CW radar sensor*. It measures continuously ground velocity in the horizontal plane and altitude above the terrain. The velocity is obtained by using the Doppler shift, and the altitude information is obtained through determination of the phase shift, induced by transit time, of the modulation frequency in the received signal. The AN/APN-187 uses a beam-lobing technique to correct variations in accuracy resulting from terrain differences over land and over water. Its antenna is stabilized in pitch and roll but is fixed to the aircraft

---

* Information disclosed of the Doppler Velocity Altimeter is the property of Singer-General Precision, Inc. It is furnished for evaluation purposes only and shall not be disclosed or used for any other purposes except as specified by contract between recipient and Singer-General Precision, Inc.

in azimuth. The antenna transmits three beams of energy in a symmetrical pattern—front left, front right, and aft—and along the heading the velocity component V_H and the cross-heading velocity component V_D are determined separately. The components are obtained from the following expressions (*McMahon*, 1970):

$$V_H = \frac{1}{k_1}(d_L - d_A) \qquad \text{and} \qquad V_D = \frac{1}{k_2}(d_L - d_R) \qquad (27.3)$$

where d_L, d_R, and d_A are the Doppler shifts in the returned energy from the left, right, and aft beams, respectively, and k_1 and k_2 are constants depending on the transmission frequency, the velocity of the electromagnetic waves, and the direction cosines of the three lines defining the beam centers. Thus with the antenna fixed in azimuth to the aircraft and maintained level in pitch and roll, the along-heading velocity can be obtained by subtracting the Doppler shift in the aft beam from the shift in the left beam and applying the proper constant of proportionality. Similarly, across-heading velocity is proportional to the difference between the left-beam Doppler shift and the right-beam Doppler shift.

Since the beams have finite widths, they overlap several equal-frequency shift lines within a narrow region, and the Doppler return consists of a spectrum of frequencies rather than a single frequency. Because the illumination of the target surface is most intense at the center of the transmitted beam, the illumination power is a function of frequency rather than of time. The spectrum is statistically equivalent to narrow-band noise, and the band center frequency is the desired mean Doppler frequency. The backscattered energy within a typical beam segment is essentially constant over land. Over water, the backscattered energy varies with the beam incidence angle and with the sea state. As a result, there is an overall attenuation of the energy in the received signal, and the center of power of the returned Doppler spectrum shifts downward in frequency, resulting in velocity measurement errors. The AN/APN-187 uses beam lobing in the γ-direction to measure the change in calibration over water. Each of the received beams is switched at a 36 Hz rate in the γ-direction, producing two overlapping Doppler spectra for each beam. If the aircraft is over land, the centers of power of the two spectra are the same, if the aircraft is over water, however, the centers of power differ. The AN/APN-187 computes the difference and adds a calibration value to the basic velocity measurement, thus providing accurate output over water. The stabilization of the antenna arrays (separate antenna arrays for transmission and reception) is established to ensure that leveling signals from the aircraft central repeater are received by control transformers in the antenna. Through separate servoamplifiers, these signals control the roll and pitch motors.

Altitude determination requires two FM frequencies—namely, 49 kHz at high altitude and 197 kHz at low altitudes—to increase the sensitivity of the measurements at low altitudes. At 49 kHz, a 360° phase change occurs every 8680 ft in altitude, and at 197 kHz the corresponding phase change occurs at every 2170 ft. The measurable altitude range is from 0 to 50,000 ft for Doppler velocity measurement and from 0 to 30,000 ft for altitude measurement. The ground speed range is from 50 to 500 knots. The pitch and roll stabilizations are maintained within ±30 and ±60°, respectively. The accuracy of the velocity determination is ±0.175% of V_g or ±0.25 knot, which ever is greater, and the accuracy of the altitude determination is ±2% ±10 ft for altitude range of 75–5000 ft, and ±5% for the altitude range of 5000–30,000 ft. The transmission carrier frequency is 13.325 GHz.

For small and medium-sized aircraft navigation, the small lightweight Doppler navigation and guidance system composed of the *SKK-1000A* Doppler and *SKQ-601* computer seems to be a promising solution. This system is manufactured by the Singer Company, Kearfott Division, Little Falls, New Jersey, and was originally developed for the Navy as a self-contained airborne navigator bearing the military designation AN/APN-153. The system has three major components: a Janus-type Doppler radar, a compass, and a computer. In this basic dead reckoning device, the control-indicator of the radar provides drift and ground speed information and the computer, when fed with these data together with the heading data from the compass, provides track information. The SKK-1000A consists of three units, the receiver-transmitter, the control-indicator, and the antenna-assembly with mount. The receiver-transmitter is of modular, solid state construction and contains 95% of the electronics for the system. The control-indicator is built for panel or console mounting, the indicator displaying drift angle to 40° left or right and the ground speed from 70 to 1000 knots. Since it can be set either for land or sea navigation, the latter arrangement compensates for the radar backscattering. The small antenna has a mercury switch that controls the motor, furnishing ±25° pitch stabilization without need for external attitude information. As an option, a roll stabilization antenna is available. The navigation control indicator of the computer has two vertical rows of counters and dials, each displaying *Desired Track, Nautical Miles to Go,* and *Nautical Miles Off Track* for each leg of the flight. While the first leg of the flight is being tracked on the left row of counters, the desired track and distance of the second leg can be inserted on the right row. When the *Nautical Miles to Go* counter reads zero, the computer will automatically switch to the right row, and the second track will be initiated. Distance and track information for the third track are then inserted into the left row. Windows above the rows of counters and dials indicate which row is providing current information and which is ready for inserts for the next leg.

The following sequence illustrates the computer programming and indicator displays at various points during navigation. At the initial point in the route, the azimuth of 058.3° to the next way point is set to the *Desired Track* display, the distance of 514.41 nmi is set to the *Nautical Miles to Go* display, and 0 is set to the *Nautical Miles Off Track* display in the computer. The Doppler control indicator is in *Land* position. For the second leg en route, the right row of displays is programmed to correspond to the route parameters as shown in Fig. 27.5. When the aircraft is passing the second way point, the computer control has automatically transferred from stage 1 to stage 2 and *Heading* has changed to 077.8°. While crossing from land to sea, the control-indicator switch has been manually turned to the *Sea* position (Fig. 27.6). It is assumed that on that

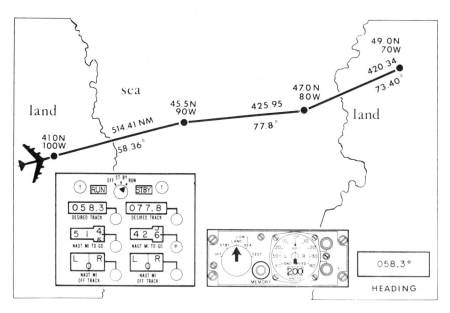

Figure 27.5 Instrument setting at initial point en route (courtesy Singer Co., Kearfott Div.).

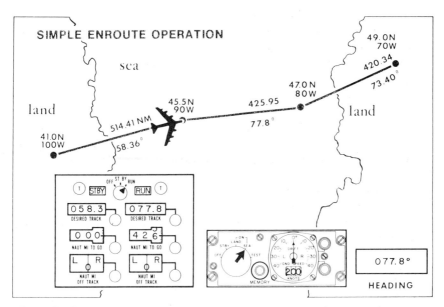

Figure 27.6 Instrument setting at first way point en route (courtesy Singer Co., Kearfott Div.).

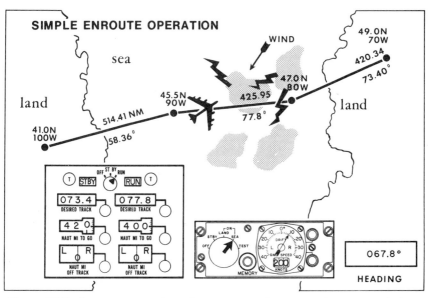

Figure 27.7 Instrument setting when encountering quartering cross wind en route (courtesy Singer Co., Kearfott Div.).

leg the aircraft encounters heavy wind quartering from left, which tends to drive the aircraft to the right. The system is designed to compensate automatically for any change in the wind direction by maintaining zero *Nautical Miles Off Track* indication. In this case the *Drift* indication is approximately 10° to the right, and the *Heading* of the aircraft has changed to 067.8° to compensate the drift. Also, the *Standby* row of displays has been programmed to represent the route parameters of the last leg (Fig. 27.7). If the pilot has to steer the aircraft off the route during the flight (e.g., because of thunderstorms), he can intercept the original route and the computer continues displaying the correct route parameters.

The SKK-1000 A weighs 48.5 lb and occupies a total of 1.2 ft³ of space. The transmitted frequency is 13.325 GHz. Within the operational altitude range of 50–50,000 ft the accuracy of the ground speed is ±0.17% and that of the drift angle is ±0.17°. The SKQ-601 weighs 23 lb and occupies 0.34 ft³ of space. The ground speed limits are 0–2000 knots, and the distances per leg are 0–999 nmi. The track angle is from 0 to 360°, and the distance off the track is from 0 to 99 nmi.

TWENTY-EIGHT

POSITIONING BY SATELLITE

Thomas A. Stansell, Jr.

28.1. INTRODUCTION

The U.S. Navy Navigation Satellite System, usually called Transit, was conceived in 1958 at the Applied Physics Laboratory of Johns Hopkins University. The technical approach grew from experiments to determine the orbit of the first artificial earth satellite, Sputnik 1, by measuring the Doppler shift of its radio signals (*Guier* and *Weiffenbach*, 1960). At the same time, there was urgent need for an accurate global navigation system for the Polaris ballistic missile submarines. A development program was begun in December 1958, and the Transit system became operational in January 1964.

In 1967 the United States government released details of the system to permit commercial manufacture and use of Transit navigation equipment. Such equipment is now available from the United States, Canada, Japan, Great Britain, France, and Holland. Applications include navigation of commercial vessels, oceanographic exploration, geodetic survey, positioning of offshore drill rigs, and offshore oil exploration. Aircraft navigation also has been successful, but system limitations prevent widespread use.

The Transit system provides intermittent position fix updates rather than continuous navigation information. However, coverage is worldwide, and system accuracy is excellent regardless of weather conditions because high precision signals are transmitted on a direct line-of-sight basis from the satellite to the navigation receiver.

463

Transit is the only operational United States navigation satellite system. (The Soviet Union also presumably operates a classified system.) Although commercial and military use of Transit continue to expand, our government also is exploring the feasibility of a new system concept called Navstar or the Global Positioning System (GPS). If the new system is implemented, it is expected to consist of 24 satellites located in three orbit planes about 20,000 km above the earth. Because users always would be within line of sight of at least four satellites, highly accurate continuous navigation in three dimensions would be possible. According to the published schedule, if the new system meets cost and performance objectives and is approved, it could become fully operational by 1984. In that event, the Transit system could be discontinued for military navigation by 1989. Until technical and economic feasibility of the GPS is demonstrated, however, use of Transit is expected to expand rapidly over the next decade, raising the possibility of its continued operation by a civilian agency if the military converts to GPS.

28.2. TRANSIT SYSTEM DESCRIPTION

28.2.1. The Satellites

At this writing there are six operational Transit satellites in orbit. Figure 28.1 illustrates their physical configuration: four panels of solar cells charge the internal batteries, and signals are transmitted to the earth by the "lampshade" antenna, which always points downward because of the gravity gradient stabilization boom. An elongated object in orbit naturally tries to align with the earth's gravity gradient. Magnetic hysteresis rods along the solar panels damp out the tendency to sway back and forth by interaction with the earth's magnetic field; that is, mechanical energy is converted to heat through magnetic hysteresis.

The six operational satellites have demonstrated exceptional longevity. Their launch dates are:

Designation	Launch Date
30120	April 14, 1967
30130	May 18, 1967
30140	September 25, 1967
30180	March 2, 1968
30190	August 27, 1970
30200	October 29, 1973

Figure 28.1 Physical configuration of Transit satellite (courtesy of the Johns Hopkins University Applied Physics Laboratory).

Also, there are a dozen or more of these spacecraft available to be launched whenever a replacement is needed. Because the satellites weigh only 135 lb and are placed in a relatively low orbit, the solid fuel Scout rocket is the launch vehicle. There should be no problem in maintaining an operational Transit system for as long as needed.

Figure 28.2 is a block diagram of the satellite electronics. The satellites transmit coherent carrier frequencies at approximately 150 and 400 MHz. Because both signals are derived by direct multiplication of the reference oscillator output frequency, the transmitted frequencies are very stable, changing no more than about one part in 10^{11} during a satellite pass. Thus they may be assumed to be constant with negligible error.

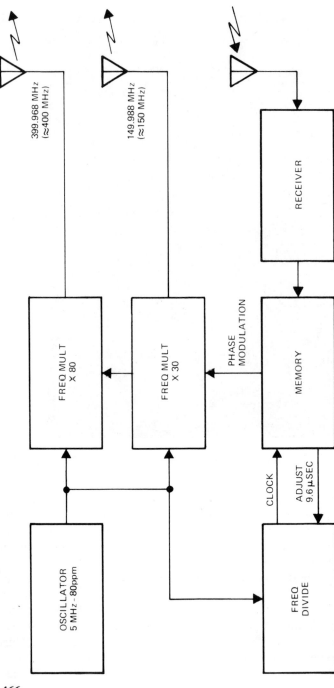

Figure 28.2 Transit satellite block diagram (courtesy of Magnavox Government and Industrial Electronics Company, Advanced Products Division, Torrance, California [Magnavox APD]).

The reference oscillator output also is divided in frequency to drive the memory system. In this way, the navigation message stored there is read out and encoded by phase modulation onto both the 150 and 400 MHz signals at a constant and carefully controlled rate. Thus the transmitted signals provide not only a constant reference frequency and a navigation message but also timing signals, because the navigation message is controlled to begin and to end at the instant of every even minute. An updated navigation message and time corrections are obtained periodically from the ground by way of the satellite's injection receiver. The time correction data are stored in the memory and applied in steps of 9.6 μsec each.

The six satellites are in circular, polar orbits about 1075 km high, circling the earth every 107 min (Fig. 28.3). This constellation of orbits forms a "birdcage" within which the earth rotates, carrying us past each

Figure 28.3 Six Transit satellites in circular, polar orbits about 1075 km above earth (courtesy Magnavox APD).

orbit in turn. Whenever a satellite passes above the horizon, we have the opportunity to obtain a position fix. The average time interval between fixes varies from about 35 to 100 min, depending on latitude.

28.2.2. Ground Support Network

Figure 28.4 is a representation of the overall Transit navigation system. The ground support network, which is operated by the U.S. Navy Astronautics Group with headquarters at Point Mugu, California, consists of four tracking stations, a computing center, and two injection stations. The four tracking stations are located at Prospect Harbor, Maine; Rosemount, Minnesota; Point Mugu, California; and Wahiawa, Hawaii. Each time a Transit satellite passes within line of sight of a tracking station, it receives the 150 and 400 MHz signals, measures the Doppler frequency shift caused by the satellite's motion, and records the Doppler frequency as a function of time. The Doppler data are then sent to the Point Mugu computing center, where they are used to determine each satellite's orbit and to project each orbit many hours into the future.

The computing center forms a navigation message from the predicted orbit, which is provided to the injection stations at Point Mugu and at Rosemount. At the next opportunity, one of the injection stations transmits the navigation message to the appropriate satellite. Each satellite receives a new message about once every 12 hours, although the memory capacity is 16 hours.

Because the ground tracking stations are located in the northern hemisphere and are spread over only 100° of longitude, it might be suspected that system accuracy would degrade outside this region. Fortunately, this is not the case. The orbit determination programs originally were developed with the aid of a worldwide tracking network called Tranet. These stations were located to provide nearly continuous coverage of each experimental satellite as it orbited the earth. Tests have been conducted to compare orbits computed from Tranet data with the same orbits computed with data from only the four operational tracking stations. The results show orbit differences on the order of one meter rms; thus it is clear that worldwide accuracy is not affected by location of the operational tracking stations.

To summarize, the Transit system is designed so that each satellite is a self-contained navigation beacon. It transmits two very stable frequencies, timing marks, and a navigation message that describes the satellite's position as a function of time. By receiving these signals during a single pass, the system user can calculate an accurate position fix.

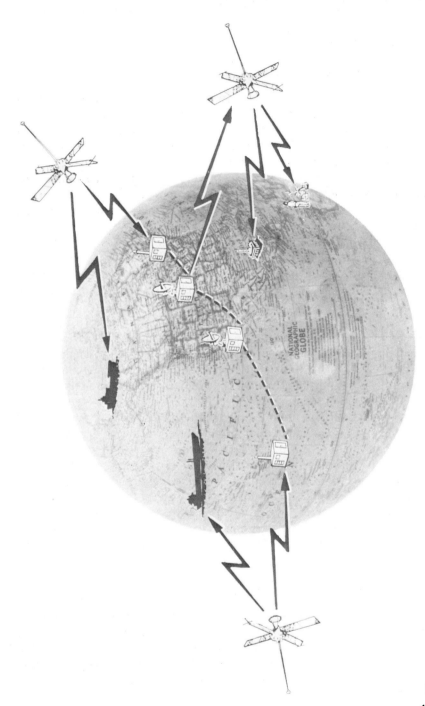

Figure 28.4 Schematic overview of the Transit navigation satellite system (courtesy Magnavox APD).

28.2.3. Obtaining a Position Fix

28.2.3.1. *The Navigation Message*

Figure 28.5 indicates how the navigation message defines the position of the satellite. Every 2 min the satellite transmits a message consisting of 6103 binary bits of data organized into six columns and 26 lines of 39-bit words, plus a final 19 bits. The message begins and ends at the instant of the even minute, which are denoted as time marks t_i and t_{i+1}. The final 25 bits of each message form a synchronization word that identifies the time mark and the start of the next two-minute message. By recognizing this word, the navigation receiver establishes time synchronization and thereafter can identify specific message words.

The orbital parameters are located in the first 22 words of column six, and those in lines nine through 22 are changed only when a new message is injected into the memory. These fixed parameters define a smooth, precessing, elliptical orbit; satellite position being a function of time since a recent time of orbit perigee.

The words in lines one through eight shift upward one place every 2 min, with a new word inserted each time in line eight. These variable parameters describe the deviation from the smooth ellipse of the actual satellite position at the indicated even minute time marks. By interpolation through the individual variable parameters, the satellite position can be defined at any time during the satellite pass.

Each binary bit is transmitted by phase modulation of the 150 and 400 MHz signals. The modulation format for a binary one is given in Fig. 28.6, and a binary zero is transmitted with the inverse pattern. As shown, this format furnishes a clock signal at twice the bit rate, which is used to synchronize the receiving equipment with the message bits.

Because the satellites transmit only a watt of power and may be thousands of kilometers away, very sensitive receivers are needed. In addition, however, the orbit parameters must be verified by comparing the redundant messages to detect and eliminate occasional errors in the received data.

28.2.3.2. *The Doppler Measurement*

By receiving the navigation message, the Transit system user learns the position of the satellite as a function of time. Thus to obtain a position fix, he must relate his position to the known satellite orbit. This relationship is established by measuring the Doppler shift, which is a unique function of the observer's position and motion relative to the known satellite orbit.

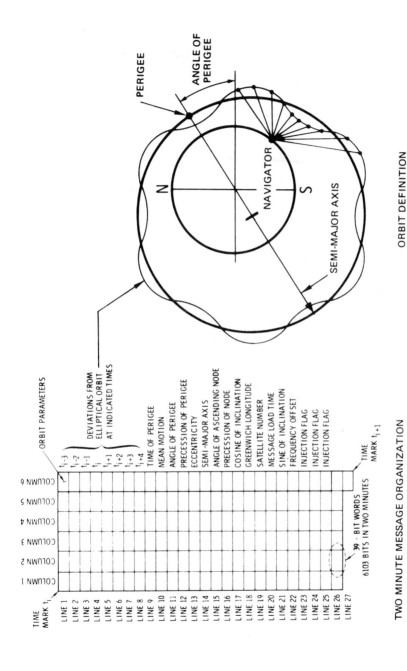

ORBIT DEFINITION

TWO MINUTE MESSAGE ORGANIZATION

Figure 28.5 Satellite message describes orbital position (courtesy Magnavox APD).

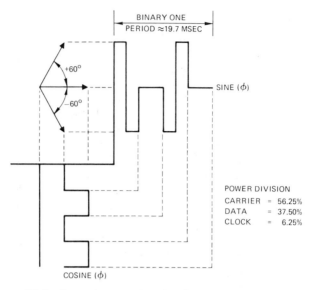

Figure 28.6 Data phase modulation (courtesy Magnavox APD).

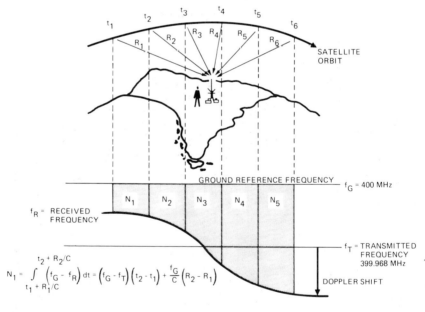

Figure 28.7 Each Doppler count measures slant range change (courtesy Magnavox APD).

472

Figure 28.7 illustrates the Doppler measurement technique. The frequency f_R being received from the satellite consists of the frequency being transmitted f_T plus a Doppler frequency shift of up to ± 8 kHz due to relative motion between the satellite and the receiver. Note that the transmitted frequency is offset low by about 80 ppm (32 kHz at 400 MHz) to prevent f_R from crossing 400 MHz.

The navigation receiver is equipped with a stable reference oscillator from which a 400 MHz ground reference frequency f_G is derived. Oscillator stability must be adequate to assume a constant frequency throughout the satellite pass. As shown by the figure, the navigation receiver forms the difference frequency $f_G - f_R$, and each Doppler measurement is a count of the number of difference frequency cycles occurring between time marks received from the satellite. Because every message bit effectively represents another time mark, the Doppler counting intervals are formed with respect to the message format of Fig. 28.5. For example, each line of the message lasts about 4.6 sec, and the commonly used Doppler count interval of 23 sec is formed by starting a new count at the end of every fifth line.

Each Doppler count is composed of two parts, the count of a constant difference frequency $f_G - f_T$ minus a count of the number of Doppler cycles received during that time interval. It is the Doppler cycle count that is physically meaningful. The count of the difference frequency is an additive constant that is eliminated by the position fix calculation.

Figure 28.7 emphasizes that the distance between the satellite and the observer changes throughout the satellite pass. It is this change, in fact, that causes the Doppler frequency shift. As the satellite moves closer, more cycles per second must be received than were transmitted, to account for the shrinking number of wavelengths along the propagation path. For each wavelength the satellite moves closer, one additional cycle must be received. Therefore, the Doppler frequency count is a direct measure of the change in distance between the receiver and the satellite over the Doppler count interval. In other words, the Doppler count is a geometric measure of the range difference between the observer and the satellite at two points in space, accurately defined by the navigation message. This is a very sensitive measure because each count represents one wavelength, which at 400 MHz is only 0.75 m.

The equation defining the Doppler count of $f_G - f_R$ is the integral of this difference frequency over the time interval between receipt of time marks from the satellite. For example,

$$N_1 = \int_{t_1 + R_1/C}^{t_2 + R_2/C} (f_G - f_R)\, dt \qquad (28.1)$$

Note that $t_1 + R_1/C$ is the time of receipt of the satellite time mark that was transmitted at time t_1. The signal is received after propagating over distance R_1 at the velocity of light C.

Equation 28.1 represents the actual measurement made by the satellite receiver, but it is helpful to expand this equation into two parts:

$$N_1 = \int_{t_1+R_1/C}^{t_2+R_2/C} f_G \, dt - \int_{t_1+R_1/C}^{t_2+R_2/C} f_R \, dt \qquad (28.2)$$

Since the first integral in Eq. 28.2 is of a constant frequency f_G, it is easy to integrate; but the second integral is of the changing frequency f_R. However, the second integral represents the number of cycles *received* between the times of *receipt* of two timing marks. By a "conservation of cycles" argument, this quantity must equal identically the number of cycles *transmitted* during the time interval between *transmission* of these time marks. Using this identity, Eq. 28.2 can be written

$$N_1 = \int_{t_1+R_1/C}^{t_2+R_2/C} f_G \, dt - \int_{t_1}^{t_2} f_T \, dt \qquad (28.3)$$

Because the frequencies f_G and f_T are assumed constant during a satellite pass, the integrals in Eq. 28.3 become trival, resulting in

$$N_1 = f_G\left[(t_2 - t_1) + \frac{1}{C}(R_2 - R_1) \right] - f_T(t_2 - t_1) \qquad (28.4)$$

Rearranging the terms in Eq. 28.4 gives

$$N_1 = (f_G - f_T)(t_2 - t_1) + \frac{f_G}{C}(R_2 - R_1) \qquad (28.5)$$

Equation 28.5 clearly shows the two parts of the Doppler count. First is the constant difference frequency multiplied by a time interval defined by the satellite clock. Second is the direct measure of slant range change measured in wavelengths of the ground reference frequency C/f_G. It happens that the wavelength of f_G is the proper scale factor because received time marks are used to start and stop the Doppler counts. If a ground clock is used to control the count intervals, the wavelength of f_T would become the appropriate scale factor.

28.2.3.3. Computing the Fix

A usable satellite pass will be above the horizon between 10 and 18 min. During this time, the number of Doppler counts acquired will be dependent on the length of each one, but typically 20 to 40 counts will be collected by modern equipment. The Doppler counts and the satellite navigation message are passed to a small digital computer for processing.

The computer first takes advantage of message redundancy to eliminate errors in the received orbital parameters. It is then able to compute the satellite's position at the beginning and end of every Doppler count. The computer also receives an estimate of the navigator's position in three dimensions, latitude, longitude, and altitude above the reference ellipsoid, from which the slant range to each satellite position can be computed. It is then possible to compare the slant range change as measured by each Doppler count with the corresponding value computed from the estimated navigator's position.

The difference between a Doppler measured slant range change and the value computed from the estimated position is called a residual e_i. The objective of the position fix calculation is to find the navigator's position that minimizes the sum of the squares of the residuals (i.e., makes the calculated slant range change values agree best with the measured values). To implement the solution, a simple, linear estimate is made of the effect each variable will have on each residual. Assuming we wish to solve for latitude (ϕ), longitude (λ), and the unknown frequency offset $F = f_G - f_T$, we can write

$$\hat{e}_i = e_i + \frac{\partial e_i}{\partial \phi} \Delta\phi + \frac{\partial e_i}{\partial \lambda} \Delta\lambda + \frac{\partial e_i}{\partial F} \Delta F \qquad (28.6)$$

This equation states that if we move the estimated position by $\Delta\phi$ and by $\Delta\lambda$ and the estimated frequency offset by ΔF, the present residual e_i will become a new value, estimated to be $\hat{e}_i$. Next we wish to minimize the sum of the squares of the estimated residuals by setting the partial derivative with respect to each variable equal to zero. This results in three equations, where the summation covers the m valid Doppler count residuals.

$$\frac{\partial}{\partial \phi} \sum_{i=1}^{m} \hat{e}_i^2 = 2 \sum_{i=1}^{m} \left(\hat{e}_i \cdot \frac{\partial \hat{e}_i}{\partial \phi} \right) = 0$$

$$\frac{\partial}{\partial \lambda} \sum_{i=1}^{m} \hat{e}_i^2 = 2 \sum_{i=1}^{m} \left(\hat{e}_i \cdot \frac{\partial \hat{e}_i}{\partial \lambda} \right) = 0 \qquad (28.7)$$

$$\frac{\partial}{\partial F} \sum_{i=1}^{m} \hat{e}_i^2 = 2 \sum_{i=1}^{m} \left(\hat{e}_i \cdot \frac{\partial \hat{e}_i}{\partial F} \right) = 0$$

Ignoring all but the first-order terms of Equations 28.7 gives three equations in the three selected variables, $\Delta\phi$, $\Delta\lambda$, and ΔF.

$$\sum_{i=1}^{m} \frac{\partial e_i}{\partial \phi} \left[e_i + \frac{\partial e_i}{\partial \phi} \Delta\phi + \frac{\partial e_i}{\partial \lambda} \Delta\lambda + \frac{\partial e_i}{\partial F} \Delta F \right] = 0$$

$$\sum_{i=1}^{m} \frac{\partial e_i}{\partial \lambda} \left[e_i + \frac{\partial e_i}{\partial \phi} \Delta\phi + \frac{\partial e_i}{\partial \lambda} \Delta\lambda + \frac{\partial e_i}{\partial F} \Delta F \right] = 0 \qquad (28.8)$$

$$\sum_{i=1}^{m} \frac{\partial e_i}{\partial F} \left[e_i + \frac{\partial e_i}{\partial \phi} \Delta\phi + \frac{\partial e_i}{\partial \lambda} \Delta\lambda + \frac{\partial e_i}{\partial F} \Delta F \right] = 0$$

Because only linear, first-order terms are used, the values of $\Delta\phi$, $\Delta\lambda$, and ΔF that satisfy these equations will be an approximation to the exact solution. Therefore, the original estimates of latitude, longitude, and frequency are adjusted in accordance with the first solution, and new slant ranges, residuals, and partial derivatives of the residuals are computed for another solution. This process is repeated, or iterated, until the computed values of $\Delta\phi$, $\Delta\lambda$, and ΔF are sufficiently small, at which point the solution is said to have converged. Normally, only two or three iterations are required, even when the initial estimate is tens of kilometers from the final solution. Note that ignoring higher order terms has no effect on final accuracy, because these terms tend to zero as the solution converges.

In summary, the Transit position fix begins with an estimated position and determines the shift in that position required to best match calculated slant range differences with those measured by the Doppler counts. The initial estimate can be in error by 200 or 300 km and the solution will converge to an accurate value.

28.2.3.4. Accounting for Motion

If the navigator is in motion during the satellite pass, the motion must be recorded before an accurate position fix can be computed. As Fig. 28.8 reveals, only if the motion is known can the calculated range differences from satellite to receiver be compared properly with the range differences measured by the Doppler counts. Automatic speed and heading inputs often are employed for this purpose. During the satellite pass the computer creates a table of the navigator's estimated latitude and longitude at the beginning and end of each Doppler count interval. As before, the fix solution provides a delta-latitude and a delta-longitude, which are added to every point in the navigator's table between iterations of the solution. Therefore, although the final position fix result may be expressed as a latitude and a longitude at one point in time, the fix solution in fact is a shift of the entire estimated track.

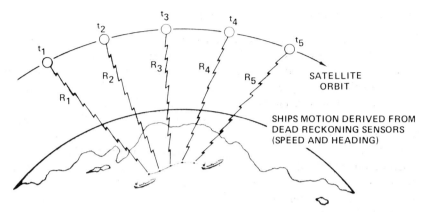

Figure 28.8 Knowledge of ship's motion is required for an accurate position fix update (adapted from *Stansell*, 1968, by permission of the Institute of Navigation).

Figure 28.9 illustrates the preferred mode of operation for a moving navigator. Between satellite fixes the computer automatically dead reckons based on inputs of speed and heading. The dead reckoning process also is used to fill the navigator's table of positions during each satellite pass. After the fix solution has converged, the delta-latitude and delta-longitude values are applied to the current position, thus correcting for the accumulated dead reckoning error.

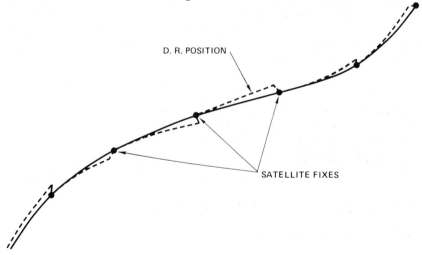

Figure 28.9 Dead reckoning allows continuous navigation between satellite fixes (courtesy Magnavox APD).

28.2.4. Frequency of Satellite Fixes

Satellite passes do not occur at a uniform rate. We saw in Fig. 28.3 that the coverage density is a minimum at the equator and increases with latitude. The figure also shows that the orbital spacing is not uniform; in fact the orbits slowly precess with respect to each other, and the spacing therefore continues to change (*Stansell*, Spring 1971, 1973).

The interval between satellite fixes also is influenced by interference between satellites. Such interference depends on time of satellite arrival at a specific location. For example, Fig. 28.10 plots the satellite passes at Washington, D.C., on December 11, 1974. The passes have been divided into three categories reflecting their maximum elevation angle above the horizon. Passes with an elevation between 10° and 70° are acceptable, but passes falling above or below these limits are more likely to have degraded accuracy and are rejected. Although the figure shows instances of two or three satellites being visible simultaneously, the receiver can operate on only one signal at a time; thus some of the passes will not be tracked. When a high or a low pass begins first, a good pass may be blocked. Furthermore, if the signals from two satellites cross each other in frequency, the receiver has a 50% probability of switching from one satellite to the other, thus denying any fix at all.

Some Transit receivers are equipped for computer control of their tuning. When properly implemented, this feature can cause the receiver to ignore undesirable passes and to track every good pass available. Most systems in use today either do not have this capability or employ a computer that has not been programmed to take advantage of it. Therefore, Fig. 28.10 lists fix interval statistics for both types of equipment. It is evident that the maximum time interval is much more sensitive to computer-aided tuning than the average interval between fixes.

In spite of all the variables, we attempt in Fig. 28.11 to describe the relationship between average fix interval and latitude. The interval decreases as latitude increases until the higher concentration of passes begins to cause more instances of cross-satellite interference. At extremely high latitudes, the elevation angle of all passes exceeds the 70° limit, and a special position fix program is needed for polar navigation.

28.3. ACCURACY CONSIDERATIONS

28.3.1. Static System Errors

Piscane et al. (1973) present an error budget for individual Transit position fixes that provides a good summary of the factors affecting

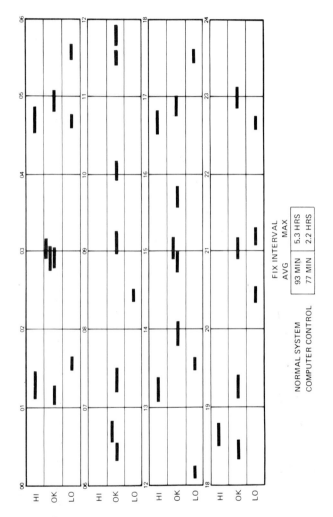

Figure 28.10 Example of daily distribution of satellite passes (courtesy Magnavox APD).

479

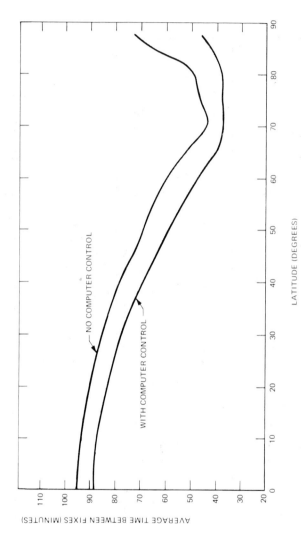

Figure 28.11 Average Interval between satellite fixes as a function of latitude (courtesy Magnavox APD).

accuracy when the navigator is not moving:

Source	Error (m)
1. Uncorrected propagation effects (ionospheric and tropospheric effects)	1–5
2. Instrumentation and measurement noise (local and satellite oscillator phase jitter, navigator's clock error)	3–6
3. Uncertainties in the geopotential model used in generating the orbit	10–20
4. Uncertainties in navigator's altitude (generally results in bias in longitude)	10
5. Unmodeled polar motion and UT1–UTC effects	0–10
6. Incorrectly modeled surface forces (drag and radiation pressure acting on the satellites during extrapolation interval)	10–25
7. Ephemeris rounding error (last digit in ephemeris is rounded)	5

Since publication of this table in 1973, the polar motion error has been modeled and is included as an adjustment to the transmitted orbit parameters. The root sum square (rss) of the remaining errors lies in the range of 18 to 35 m, which we believe is slightly optimistic due to the laboratory standards and the sophisticated refraction correction models employed by the Applied Physics Laboratory. Field results usually lie in the range of 27 to 37 m rss. Fig. 28.12 is a typical set of stationary fix results. The maximum error is 77 m, and the rss radial error is 32 m for all 69 points.

28.3.1.1. *Refraction Error*

There are two sources of refraction error; the larger one is due to the ionosphere. As illustrated by Fig. 28.13, as the 150 and 400 MHz signals pass through the ionosphere, their wavelengths are stretched because of interaction with free electrons and ions. This stretching represents a *phase velocity* greater than the speed of light, which is characteristic of a

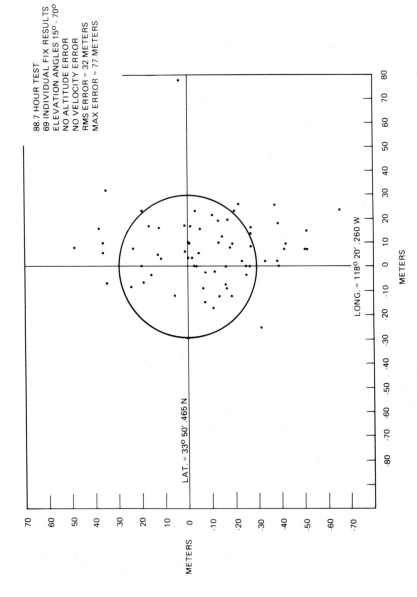

Figure 28.12 Typical dual-channel position fix results (courtesy Magnavox APD).

88.7 HOUR TEST
69 INDIVIDUAL FIX RESULTS
ELEVATION ANGLES 15° - 70°
NO ALTITUDE ERROR
NO VELOCITY ERROR
RMS ERROR = 32 METERS
MAX ERROR = 77 METERS

LAT. = 33° 50′ .465 N

LONG. = 118° 20′ .260 W

METERS

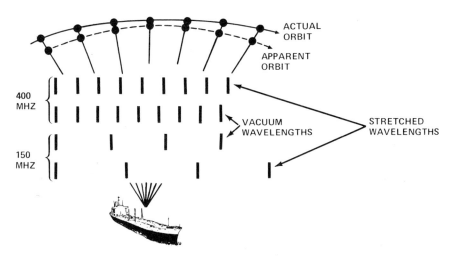

Figure 28.13 Ionospheric refraction stretches signal wavelength, causing greater apparent orbit curvature (courtesy Magnavox APD).

dispersive medium. To a close first-order approximation, the wavelength stretch is inversely proportional to the square of transmitted frequency. Because satellite motion changes the path length through the ionosphere, the rate of change of this stretch causes an ionospheric refraction error frequency shift in the received signal. *Guier* and *Weiffenbach* (1960) showed that an excellent refraction correction could be obtained by combining the Doppler measurements made at two different frequencies, and this is why Transit satellites transmit both 150 and 400 MHz signals.

For applications not requiring the ultimate system accuracy, 400 MHz single-channel receiving equipment can be used. Figure 28.13 demonstrates that because of wavelength stretching, the satellite will appear to follow a path with greater curvature about the navigator. The effect is to reduce the total Doppler shift somewhat, pushing the position fix solution away from the satellite orbit to explain the lower Doppler slope. Because the satellites move primarily along north-south lines, the resultant navigation errors are mostly in longitude. The magnitude of these errors varies with density of the ionosphere from very small at night to peaks of 200 to 500 m in daylight, depending on sunspot activity and location with respect to the magnetic equator, where the ionosphere is most dense. Figure 28.14 is a plot of typical single-channel results containing both daytime and nighttime fixes in which the maximum error is 242 m and the rss error is 88 m.

The second source of refraction error is the troposphere. In this case,

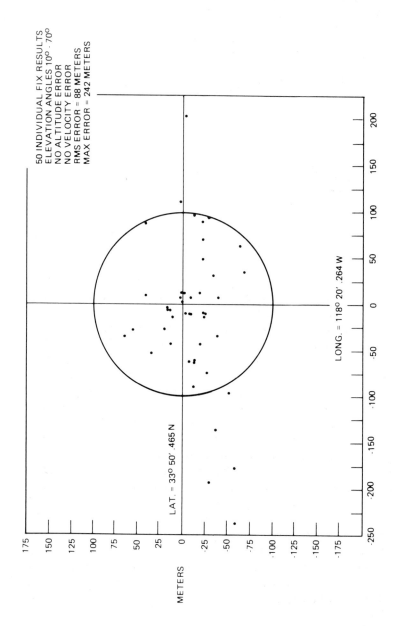

Figure 28.14 Typical single channel position fix results (courtesy Magnavox APD).

propagation speed is slowed as the signal passes through the earth's atmosphere, which compresses the signal wavelength. The effect is directly proportional to transmitted frequency, as is the Doppler shift, and therefore it cannot be detected like ionospheric refraction. There are only two ways to reduce the effect of tropospheric refraction. First is by modeling its effect on the Doppler counts. Very sophisticated models employing measurements of temperature, pressure, and humidity have been published for this purpose, but less sophisticated models are usually sufficient (*Moffet*, 1973). This is especially true in conjunction with the second technique, which is to delete Doppler data taken close to the horizon where the tropospheric refraction error is greatest. Above 5° to 10° of elevation, the tropospheric error is many times smaller than at the horizon.

28.3.1.2. *Altitude Error*

The specific Doppler curve obtained as a satellite passes is predominantly a function of the navigator's position along the line of satellite motion and his distance from the orbit plane. Because Transit satellites are in polar orbits, the along-track position closely relates to latitude and the cross-track distance is a combination of longitude and altitude.

Figure 28.15 is the cross section of a pass where the satellite is moving in its orbital plane perpendicular to the page. It has just reached the center of pass with respect to stations *X*, *Y*, and *Z*. The figure illustrates

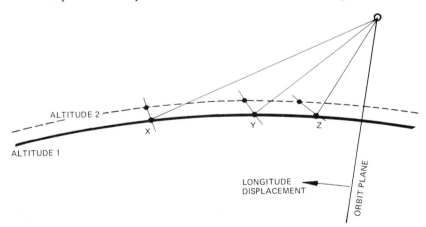

Figure 28.15 Effect of altitude estimate on position fix (adapted from *Stansell*, April 1971, by permission of the Institute of Electrical and Electronics Engineers, Inc.).

how the cross-track distance is a function of both longitude and altitude, which affect the Doppler curve in similar ways. To compute an accurate fix, therefore, it is necessary to have a priori knowledge of altitude. Figure 28.16 shows the sensitivity of fix error to altitude error as a function of maximum satellite pass elevation angle. The elevation angle is plotted on a scale that is uniform in probability of satellite pass occurrence. In other words, more passes fall between 10° and 20° than between 70° and 80°, except at very high latitudes.

For satellite navigation, "altitude" means height above or below the reference spheroid (the reference ellipsoid or satellite datum). This surface is chosen to be a worldwide best fit to mean sea level, which is the true geoid. Figure 28.17 illustrates the differences between the geoid, the spheroid, and topography. Therefore, knowing height above mean sea level is not sufficient for an accurate position fix. One also must know the local geoidal height, which is the deviation between the geoid and the spheroid. Figure 28.18 is a geoidal height map indicating that these deviations reach nearly 100 m. The ellipsoidal reference for Fig. 28.18

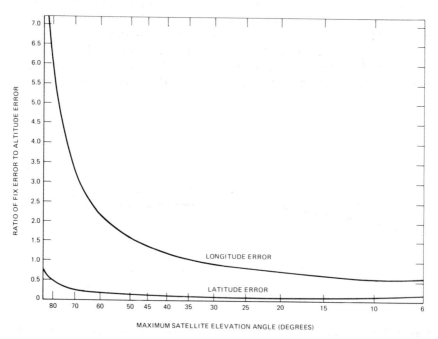

Figure 28.16 Sensitivity of satellite fix to altitude estimate error (courtesy Magnavox APD).

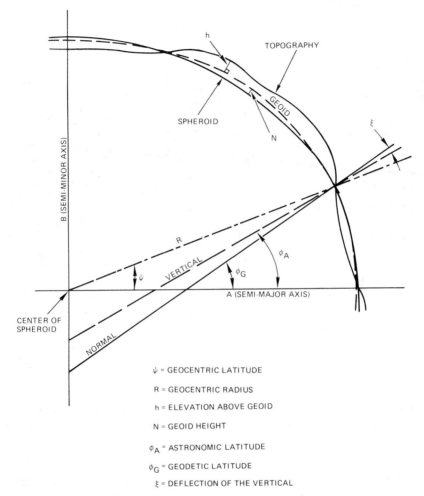

Figure 28.17 Relationships of geodetic surfaces (from *NASA Directory of Observation Station Locations*, 2nd ed., Vol. I, November 1971, Goddard Space Flight Center).

has a semimajor axis (equatorial radius) of 6,378,137 m and a flattening coefficient of 1/298.25 (flattening coefficient is the difference between the major and minor axes divided by the major axis).

The geoidal height chart of Fig. 28.18 was developed by observing the influence of the earth's gravity field on satellite orbits. As a result, extremely fine grain structure cannot be detected, and the map is known to be in error by ±20 m in many places. Since the map was first published

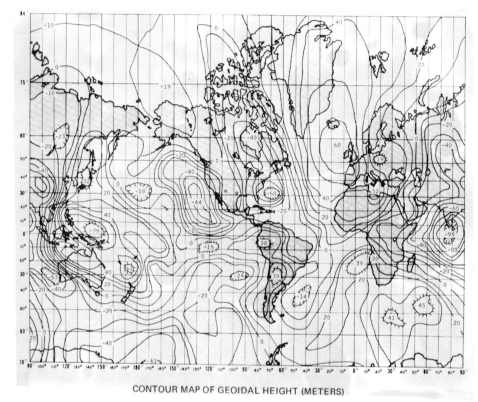

CONTOUR MAP OF GEOIDAL HEIGHT (METERS)

Figure 28.18 Geoidal height chart obtained from model of earth's gravity field. Dimensions are meters of mean sea level above the reference spheroid (from *Moffett*, 1973).

in 1967, refinements have been made but have not been released because of military classification. Thus for maximum accuracy it is better to determine local geoidal height by the fixed site survey techniques described in Section 28.4.3.

28.3.2. Accuracy Underway

All discussion of navigation error for a stationary receiver applies equally well to a receiver that is moving, as long as that motion is precisely known and secondary system errors are not introduced. If the motion is not known accurately, however, additional position fix error will occur.

Figure 28.19 is a useful way to visualize the effect of velocity error. The ellipse is fitted through eight position fixes, with a one-knot velocity error in each of eight compass directions. Note that fix error is greater when the velocity error is in a north-south direction, and the fix error direction depends on direction of satellite motion and on whether the pass is to the east or to the west of the observer.

Whereas Fig. 28.19 is for a single position fix, Figs. 28.20 and 28.21 show the errors caused by a one-knot velocity error north and east, respectively, as a function of maximum pass elevation angle. As with Fig. 28.16, elevation angle is plotted to represent a uniform probability of pass occurrence.

Figure 28.22 represents an attempt to express overall Transit system accuracy in a single set of curves. The result is overly simplified and somewhat conservative, but it does indicate realistic rss performance levels. One can see that a dual-channel system provides maximum benefit when there is an accurate source of velocity. The other benefit of the

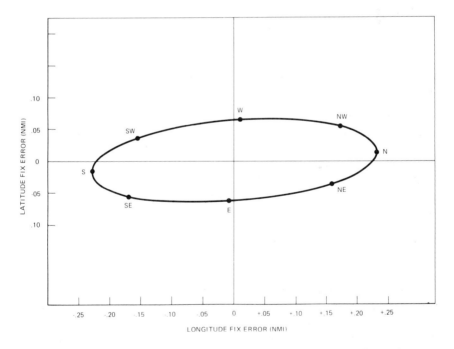

Figure 28.19 Effect of a one-knot velocity error on fix from a 31° satellite pass. Direction of velocity error is noted beside each of eight fix results. Satellite was east of receiver and heading north (courtesy Magnavox APD).

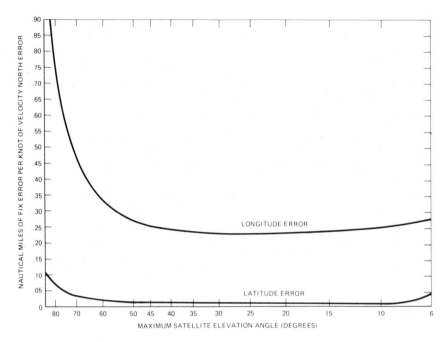

Figure 28.20 Sensitivity of satellite fix to a one-knot velocity north estimate error (courtesy Magnavox APD).

dual-channel system is to eliminate the peak 200–500 m errors that occur with single-channel equipment during the day, dependent on sunspot activity. The error curves begin at a point above the fixed site accuracy levels. This adjustment is intended to account for such secondary effects as the impact of pitch and roll on antenna position and on reference oscillator stability, and the likelihood of additional error in knowledge of antenna altitude.

28.3.3.   Velocity Solution

The normal position fix solution determines latitude, longitude, and frequency offset by means of Eqs. 28.8. These equations easily could be expanded to include other system variables such as velocity north, velocity east, altitude, and even acceleration. With every new variable, however, accuracy would become more and more sensitive to system noise. In fact, studies have shown that velocity north is the only parameter that can be added without creating intolerable noise sensitivity; that is, it is the only

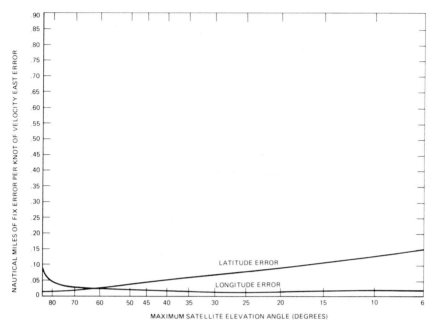

Figure 28.21 Sensitivity of satellite fix to a one-knot velocity east estimate error (courtesy Magnavox APD).

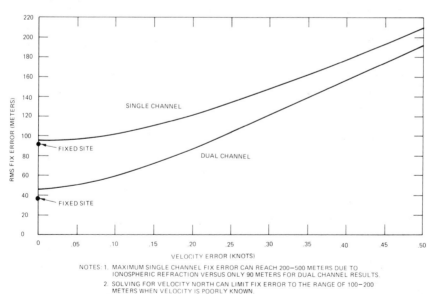

NOTES: 1. MAXIMUM SINGLE CHANNEL FIX ERROR CAN REACH 200–500 METERS DUE TO IONOSPHERIC REFRACTION VERSUS ONLY 90 METERS FOR DUAL CHANNEL RESULTS.

2. SOLVING FOR VELOCITY NORTH CAN LIMIT FIX ERROR TO THE RANGE OF 100–200 METERS WHEN VELOCITY IS POORLY KNOWN.

Figure 28.22 Satellite fix accuracy as a function of velocity error (courtesy Magnavox APD).

other variable that affects Doppler curve shape in a way that can be discerned clearly from the effects of latitude, longitude, or frequency. To be precise, the added variable should be velocity parallel to satellite motion, but velocity north is an adequate approximation at most latitudes because the satellites are in polar orbits.

Solving for velocity north increases position fix error when ship's motion is accurately known. Therefore it should be attempted only when velocity errors are likely to exceed about 0.4 knot, in which case the expanded solution should be better. Qualifications are expressed because the expanded solution is more sensitive to other sources of system noise, such as asymmetric Doppler data, and it does not work well for pass elevation angles below 20°. Finally, the velocity north result becomes the scapegoat for other system errors and is not a dependable measure of velocity north error; it simply allows the latitude and longitude to be more accurate in the face of large velocity errors.

28.3.4. Reference Datum

A local survey is always performed with reference to previously established benchmarks, which in turn are defined with respect to the local datum. For example, the reference datum for the United States is the North American Datum of 1927 (NAD-27), with a defined origin at Meades Ranch in Kansas. Figure 28.23 lists nine of the most widely used reference datums.

Because a satellite system operates on a global basis, it cannot provide fix results on each local datum. Thus a global datum is adopted, and position fixes must be transformed from the global datum to the appropriate local datum. To pursue this subject, we must understand what is meant by "position."

Assume the earth to be an oblate spheroid (ellipsoid) having X-, Y-, and Z-axes with their origin at the center of the spheroid. The X- and Y-axes lie in the equatorial plane, with X pointing toward the prime meridian (Greenwich) and Y pointing toward 90° longitude. The Z-axis points north and nominally is the rotational axis of the earth: for Transit, it is defined as the mean position of the rotational axis for the years 1900–1905, the "mean pole, 1900–1905," or the "Conventional International Origin" (CIO) (*Piscane et al.*, 1973). Any point on the earth's surface can be defined by coordinates x, y, and z. The longitude of a point is the angle between the X-axis and the point x, y projected onto the equatorial plane.

Figure 28.23 Transformation constants between world geodetic systems (adapted from *Stansell*, 1973, by permission of the Society of Exploration Geophysicists, with additions from *Seppelin*, 1974).

DATUM	SPHEROID	SEMI-MAJOR AXIS (METERS)	RECIPROCAL FLATTENING	SHIFT TO NWL-8D a = 6378145 1/f = 298.25			SHIFT TO SAO-C7 a = 6378142 1/f = 298.255			SHIFT TO WGS-72 a = 6378135 1/f = 298.26		
				ΔX	ΔY	ΔZ	ΔX	ΔY	ΔZ	ΔX	ΔY	ΔZ
NAD 1927	CLARKE 1866	6378206	294.98	-23	159	185	-26	155	185	-22*	157*	176*
EUROPEAN	INTERNATIONAL	6378388	297.00	-81	-99	-118	-93	-132	-143	-84	-103	-127
TOKYO	BESSEL	6377397	299.15	-147	530	676	-140	510	689	-140	516	673
INDIAN	EVEREST	6377276	300.80				293	697	228			
AUSTRALIAN NATIONAL	REFERENCE ELLIPSOID 1967	6378160	298.25				-88	-36	86	-122	-41	146
OLD HAWAIIAN	CLARKE 1866	6378206	294.98	52	-262	-183	59	-263	-203			
MAUI										65	-272	-197
OAHU										56	-268	-187
KAUAI										46	-271	-181
CAPE (ARC)	CLARKE 1880 (MOD)	6378249	293.47				-130	-147	-347	-129	-131	-282
SOUTH AMERICAN	REFERENCE ELLIPSOID 1967	6378160	298.25				-284	116	-410	-77	3	-45
ARGENTINE	INTERNATIONAL	6378388	297.00				-167	128	25			

*VALUES OF -9, 139, AND 173 SHOULD BE USED FOR ALASKA AND CANADA

Latitude of point x, y, z is more complex, being *geodetic* latitude rather than geocentric latitude. The geodetic latitude is the angle between the equatorial plane and a line through x, y, z which is normal to the spheroid at that point, as illustrated by Fig. 28.17. Thus latitude is influenced by the shape of the spheroid.

To complete the picture, altitude is normally surveyed with respect to mean sea level, and the "geoid" is the surface that describes mean sea level over the entire earth. A spheroid is derived to be a "best fit" in some way to the true geoid. Nevertheless, the perfectly smooth spheroid and the actual, rather lumpy geoid deviate from each other, as shown previously by Figs. 28.17 and 28.18. The difference is known as geoidal height.

It would be convenient if all the world used the same spheroid for mapping. Unfortunately, many different spheroids are in use, each one having been selected to "best fit" the geoid in that particular area. Furthermore, regional surveys are conducted with respect to a local datum, such as NAD-27. A datum is established based on many points, surveyed with respect to each other and also positioned independently by astronomic observations. All data are combined to establish a best fit datum having an origin point usually represented by a physical marker for which there are defined values of latitude, longitude, height, and deflection of the vertical (which Fig. 28.17 shows to be the angular difference at that point between the normal to the spheroid and local vertical).

Establishing a datum has the effect of defining the center of the spheroid. Thus if we could draw two spheroids associated with different datums, not only would the semimajor axes and flattening coefficients likely be different for the two spheroids, but their centers also would be at different positions. The net result of these considerations is that the coordinates of a point are dependent on which datum is being used. The same point surveyed in two different datums will have two different latitudes, longitudes, and heights above the spheroid, although height above mean sea level obviously will be the same.

By means of satellite measurements (satellite geodesy), the offset of one datum with respect to all others can be measured. The worldwide datum now being used by the Transit system is quite similar to the NWL-8D Datum, developed by the Naval Weapons Laboratory, and also similar to the SAO-C7 Datum, developed by the Smithsonian Astrophysical Observatory. Figure 28.23 tabulates nine datums in common use around the world, defines their spheroidal constants, and gives the origin shifts required to translate positions from these local datums to either the NWL-8D or the SAO-C7 datum. The shift parameters are known to a radial accuracy of 10 to 30 m, and as satellite geodesy progresses, accuracy will improve.

Given the information in Fig. 28.23, it is possible to calculate the shift in latitude and longitude required to transform positions from one datum to another (*Moffett*, 1973; *Seppelin*, 1974). For example, the shift to satellite coordinates at the origin of the Tokyo Datum is 347 m north and 309 m west. At the origin of the European Datum, the shift is 80 m south and 78 m west. A chart covering the United States, Mexico, and Central America (Fig. 28.24) emphasizes that the shift between the North American Datum and the NWL-8D Satellite Datum, for example, is not a constant offset but is a function of position.

When comparing local survey data with satellite fix results, the effect of datum shift naturally must be considered, but when there is a difference it may not be due totally to datum shift. Too often, after a difference is found, the local survey is checked and found to be in error. A local

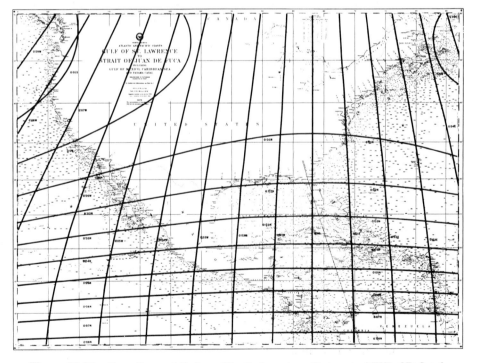

Figure 28.24 Coordinate shift from North American Datum to NWL-8D Satellite Datum (from *Stansell*, 1973, by permission of the Society of Exploration Geophysicists).

system may show excellent precision and consistency of results and still be quite wrong because of survey error. If both a local system and a satellite system are to be used, a calibration of position difference before the survey begins is extremely wise. The offset is a measure both of datum shift and of local survey error, and it can be applied as a bias throughout the survey.

Sometime during 1975 it is planned that the Naval Astronautics Group will convert the Transit system to a more recent model of the earth's gravity field which is consistent with WGS-72 (World Geodetic System, 1972: *Seppelin*, 1974). The change should result in improved orbit determination accuracy, but a number of problems will be encountered. First, the WGS-72 spheroid constants are $a = 6378135$ m and $f = 1/298.26$, which are different from the previously used NWL-8D constants, and position fix results may shift by up to 15 m because of the new geoid definition. It is hoped that the government will release a new and more accurate geoidal height chart consistent with the new datum to replace the map of Fig. 28.18. *Seppelin* (1974) lists datum shift parameters which were developed as part of the WGS-72 effort. Some of these have been included in Fig. 28.23 and are considered superior to previous values. (Note: the change to WGS-72 was implemented in late 1975.)

28.4. SYSTEM APPLICATIONS

28.4.1. Commercial Navigation

Transit is the only global navigation system at this writing. It is an extremely reliable source of accurate position information, since it is unaffected by weather conditions and does not have the lane slip or sky wave problems that plague terrestrial systems. For these reasons, the use of Transit for commercial navigation has been increasing steadily since 1971 when lower cost, single-channel navigation sets first became available. As demand has increased, the cost of the equipment has decreased, making such devices available for a wider range of applications. Today, tankers, freighters, container ships, fishing vessels, and even major yachts are employing satellite navigation.

The elements of every installation are:

1. Small digital computer.
2. Computer program.
3. Satellite receiver.
4. Operator control and display.

These are depicted in Fig. 28.25, which shows typical single-channel satellite navigation equipment. In this instance, the computer program is contained on a Philips-type magnetic tape cassette that is loaded into the computer by the reader located in the operator's console. The computer is equipped with a three-button front panel for ease of operation. The satellite receiver consists of the antenna/preamplifier unit that mounts on the ship's mast, two small modules that plug inside the computer, and a reference oscillator attached to the computer. Speed and heading signals are obtained from conventional speed log and gyrocompass units and are connected by an interface module, also plugged inside the computer. A fourth computer module is the link to the operator console, which is usually mounted near the chart table or on the ship's bridge.

By automatic dead reckoning, the system in Fig. 28.25 constantly provides Greenwich time, latitude, and longitude. After every acceptable satellite pass, the time and position information are updated automatically. The system also is programmed to display the great circle range and bearing to any selected way point. Simplicity of operation is stressed, and a navigator can learn to use the system in a half-hour or less.

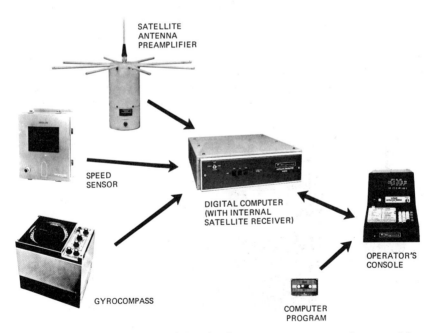

Figure 28.25 Typical commercial navigation system components (courtesy Magnavox APD).

28.4.2. Oceanographic Exploration

The first application of Transit navigation beyond its original military objectives was for oceanographic exploration. For the first time, mid-ocean scientific measurements could be tied to their geographic origin with high accuracy. Figure 28.26 presents the type of dual-channel Transit equipment often used for oceanographic exploration. Comparing it with the single-channel set of Fig. 28.25, we see that there is a larger computer with complete front panel controls. The dual-channel receiver, being more complex, is separate from the computer, and the antenna/preamplifier unit is larger. Speed and heading signals are connected to the computer by an interface module, as with the commercial equipment. The computer program is provided on punched paper tape and is loaded by means of a photoreader. Operator interface may be a Teletype or an electronic printer and keyboard, with a remote video display for instant readout.

In addition to the capabilities provided by the commercial single-channel system, the dual-channel equipment gives high accuracy position fixes that are unaffected by variations in ionospheric refraction. The printed output (Fig. 28.27) automatically records the dead reckoned position at selected time intervals, as well as every satellite fix with such associated parameters as course and speed, elevation angle of the satellite pass, and the estimated error in latitude (S-LA) and in longitude (S-LO). Based on the magnitude of the position update since the last satellite fix, the program automatically computes the set and drift affecting the dead

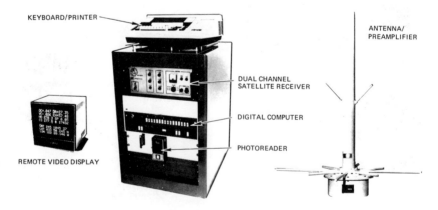

Figure 28.26 Dual-channel Transit navigation system (courtesy Magnavox APD).

```
RC   FIX MAP-VN-75090
111111111111111111111111111111111111111000000
            ↑↑
DAY        TIME       LAT        LON        ANT        CRSE      GSPD      NVEL
243    151000    033 50·476  N  118 20·276  W -12·50     ·0000     ·0000     ·0000
ITER       ELEV      GEOM       SAT     S-LA       S-LO     S-VN      RMS       FREQ
03          49·       S-E     30180    ·0075     ·0138    ·0000     ·0078    4·6973

S--UPDAT      ·01313354     --·01027032      ·01667242
C-WHDG   C-WSPD
 359·4   ·0023
N--UPDAT


RC   VNFIX MAP-VN-75090
111111111111111111111111111111111111111000000
            ↑↑
DAY        TIME       LAT        LON        ANT        CRSE      GSPD      NVEL
243    151000    033 50·478  N  118 20·304  W -12·50     ·0000     ·0000     ·0926
ITER       ELEV      GEOM       SAT     S-LA       S-LO     S-VN      RMS       FREQ
02          49·       S-E     30180    ·0230     ·0632    ·1936     ·0078    4·6387

S--UPDAT      ·01560802     --·03463127      ·03798598
C-WHDG   C-WSPD
 358·0   ·0027
2D-UPDAT
```

Figure 28.27   Printout of satellite position fix (courtesy Magnavox APD).

reckoning (C-WHDG and C-WSPD), and these values can be used to improve future dead reckoning accuracy. The printout of Fig. 28.27 is from a program with the velocity north calculation feature. It is employed only when the satellite pass is above 20° maximum elevation, and then only if the Doppler data are quite complete and symmetric about the center of the pass. The printout shows the extent and symmetry of Doppler data by the line of zeros and ones with two arrows underneath. Each character represents a potential Doppler count, and the ones show which counts were employed in the fix process.

28.4.3.   Fixed Point Survey

Transit equipment can be used to bring precise geodetic control to areas that are difficult to reach and where conventional survey techniques are too slow and too costly. Examples are positioning of offshore drill rigs and providing control points for regions such as the Amazon Basin in Brazil and the north slope of Alaska. The time required to establish a control point with Transit can be measured in hours rather than in weeks or months if conventional land survey techniques are employed over great distances or under adverse conditions. One very powerful and effective technique is to combine aerophotogrammetry with Transit position fixes.

Visible markers are placed at suitable intervals throughout the area to be mapped, and Transit equipment is used to establish the position of each marker. Transportation to these sites is often by helicopter. When the aerial photographs are converted into maps, the visible markers provide the necessary geodetic reference.

A single Transit position fix is not sufficiently accurate to satisfy most survey requirements. Therefore, special techniques have been developed for this application.

28.4.3.1. *Three-Dimensional Point Survey*

Although one satellite position fix is not sufficiently accurate, the average of many fixes taken at one location can be quite accurate. Here we must distinguish between accuracy and repeatability. Positions that are repeatable to one meter can be established and tied to the local reference datum. However "accuracy" would imply no geodetic bias in the satellite orbit parameters or in the measurement techniques. Such biases are believed to be on the order of 10 m or less, and only future improvements in orbital determination techniques will detect and eliminate these small biases. In other words, if a point is 10 m off on a worldwide grid, no other survey technique can prove it.

Although averaging results from many fixes eventually will yield a satisfactory answer, this process presents two questions:

1. Is the altitude estimate correct?
2. How should the individual fixes be weighted in reaching an average?

Fortunately, the questions can be avoided by using a special computer program that determines the one three-dimensional position that best fits all the Doppler measurements obtained from all the satellite passes taken at that location (*Stansell*, April 1971). Figure 28.28, as compared with Fig. 28.15, provides an intuitive grasp of the process. Altitude error affects the calculated longitude of individual position fixes. If measurements from several satellite passes are used to calculate a single position, however, both longitude and altitude can be determined.

Figure 28.29 shows how a typical three-dimensional survey converges to the final answer. Each time another satellite is tracked, its data are combined with all previous data and a new solution is computed. The figure indicates that with each successive satellite pass, the latitude, longitude, and altitude parameters converge toward the final solution.

Other tests have been conducted to measure the repeatability of the

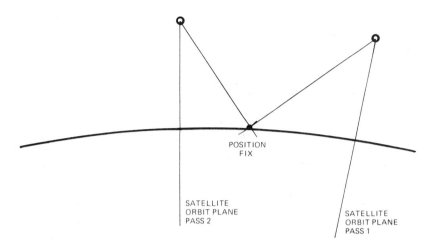

Figure 28.28 Multiple satellite passes can define latitude, longitude, and altitude (from *Stansell,* April 1971, by permission of the Institute of Electrical and Electronics Engineers, Inc.).

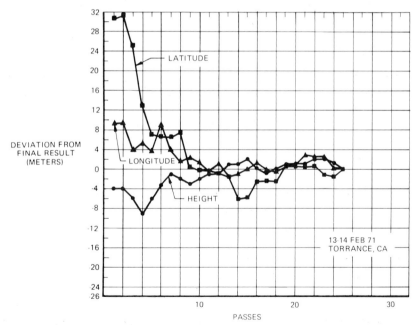

Figure 28.29 Convergence of three-dimensional fix solution (adapted from *Stansell,* April 1971, by permission of the Institute of Electrical and Electronics Engineers, Inc.).

three-dimensional solutions. For example, 11 independent solutions each employing N different satellite passes have been compared with the following results.

Comparison of 11 Independent Station Solutions (m)

Passes per Solution	Sigma Latitude	Sigma Longitude	Sigma Altitude	Sigma Radial	Maximum Radial
10	6.3	6.0	3.2	9.3	18.6
25	4.1	3.8	2.7	6.3	10.4
50	2.6	3.6	2.0	4.9	9.6

These data show the level of repeatability to be expected depending on the number of satellite passes processed at each site. (The values may be somewhat pessimistic, because the data were taken over the course of a year prior to the polar wander correction (*Piscane et al.*, 1973), instrumentation error was not minimized with a clock option, and only single precision computer program results are presented.)

28.4.3.2. Precise Ephemeris

The results just described are available to every system user with the necessary equipment and computer program. The principal source of error is misknowledge of satellite orbital position. *Piscane et al.* (1973) list three reasons for error in the orbit parameters transmitted by Transit satellites:

1. Uncertainties in geopotential model used in generating orbit — 10–20 m
2. Incorrectly modeled surface forces (drag and radiation pressure acting on satellites during extrapolation interval) — 10–25 m
3. Ephemeris rounding error (last digit in ephemeris is rounded) — 5 m

    rss — 15–32 m

The orbit parameters in the satellite memory are a prediction of its position based on past tracking data. The prediction is obtained by numerical integration of the equations of motion, taking into account all known forces acting on the satellite, such as the gravity fields of the earth, sun, and moon, plus drag and radiation pressures. To the extent that these forces are not known precisely, the predicted orbit will deviate from the actual orbit. As listed, these differences account for most of the 27–37 m rss error in individual Transit position fixes.

If the orbit did not have to be predicted into the future, a more precise determination could be made, and the U.S. Defense Mapping Agency (DMA) employs this technique in reducing satellite Doppler data from survey receivers such as the AN/PRR-14 Geoceiver. Field data are recorded on punched paper tape and returned to a computing center for evaluation. There the Doppler data are combined with a precise ephemeris of satellite positions based on tracked rather than predicted orbits; thus individual position fixes have a typical scatter of only 6.3 m rss. Naturally a three-dimensional, multipass solution converges to the required accuracy much faster with this technique than when using predicted orbit parameters from the satellite. However, DMA seldom computes a precise ephemeris for more than one or two satellites at a time and immediate results cannot be obtained in the field, offsetting slightly the advantage just described. Even so, equipment using the predicted orbits must remain on station from 4 to 10 times longer than equipment using the precise ephemeris for equivalent accuracy results.

Precise ephemeris information is not available for commercial use. However, there is precedent for the DMA to supply this information to other nations based on cooperative international survey agreements.

28.4.3.3. *Translocation*

Although precise ephemeris data are not available commercially, another technique called translocation can yield equivalent results (*Stansell*, April 1971; *Krakiwsky et al.*, 1973; *Gloeckler et al.*, 1974). Advantage is taken of the fact that almost all the error in a position fix is caused by factors external to the satellite receiver. Thus two receivers tracking the same satellite pass at the same location should produce nearly the same result (i.e., the errors are strongly correlated). Experience has shown that the correlation is quite effective for interstation separations of 200 km or more (*Krakiwsky et al.*, 1973). As a result, two or more stations can be located with respect to each other with an accuracy of 2 m or better over very considerable distances.

If altitude is known at both stations, a single satellite pass can determine relative horizontal position to within 3 m rss, assuming precise

Doppler measurements are made and the calculations are carefully performed. With as few as three satellite passes, the relative three-dimensional position can be determined to about 2 m rss in the horizontal plane. Thus translocation, if done properly, can yield very accurate results with only a few satellite passes.

One method of using translocation is to establish a base station that collects data from all available satellite passes for days or weeks. When fed to the three-dimensional point positioning program, these data will yield an excellent absolute position determination. In the meantime, one or more portable receivers move from one location to another, gathering three to ten passes at each site. These data are then processed by translocation to transfer the accurate base station position to each of the remote locations.

To maximize the error correlation potential of translocation, it is important that only common data be used. Only one set of orbit positions should be computed per satellite pass, and only Doppler counts taken over identical time intervals should be used. If one receiver misses a Doppler count, the corresponding count from the second receiver should be discarded. For this reason it is best to record relatively short Doppler counts to minimize such data loss. (This problem was quite serious in previous years when 120 sec Doppler counts were commonly used, giving only four to eight counts per pass. Modern practice provides a count interval of 24 sec or less; thus 20 to 40 counts are recorded per pass.)

With the higher precision available through translocation or with precise ephemeris data, fix results become more dependent on satellite receiver instrumentation accuracy. Depending on the quality of the instrument, single-pass translocation results can range from 3 to 14 m rss or worse.

28.4.3.4. Land Survey Equipment

Figure 28.30 shows a typical complement of satellite land survey equipment. Rugged carrying cases are important to withstand transport by helicopter, jeep, or pack animal. The equipment consists of a collapsible antenna/preamplifier unit, the satellite receiver, and a paper tape recorder. Each satellite pass is tracked automatically, and the raw data are punched on tape for later entry into the computer. The recorded data include the satellite orbit parameters, Doppler counts, and identification and timing marks. Power sources range from batteries for brief operating periods to portable generators.

Instrumentation accuracy becomes an important factor whenever orbit errors do not dominate (e.g., translocation or surveys using precise

Figure 28.30 Land survey equipment (courtesy Magnavox APD).

TABLE 28.1 TRANSLOCATION AND POINT POSITIONING RESULTS: VARIOUS TYPES OF EQUIPMENT AND DATA PROCESSING TECHNIQUES[a]

Clocked Doppler Counts	Doppler Length (sec)	Number of Passes	Horizontal Error (m rss)
Two-dimensional translocation (37–57 km)			
No	120	1	12.3
No	30	1	13.2
Yes	30	1	2.9
Three-dimensional translocation (37 km)			
Yes	30	2	6.0
Yes	30	2	4.9 (selected geometry)
Yes	30	3	2.1 (selected geometry)
Yes	30	4	1.5 (selected geometry)
Yes	30	6	0.9 (selected geometry)
Yes	30	8	1.1 (selected geometry)
Point positioning (individual fix statistics)			
No	120		41.7
No	24		32.0
Yes	30		27.0
Yes	30		6.3 (precise ephemeris)

[a] Largely based on data from F. M. Gloeckler et al., *Doppler Translocation Test Program*, U.S. Army Engineer Topographic Laboratories, Fort Belvoir, Virginia, Report ETL-ETR-74-5, December 1974.

ephemeris data). Table 28.1 summarizes typical results obtained under such conditions as a function of Doppler count length and whether the receiver is equipped with a Doppler clock.

The primary source of Doppler error in the receiver is uncertainty in the start and stop time of each count. As described in Section 28.2.3, a count starts or ends on receipt of a time mark from the satellite. From Fig. 28.6 we know that only 6.25% of the signal power is devoted to the clocking signal, which has a period of nearly 10,000 μsec. Thus it is not surprising that clock errors of 50 to 150 μsec rms are typical. Since the Doppler frequency being counted has a period of 25–40 μsec, time recovery jitter can cause a count to begin several cycles early or late, where each count represents 0.75 m. Although such errors are relatively small compared with error in the predicted orbit, they are substantial for translocation.

The best technique for removing the effect of time recovery jitter is a precise clock that measures the time interval of each Doppler count, using the reference oscillator as a standard. The position fix equations must be modified slightly to accept the measured time interval information, but the result is a substantial improvement in translocation accuracy, such as the change from 12.3 to 2.9 m rss per pass in Table 28.1.

Only after the time recovery jitter problem has been solved is it necessary to turn to secondary sources of error. For example, typical dual-channel receivers employ reference oscillators with a rated stability of 2×10^{-11} per Doppler count. For 24 sec counts at 400 MHz, this translates into a stability of 0.2 count, or an error of about 15 cm. Better oscillators are available which reduce even this error by a factor of 2.

Another small error is delay through the receiver in recognizing the satellite time marks. A typical delay of 600 μsec causes the position fix to shift about 3.7 m opposite to the direction of satellite motion. In translocation, if both receivers have the same delay, the effect cancels. Also, using an equal number of satellite passes in opposite directions will tend to cancel residual error. Nevertheless, a receiver delay calibration is desirable when precise results are wanted.

28.4.4. Survey Navigation

The first commercial application of the Transit system was for offshore oil exploration. Survey vessels towing seismic streamers several kilometers long record signals from 24, 48, or 96 groups of hydrophones spaced along the streamer each time an acoustic source, such as a set of air guns, is actuated. Sophisticated processing of these signals can produce vertical section maps showing geological features of the earth below the sea. Before 1968, navigation of these survey vessels was performed only by shore-based radiolocation systems such as Decca, Raydist, and Shoran. Establishing accurate navigation in a remote area was difficult, time-consuming, and often very expensive. In addition, long-range systems did not offer adequate accuracy, and the accurate systems had limited range as well as lane slip problems.

The oil industry saw Transit as a way to obtain accurate position, anywhere in the world, without the need for expensive shore installations. There was an initial rush of enthusiasm followed by some disappointment when the system limitations were discovered. Exploration required continuous navigation to high accuracy, not intermittent fixes with degraded accuracy due to velocity error.

At about the same time, another new product was being tested for oil exploration. This was Doppler sonar, which transmits pulses of high frequency sound in four narrow beams toward the ocean bottom. Signals reflected from the bottom, or from scattering particles in the water in deep areas, are received, amplified, and tracked in frequency. The Doppler frequency shift of the returned signals is a direct measure of ship's velocity. By means of a gyrocompass, the sonar velocity signals can be resolved into north and east coordinates for continuous dead reckoning.

By combining Transit position fixes with the Doppler sonar equipment, it appeared that an accurate, continuous navigation capability could be achieved, with no dependence on shore stations. It was soon learned that the two systems could not operate independently if the best results were wanted, and this recognition spawned the development of several varieties of integrated navigation system.

Figure 28.31 shows the elements of a modern integrated navigation system (*Stansell*, April 1971, 1973, 1974). All components are connected to the system computer which is programmed to perform the following functions:

1. Navigation.
2. Control of the survey.
3. Automatic data logging.
4. Collection of auxiliary geophysical data.

28.4.4.1. *Navigation*

The Doppler sonar transducer protrudes beneath the vessel to operate below the aerated water and bubbles next to the hull. In addition to the four sonar transducer crystals, a thermistor measures water temperature for speed of sound calibration and a two-axis inclinometer assembly at the top permits compensation for pitch and roll effects. The transducer is mounted through a "gate valve" for service removal without requiring the vessel to enter dry dock.

Signals from the sonar transducer are sent to the velocity tracker electronics, where the received pulses are converted into continuous,

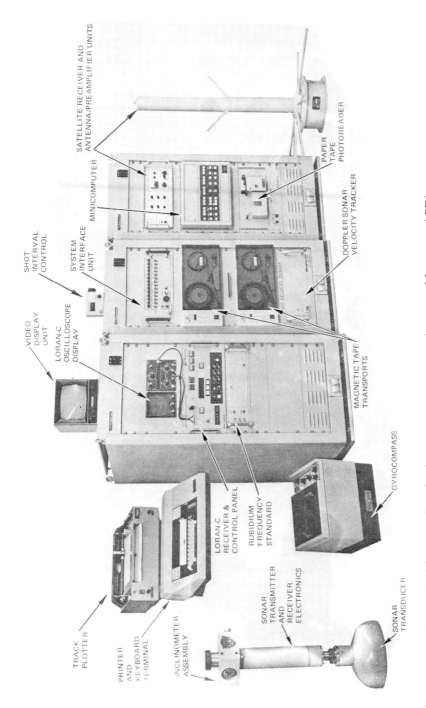

VIDEO
DISPLAY
UNIT

SHOT
INTERVAL
CONTROL

SYSTEM
INTERFACE
UNIT

LORAN-C
OSCILLOSCOPE
DISPLAY

SATELLITE RECEIVER AND
ANTENNA/PREAMPLIFIER UNITS

MINICOMPUTER

PAPER
TAPE
PHOTOREADER

DOPPLER SONAR
VELOCITY TRACKER

MAGNETIC TAPE
TRANSPORTS

GYROCOMPASS

TRACK
PLOTTER

PRINTER
AND
KEYBOARD
TERMINAL

INCLINOMETER
ASSEMBLY

LORAN-C
RECEIVER &
CONTROL PANEL

RUBIDIUM
FREQUENCY
STANDARD

SONAR
TRANSMITTER
AND
RECEIVER
ELECTRONICS

SONAR
TRANSDUCER

Figure 28.31 Typical integrated navigation system components (courtesy Magnavox APD).

509

two-axis velocity information. The Doppler velocity, the thermistor temperature reading, and the inclinometer outputs are all routed through the system interface unit to the computer.

The computer also receives heading information from the gyrocompass. With this, the sonar velocity and calibration data are resolved into velocity north and velocity east components for continuous dead reckoning. All gyrocompasses suffer from errors introduced by ship's velocity, acceleration, and latitude. One of the synergistic advantages of an integrated system is that by computing and feeding back torquing signals to the gyrocompass, these errors can be greatly reduced. Perhaps more important, the need for manual adjustment is eliminated, along with the opportunity for human error.

During a satellite pass, the computer employs the automatic dead reckoning process to describe ship's motion while it collects the satellite message and Doppler counts. Because the sonar accurately measures velocity with respect to the bottom, velocity error is minimized and an accurate fix can be computed even when the vessel maneuvers during the satellite pass. Once computed, the delta-latitude and delta-longitude results are added to the current dead reckoned position (i.e., an update is applied if all automatic acceptance criteria are met). Actually, less than the entire update is applied—the percentage is a function of the expected error in the satellite fix and in the dead reckoned position. The result is a position with the minimum expected error.

When the survey vessel is in water too deep for the sonar to reach the bottom, unknown water current becomes a velocity error. Then it is desirable to employ radio sensors such as the Loran-C system (Fig. 28.31). In this case, an atomic frequency standard permits the Loran signals to be used in the direct ranging mode rather than the conventional hyperbolic or range difference mode. At every satellite fix, the computer applies a bias to the Loran range readings to adjust for propagation and time drift errors. Thereafter, the computer causes the navigated position to agree with the measured Loran ranges. This process is highly filtered to remove position jitter, but the sonar and gyrocompass inputs enable the computer to respond rapidly to turns or other accelerations.

Navigational accuracy is dependent on a number of factors (*Stansell*, 1973), including complement of equipment, adequacy of calibration, water depth, and sea state. Figure 28.32 shows how position error grows with time since the last satellite fix for a rather complete system and for an austere integrated system under both good and poor conditions. The figure also reveals how error increases much more rapidly when the sonar cannot reach bottom, unless a radio aid such as Loran-C is available.

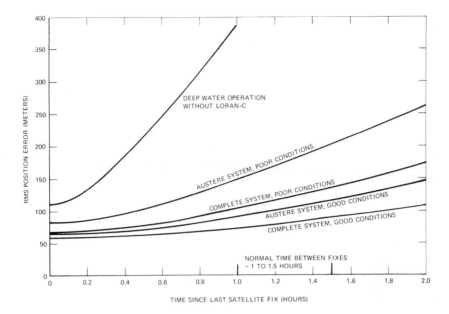

Figure 28.32 Integrated system error as function of time since the last satellite fix update (courtesy Magnavox APD).

The search for oil in ever deeper water is spurring development of sonar equipment that can reach beyond the typical 300 m depth limit to the range of 1200–1800 m. In addition, development of other radio navigation interfaces has accelerated.

For surveys in areas already well covered by high precision radiolocation systems such as Hi-Fix, Raydist, and Lorac, a simpler form of satellite navigation equipment has been used. These radio systems are very precise, but loss of lane count is a serious problem, and reacquisition of lane count can be costly to the survey. In such cases, satellite fixes are being used to verify or to establish lane count, and ship's motion for the fix is being obtained from the radio system.

28.4.4.2. Control of the Survey

An integrated system such as the one in Fig. 28.31 helps control the survey in two ways. First, it allows the vessel to follow predefined survey lines. The latitude and longitude of the start point and of the end point of each line are entered into the computer. The computer is then able to display the vessel's position relative to the survey line in along-track and cross-track coordinates. If the vessel is steered to minimize the cross-track component, it will follow the desired line. Displays can be by means of the video monitor, track plotter, or the printer and keyboard terminal.

Some integrated systems include the ability to steer the survey vessel. The automatic steering capability not only keeps the vessel on the desired line, but on completion it turns the vessel and brings it onto the next survey line.

In addition to guiding the vessel, the system usually actuates the seismic shots. These are fired at uniform time intervals or at uniform increments of distance traveled, as selected.

28.4.4.3. *Data Logging*

The computer not only prints navigational information on the output terminal and drives the track plotter, it records most aspects of the survey on magnetic tape. Data records are made at every shot point, every satellite fix, and to record all manual inputs. Often two tape transports are used to assure that all data are recorded without interruption.

Because an integrated system contains a programmable digital computer that drives magnetic tape recorders, many other types of data can be fed to the computer and recorded along with the navigational information. Typical inputs include those from depth sounders, magnetometers, and gravity meters.

28.4.5. Aircraft Navigation

The Transit system can be used very effectively for aircraft navigation when the flight duration exceeds 4 or 5 hours and when primary navigation is by dead reckoning. For shorter flights, there would be too few satellite fixes to justify carrying the equipment. For long flights, .dead reckoning instruments such as inertial navigation, Doppler radar, or air data computers have decreasing accuracy with time and distance. Thus an occasional position fix update can be quite valuable. Examples are the Project Magnet aircraft, which performs worldwide magnetic surveys, and the P-3 ASW aircraft (*Merkel*, 1973), which spend many hours on submarine patrol. In implementing a Transit system for aircraft navigation, integration with the velocity and altitude sensors is mandatory for reasonable accuracy.

28.5. FUTURE DEVELOPMENTS

A new generation of Transit satellite called TIPS (Transit Improvement Program Satellite) has been developed. The TIPS-I satellite has been launched but will not be used for navigation. The TIPS-II is scheduled for launch in 1975, and after experimental evaluation it may be used for navigation. These new spacecraft contain a number of interesting new features that will improve accuracy and add utility.

A major source of error in the Transit system is due to unmodeled drag forces that act on the satellite, causing it to speed up or slow down with respect to its predicted path. A displacement compensation system (DISCOS) has been added to TIPS-I and II to eliminate this problem. The DISCOS consists of a proof mass that flies within a container, protecting it from all drag forces. Small thrusters on the spacecraft are used to center the container around the proof mass; and because the proof mass is in a drag-free orbit, the satellite by definition will follow the same orbit. The thrusters have enough fuel for a 5-year life, and the results from TIPS-I were considered outstanding.

The TIPS satellite includes a powerful digital computer. Thus by sending programs from the ground, investigators can program the satellite to perform experiments, store and process data, switch out a faulty module, or act like a standard navigation satellite.

Finally, the TIPS provides very precise timing signals, allowing range measurements in addition to the standard Doppler measurements. A pseudorandom code will be transmitted at a high chip rate to permit precise time and range measurements. Oscillator frequency steering gives accurate control of time epoch. A very interesting result from TIPS-I was the discovery that ionospheric refraction correction could be accomplished with only one channel of data by combining Doppler and range measurements. The surprising effect occurs because ionospheric refraction acts to delay the pseudorandom modulation (group delay), whereas it acts to speed up (negative phase delay) the carrier signal in an equal but opposite way.

It is expected that only one or two experimental TIPS will be launched as long as the Global Positioning System continues its successful development. The improvements are only a proof of technology and are available on a contingency basis should GPS be delayed severely.

TWENTY–NINE

ADDITIONAL ELECTRONIC NAVIGATION SYSTEMS

This chapter is a condensed summary of the systems that were not analyzed in the previous chapters but are important because of their significance in the development of the present systems or because they are products of the latest designs not well known to readers.

Delrac (for *Dec*ca *l*ong-*r*ange *a*rea *c*overage) is based on the use of VLF radio waves that can reach long distances along the globe when reflected from the ionosphere. Delrac resembles the standard hyperbolic Decca Navigator System; instead of using the usual star formation, however, the stations work in pairs, one master to each slave, on very long baselines of about 1300–1600 km. The system uses the common transmission frequency of about 10–14 kHz on a time-sharing basis. Each pair of stations has an estimated range of about 5000 km with an accuracy of about ±15 km. Delrac was an experimental system developed by the Decca Navigator Company in the late 1950s; it never became operational, probably because of the successful introduction of the Omega system.

EOLE, a cooperative application satellite, was launched on August 16, 1971, from the NASA Wallops Island station in Virginia. Its orbit is 690 km perigee, 900 km apogee, and 50° inclination. The EOLE system was originally designed for data collection and positioning of meteorological balloons flying at constant altitude in the southern hemisphere; ground-based transponders were also intended to benefit. In addition to its original purpose, EOLE has been used successfully for tracking and positioning drifting objects such as buoys and icebergs within an accuracy of about ±1 km. The interrogation program is prepared and transmitted to the special memory circuit of the satellite by way of telecommand stations on the ground. According to the program, the satellite interrogates each transponder in sequence at a frequency of 460 MHz; upon recognition of its own code number, the transponder answers back at a frequency of 400 MHz. The range and range rate measured in the satellite, together with the memorized data from that particular transponder, are transmitted to telemetry stations on the ground. All data are finally transmitted to the computing center of the Centre National d'Etudes Spatiales (CNES) in Bretigny, near Paris, for processing (*Brachet et al.*, 1974).

GEOLE (geodetic EOLE) is a proposed system whose development has been undertaken by the CNES to create a high accuracy location method based on the EOLE principle. The goals are to achieve an accuracy of about 1 or 2 m within a few hours or about 10–20 m within a few minutes in the location of isolated fixed or slightly mobile points. The system consists of fully automatic, self-contained beacons whose locations will be fixed, and one or more satellites that interrogate these beacons, measuring the ranges and range rates. The data from the beacons will be sent over a conventional telemetry link to the central station. The carrier frequencies involved would be in the region of 400 MHz and 1.5–2.0 GHz (*Guerit*, 1974).

INS (inertial navigation systems) are basically comparable to the conventional, two-dimensional dead reckoning systems, where the position of a new way point is found by the aid of the direction and distance of the vehicle from the previous point. INS instruments are truly self-contained; they operate anywhere in the world independent of the environment (e.g., terrain or atmosphere), and in certain applications (guidance) they eliminate the necessity for man to be involved in the navigation. To determine the attitude and position of a vehicle with respect to the previous situation, a fixed reference coordinate system is needed. This is provided by three gyros whose axes are perpendicular to each other; the gyros are mounted on gimbals to keep the platform coordinate system steady and independent of the movements of the vehicle. The gyros sense the angular changes of the vehicle, while three mutually orthogonal accelerometers mounted on the platform sense the relative changes in acceleration. The first integration of the acceleration yields the velocity of the vehicle, and the second integration yields the distance from the previous way point. When fed to a computer, the acceleration and angular data give the position of the new point in three-dimensional Cartesian coordinates of the accelerometer reference frame. Inertial systems installed on shipboard are designated SINS.

Loran-B, an experimental system developed in the late 1950s, was an intermediate step between Loran-A and Loran-C. It employed medium-long waves at a frequency band of about 1850–1950 kHz similar to Loran-A; but instead of using pulse-matching technique for phase measurement like Loran-A, it used both pulse- and phase-matching techniques, as in the present Loran-C.

Navarho was an experimental system developed in the mid-1950s to be used as a long-range Rho-Theta navigation system providing simultaneous distance and azimuth information. The design characteristics of the system were: a 90–110 MHz carrier frequency band, and an expected range of 2000 nm over land and 2600 nm over water. The antennas were arranged in the form of an equilateral triangle at 0.4 wavelength apart. The system was designed to use amplitude comparison for azimuth determination and phase comparison for distance determination. The system was never implemented, but many of the results obtained through extensive testing were applied to Omega.

SPOT (speed, position, and track) is a proposed system that uses synchronous satellites as space platforms from which information originating at a ground-based center station can be relayed to users of the navigation system. SPOT's phase navigation principle is summarized here. The center station transmits a carrier wave modulated by a low frequency modulation wave. This signal, which is called the field signal, is received in a synchronous satellite and transmitted back to earth to the users. When the signal arrives on earth, the same phase of tone is experienced at all points having the same slant distance to the satellite, and the points are located in a constant-phase locus that is a circle. A set of such circular LOPs can be visualized to exist on the earth's surface (and drawn on a map). To find which of the lines he is occupying, the user transponds the signal back to the center station by way of the satellite. The phase comparison between the returned signal and the generated signal in the center station determines the particular LOP (*Breckman* and *Barnla*, 1970).

VLF transmission has been gaining more popularity as an aid to navigation in recent years. Early in 1973 three transmitters in the U.S.S.R. were transmitting navigational signals at frequencies of 11.905,

12.649, and 14.881 kHz. The central station of the system is located in Novosibirsk, the other two stations being near the east and west boundaries of the country. Each station transmits sequentially on all frequencies a burst of signal 0.4 sec long, having a repetition rate of 3.6 sec.

The group of U.S. Navy transmitters occupying the 15–25 kHz spectrum are used primarily for communication purposes. Since most of the transmissions use cesium beam standards as their master frequency source, thus stabilizing the frequency and the phase, they can serve for navigation and tracking. The signal transmission is continuous, with frequency shift keying as the usual modulation. The transmitters have the highest power in the VLF spectrum, from 100 kW up to 1 MW (*Beukers,* 1973).

VOR/DME is an acronym representing two systems: VHF omnidirectional range/distance measurement equipment; thus it is a Theta-Rho system. It is widely used for civilian air traffic around the world. The VOR operates in the 108–118 MHz band with the channels spaced 100 kHz apart, preventing a given station from interfering with another station. The ground station radiates a cardioid pattern that rotates at 30 Hz, generating an AM 30 Hz modulation signal. The ground station also radiates an omnidirectional signal that is FM modulated with a fixed 30 Hz reference tone. The phase between the two 30 Hz tones varies according to the azimuth of the aircraft's track. The phase difference is sensed by a phasemeter of the aircraft's receiver and is read by the pilot as a direct visual readout. The DME, which is used in conjunction with the VOR to obtain a fix, operates in the 960–1215 MHz band and gives the distance from the aircraft to the ground station up to 150 km. In this pulse-type system, the ground-based transponder returns the pulse to the airborne receiver upon the triggering of the airborne interrogator pulse. The receiver unit provides the pilot a digital readout in nautical miles.

D-VOR (Doppler VOR) is an improved version of the conventional VOR and was developed in the 1960s in West Germany (*Kramer,* 1970). The largest source of errors in the conventional VOR is the reflection of radio waves from obstructions, which can cause navigation accuracy to drop below the required standard when an aircraft is flying over hilly terrain. The reduction factor of error of D-VOR compared to that of VOR can be of the order of 1:10. Improved accuracy is achieved by using a wide base circular antenna array and taking advantage of the Doppler principle to frequency modulate the azimuth-dependent 30 Hz signal. When the antenna circle is electrically rotated, the distance between the transmitter and the receiver is periodically varied, introducing a

Doppler effect in the transmitted frequency. Whereas in the conventional VOR the azimuth-dependent signal is amplitude modulated and the reference signal is frequency modulated, in D-VOR the reference signal is transmitted as an amplitude modulation of the carrier and the azimuth-dependent signal is transmitted as a frequency modulation of the sideband.

REFERENCES

American Ephemeris and Nautical Almanac for the Year 1976. Government Printing Office, Washington, D.C., 1974.

Aslakson, C. I. "New Determination of the Velocity of Radio Waves." *Trans. Am. Geophys. Union*, Vol. 32, No. 6, 1951.

Bakkelid, S. "Determination of Zero Correction for Tellurometers on Trandum Base, 1963." Unpublished report of the Geographical Survey of Norway, Oslo, 1963.

Barger, R. L. and J. L. Hall. "Wavelength of the 3.39 μm Laser-Saturated Absorption Line of Methane." *Appl. Phys. Lett.*, Vol. 22, No. 4, 1973.

Barrell, H. and J. E. Sears. "The Refraction and Dispersion of Air for the Visible Spectrum." *Philos. Trans. R. Soc. Lond.*, Series A, Vol. 238, 1939.

Baumgartner, W. S. and M. F. Easterling. "A World-Wide Lunar Radar Time Synchronization System." *AGARD Conference Proceedings* No. 28, *Advanced Navigational Techniques*, Ed. W. T. Blackband. Technivision Services, Slough, England, 1970.

Bay, Z., G. G. Luther, and J. A. White. *Phys. Rev. Lett.*, Vol. 29, 1972, p. 189.

Bean, B. R. "Atmospheric Bending of Radio Waves." In *Electromagnetic Wave Propagation*. Eds. M. Desirant and J. L. Michiels, Academic Press, 1960.

Bedsted, O. "A Note of the Cyclic Error." Paper presented at the Tellurometer Symposium, London, 1962.

Bender, P. L. "Prospects for Rapid Realization of a Quasi-Earth-Fixed Coordinate System." IAU Colloquium No. 26, Torun, Poland, August 26–31, 1974.

Bender, P. L. *et al.* "The Lunar Laser Ranging Experiment." *Science*, Vol. 182, October 1973.

Bender, P. L. and E. C. Silverberg. "Present Tectonic Plate Motions from Lunar

Ranging." *Proceedings of the International Symposium on Recent Crustal Movements,* Zürich, Switzerland, 1974.

Bergstrand, E. "A Determination of the Velocity of Light." Arkiv för Fysik, No. 15, Stockholm, 1950.

Beuckers, John M. "A Review and Applications of VLF and LF Transmissions for Navigation and Tracking." *Proceedings of the National Radio Navigation Symposium.* The Institute of Navigation, Washington, D.C., November 13–15, 1973.

Bomford, G. *Geodesy.* 2nd ed., Clarendon Press, 1962.

Bossler, J. D. "A Study of the Zero Error in the Model MRA-3 Tellurometer." Thesis, Department of Geodetic Science, Ohio State University, Columbus, 1964.

Bossler, J. D. and S. Laurila. "Zero Error of MRA-3 Tellurometer." *Bull. Geod.,* No. 76, June 1965.

Brachet, G., M. Vincent, and Y. Kakinuma. "The French EOLE Positioning System Application to Drifting Buoy Tracking." International Symposium on the Applications of Marine Geodesy. Battelle Memorial Institute, Columbus, Ohio, 1974.

Breckman, J. and J. D. Barnla. "The SPOT Navigation Satellite System." AGARD Conference Proceeding No. 28, *Advanced Navigational Techniques.* Ed. W. T. Blackband. Technivision Services, Slough, England, 1970.

Bremmer, H. *Terrestrial Radio Waves.* American Elsevier, 1949.

Borodulin, G. I., "Field Tests of the DST 2 Geodimeter Type Instrument." *Geod. Cartogr.* (American Geophysical Union transl.), No. 11–12, November–December, 1961.

Brook, I. "Some Preliminary Results of Measurements Carried out by the Geodetic Division of the Geographical Survey to Determine the Index Error in Tellurometers MRA-2 #462 and #444." Unpublished report of the Swedish Geographic Survey, Stockholm, 1962.

Budden, K. G. *The Wave-Guide Mode of Wave Propagation.* Logos Press, London, 1961.

Burger, W. H. "A Study of Measurements Made with the Model MRA-1 Tellurometer." Thesis, Department of Geodetic Science, Ohio State University, Columbus, 1965.

Burnside, C. D. *Electromagnetic Distance Measurement.* Crosby Lockwood & Son Ltd., 1971.

Calvo, A. B. "Evaluating the Accuracy of Omega Predicted Propagation Corrections." *Navigation,* Vol. 21, No. 2, Summer 1974.

Carter, W. E. "University of Hawaii LURE Observatory." *Proceedings of the Symposium on Earth's Gravitational Field and Secular Variations in Position,* Sydney, Australia, 1973.

Cha, Milton Y. and Peter B. Morris, "Omega Propagation Corrections: Background and Computational Algorithms", U.S. Dept. of Transportation Report Number: ONSOD-01-74, Washington, D.C.

Cole, A. E., A. Court, and A. J. Kantor. "Model Atmospheres." In *Handbook of Geophysics and Space Environments*. Ed. Shea Valley, McGraw-Hill, 1965, Chapter 2.

Cole, A. E. and A. J. Kantor. "Atmospheric Temperature, Density, Pressure, and Moisture." In *Handbook of Geophysics and Space Environments*. Ed. Shea Valley, McGraw-Hill, 1965, Chapter 2.

Connell, D. V. "NPL–Hilger & Watts Mekometer." *Symposium on Electromagnetic Distance Measurement*, Oxford, September 1965. Hilger-Watts Ltd. London.

Cordova, William R. "Secor (Sequential Collation of Range)." Thesis, Department of Geodetic Science, Ohio State University, Columbus, 1965.

Cubic Corporation. "Aeris, Airborne Electronic Ranging Instrumentation System." Report, San Diego, Calif., March 26, 1962.

Cubic Corporation. "AN/USQ-32 Microwave, Geodetic Survey System (Shiran)." ASD Technical Documentary Report ASD-TDR-62-872, San Diego, Calif., 1963.

Davies, K. "Ionospheric Radio Propagation," NBS Monograph 80. Government Printing Office, Washington, D.C., 1965.

Day, Richard Lee. "Description and Comparison of Selected Electronic Navigation Systems." Thesis, Department of Geodetic Science, Ohio State University, Columbus, 1966.

Dean, W. and R. Frank. "Final Engineering Report of CYTAC Long Range Tactical Bombing System." Sperry Engineering Report 5223-1307-14, Great Neck, New York, March 1956.

[The] Decca Navigator Company, Ltd. *Decca Navigator System*. London, 1962.

Denison, E. W. "Electromagnetic Distance Measurement." *Report of IAG Special Study Group* No. 19, 1963–1967.

Dines, L. H. G. "A Comparison Between the Geostrophic Wind, the Surface Wind, and the upper Winds." Meteorological Office 420, London, 1938.

Evenson, K. M., J. S. Wells, F. R. Petersen, B. L. Danielson, G. W. Day, R. L. Barger, and J. L. Hall. *Phys. Rev. Lett.* Vol. 29, 1972, p. 1346.

Evenson, K. M., J. S. Wells, F. R. Petersen, B. L. Danielson, and A. W. Day. *Appl. Phys. Lett.*, Vol. 22, 1973, p. 192.

Ewing, C. "The Parallel Radius Method of Solving the Inverse Shoran Problem." Dissertation, Department of Geodetic Science, Ohio State University, Columbus, 1955.

Faller, J. "The Apollo Retroreflector Arrays and a New Multilensed Receiver Telescope." *Space Research XII*. Academie-Verlag, Berlin, 1972.

Faller, J. E., *et al*. "Geodesy Results Obtainable with Lunar Retroreflectors." *Geophysical Monograph Series*, Vol. 15, American Geophysical Union, Washington, D.C., 1971.

Frank, R. L. "Current Developments in Loran-D." *Navigation, J. Inst. Nav.*, Vol. 21, November 1974.

Franssila, M. "Eine graphische Method zur Bestimmung des Geostrophischen Windes." *Mitt. Meteorol. Zentralanst.*, Helsinki, 1944.

French, Stan. "Old Navaid Passing, LSC Introduces New Loran-C." *Seis. News*, March–April, 1973.

Froome, K. D. and R. H. Bradsell. "Distance Measurements by Means of a Modulated Light Beam yet Independent of the Speed of Light." *Symposium on Electronic Distance Measurement*, Oxford, September 1965. Hilger-Watts Ltd. London.

Furry, R. E. "An Examination of the Errors in Short Line Measurement with the MRA-2 Tellurometer." Thesis, Department of Geodetic Science, Ohio State University, Columbus, 1965.

Galejs, J. *Terrestrial Propagation of Long Electromagnetic Waves*. Pergamon Press, New York, 1972.

Goddard, Robert B. "Differential Loran-C Time Stability Study." U.S. Department of Transportation, Report CG-D-80-74, November 1973.

Guerit, P. "GEOLE." International Symposium of Applications of Marine Geodesy, Battelle Memorial Institute, Columbus, Ohio, 1974.

Guier, W. H. and G. C. Weiffenbach. "A Satellite Doppler Navigation System." *Proc. IRE*, Vol. 48, No. 4, April 1960.

Gunn, R. "Investigation of Ground Reflection by Using the MRA-1 Tellurometer." Unpublished notes, Department of Geodetic Science, Ohio State University, Columbus, 1963.

Gunn, R. "Some Aspects of Ground and Side Reflections in the Microwave Distance Measuring Instruments." Unpublished notes, Department of Geodetic Science, Ohio State University, Columbus, 1964.

Hall, John S., Ed. *Radar Aids to Navigation*, Radiation Laboratory Series, Vol. 2, 1st ed., McGraw-Hill, 1947.

Hallsten, Gunnar. "Test Measurements with AGA Geodimeter 700." Report, National Road Administration, Stockholm, Sweden, 1971.

Hart, C. A. "Surveying from Air Photographs Fixed by Remote Radar Control." *R. Soc. Empire Sci.*, Chart. 7/46, London, 1946.

Hart, C. A. "Some Aspects of the Influence on Geodesy Reference to Radar Techniques." *Bull. Geod.*, October 1948.

Hassard, R. W. "Low Cost Loran-C Receivers." *Coast Guard Eng. Dig.*, July–August–September, 1973.

Hastings, C. E. "Raydist—A Radio Navigation and Tracking System," *Tele-Tech*, Philadelphia, Pennsylvania, June 1947.

Hastings, Charles E. "Electronic Systems for Precision Surveying and Navigation." *Undersea Technol.*, December 1971.

Hastings, Charles E. and Allen L. Comstock. "Pinpoint Positioning of Surface Vessels beyond Line-Of-Site." Paper presented at the National Marine Navigation Meeting of the Institute of Navigation, San Diego, Calif., November 1969.

Hatch, H. "A Study of the Tellurometer Cyclic Zero Error." Thesis, Department of Geodetic Science, Ohio State University, Columbus, 1964.

Hefley, Gifford. "The Development of Loran-C Navigation and Timing." National Bureau of Standards (NBS) Monograph 129, October 1972.

Hirvonen, R. A. Private communications. Helsinki Institute of Technology, Helsinki, Finland, 1961.

Horton, J. W. *Fundamentals of Sonar.* United States Naval Institute, Annapolis, Md., 1965.

Huggett, G. R. and L. E. Slater. "Precision Electromagnetic Distance-Measuring Instrument for Determining Secular Strain and Fault Movements." Paper presented at the International Symposium on Recent Crustal Movements, Zürich, Switzerland, 1974a.

Huggett, G. R. and L. E. Slater. "Electromagnetic Distance-Measuring Instrument Accurate to 1×10^{-7} without Meteorological Corrections." Paper presented at the International Symposium on Terrestrial Electromagnetic Distance Measurements, Stockholm, Sweden, 1974b.

Humphreys, W. J. *Physics of the Air.* McGraw-Hill, 1940.

I. A. G. General Assembly Report of Special Study Group No. 19, International Union of Geodesy and Geophysics Meeting, Berkeley, Calif., 1963.

Ingham, A. E. *Hydrography for the Surveyor and Engineer.* A Halsted Press Book, Wiley, 1974.

International Astronomical Union. Information Bulletin No. 31, January 1974.

Jacobsen, C. E. "High-Precision Shoran Test, Phase I." Air Force Technical Report No. 6611, Wright Air Development Center, Dayton, Ohio, 1951.

Jansky and Bailey, Inc. "The Loran-C System of Navigation," Atlantic Research Corporation report, Arlington, Va., February 1962.

Johler, J. R. "On the Analysis of LF Radio Propagation Phenomena." *NBS J. Res.*, Vol. 65D, 1961.

Johler, J. R. "The Propagation Time of a Radio Pulse." *IEEE Trans. Antenna Propag.*, Vol. AP-11, No. 6, 1963.

Johler, J. R. "Spherical Wave Theory for MF, LF, and VLF Propagation." *Rad. Sci.*, Vol. 5, No. 12, 1970.

Johler, J. R. Private communication, March 1975.

Johler, J. R., W. J. Keller, and L. C. Walters. "Phase of the Low Radio Frequency Groundwave." *NBS Circular 573*, 1956.

Kasper, J. F. "A Skywave Correction Adjustment Procedure for Improved Omega Accuracy." *Proceedings of the ION Maritime Meeting*, Connecticut, October 1970.

Kasper, J. F. "Omega Utilization for Non-Military Subscribers." U.S. Department of Transportation Report 27, July 1971.

Kelly, I. J. "Report on Trials of Decca/Dectra in Boeing Aircraft EI-ALA, June 18th–August 24th, 1963." Electronics Development Section, Aer Lingus–Irish International Airlines, 1963.

Kondraschkow, A. W. *Electrooptische Entfernungsmessung.* VEB Verlag für Bauwesen, Berlin, 1961.

Kramer, E. L. "The Doppler VOR—Its Foundation and Results." *AGARD Conference Proceedings* No. 28, *Advanced Navigational Techniques*, Ed. W. T. Blackband. Technivision Services, Slough, England, 1970.

Krakiwsky *et al.* "Geodetic Control from Doppler Satellite Observations for Lines under 200 km." *Can. Surv.*, Vol. 27, No. 2, June 1973.

Larsson, H. "Investigation of the Accuracy Obtained with the Decca System for Survey in the Southern Baltic." *Int. Hydrogr. Rev.*, May 1954.

Laurila, S. "On the Application of Barometric Air Levelling and Decca Radio Position Fixing in the Formation of Geodetic Networks for Small-Scale Maps." Dissertation, Finland's Institute of Technology, Helsinki, 1953.

Laurila, S. "An Investigation of Shoran-Photogrammetric Position Fixing." Mapping and Charting Research Laboratory Technical Paper No. 188, Ohio State University, Columbus, 1954.

Laurila, S. "Decca in Off-Shore Survey." *Bull. Geod.* No. 39, 1956.

Laurila, S. *Electronic Surveying and Mapping.* Ohio State University Press, Columbus, 1960.

Laurila, S. "Accuracy of Tellurometer over Short Distances." *J. Surv. Mapping Div., Proc. ASCE*, Vol. SU 3, October 1963.

Laurila, S. "Expected Instrument Accuracy of Microwave Distancers." Department of Geodetic Science, Ohio State University, Report No. 64, December 1965.

Laurila, S. H. "Refraction Effect in Satellite Tracking." Technical Reports of The Hawaii Institute of Geophysics, No. 68–19, Honolulu 1968.

Laurila, S. "Statistical Analysis of Refractive Index Through the Troposphere and the Stratosphere." *Bull. Geod.* No. 92, 1969.

Lehr, C. G., M. R. Pearlman, H. M. Salisbury, and T. F. Butler, Jr. "The Laser System at the Mount Hopkins Observatory." Paper presented at the International Symposium on Electromagnetic Distance Measurement and Atmospheric Refraction, Boulder, Col., 1969.

Leitz, Helmut. "Two Electronic Tacheometers by Zeiss." Transl. of *Allgemeine Vermessungs-Nachrichten*, 76th ed., Herbert Wichman Verlag GmbH, Karlsruhe, 1969.

Liebe, H. J. "Calculated Tropospheric Dispersion and Absorption due to the 22-GHz Water Vapor Line." *IEEE Trans. Antenna Propag.*, Vol. AP-17, No. 5, 1969.

Liebe, H. J. and W. M. Welch. "Molecular Attenuation and Phase Dispersion between 40 and 140 GHz for Slant Path Models from Different Altitudes." Telecommunications Research and Engineering Report, U.S. Department of Commerce, Boulder, Colorado, 1973.

Lilly, J. E. "Tellurometer Cyclic Zero Error." Paper presented at the Thirteenth General Assembly of the International Union of Geodesy and Geophysics, Berkeley, Calif., 1963.

List, Robert, J. *Smithsonian Meteorological Tables*, The Smithsonian Institution, Washington, D.C., 1966.

Macomber, M. "Computation of Super Long Distances with the Chord Method." Unpublished report, Department of Geodetic Science, Ohio State University, Columbus, 1957.

Marshall, A. G. "Development of the MRB 201/301." Research and Development, Tellurometer Division of Plessey, Farmingdale, N.Y., 1971.

McCullough, William R. "The Measurement of Distance Using Light." Paper presented at the Fifteenth Survey Congress in Newcastle, N.S.W., Australia, 1972.

McGoogan, J. T. "Precision Satellite Altimetry." NASA Wallops Station, Wallops Island, Va., 1974.

McGoogan, J. T., C. D. Leitao, L. S. Miller, and W. T. Wells. "Skylab S-193 Altimeter Experiment, Performance, Results and Applications." Paper presented at the International Symposium on Applications of Marine Geodesy, Battelle Memorial Institute, Columbus, Ohio, 1974a.

McGoogan, J. T., C. D. Leitao, W. T. Wells, L. S. Miller, and G. S. Brown. "Skylab Altimeter Applications and Scientific Results." Paper presented at the American Institute of Aeronautics and Astronautics/American Geophysical Union Conference on Scientific Experiments of Skylab, Huntsville, Ala., 1974b.

McKay, M. W. *The AN/APN-96 Doppler Radar Set.* General Precision Laboratory, Technical Series, Vol. 8, Pleasantville, N.Y., 1958.

McMahon, F. A. *The Radan Model PC-221A.* General Pecision Laboratory, Technical Series, Vol. 4, Pleasantville, N.Y., 1957.

McMahon, F. A. "AN/APN-187 Doppler Velocity Altimeter Radar Set." *Digest of U.S. Naval Aviation Weapons Systems*, Avionics Edition, Navair 08-1-503, Indianapolis, Indiana, May 1970.

Meade, B. K. "Correction for Refractive Index as Applied to Electro-Optical Distance Measurements." Paper presented at the Symposium on Electromagnetic Distance Measurement and Atmospheric Refraction, Boulder, Colo., 1969.

Mendoza, E. R. "A Method of Determining the Velocity of Radio Waves over Land on Frequencies near 100 kHz." *J. IEE*, Vol. 94, No. 32, 1947.

Merkel, T. B. "Military Application of the TRANSIT Satellite Navigation System in the P-3C ASW Aircraft." *Navigation*, Vol. 20, No. 3, Fall 1973.

Miller, L. S., G. S. Brown, and G. S. Hayne. "Engineering Studies Related to Geodetic and Oceanographic Remote Sensing Using Short Pulse Technique." Research Triangle Institute, Park, N.C., final report, February 1973.

Mofenson, J. "Radar Echoes from the Moon." *Electronics*, April 1946.

Moffett, J. B. "Program Requirements for Two-Minute Integrated Doppler Satellite Navigation Solution." Johns Hopkins University Applied Physics Laboratory Report TG 819-1 (Rev. 2), Baltimore, Md., 1973.

Mulholland, J. D. and E. C. Silverberg. "Measurement of Physical Librations Using Laser Retroreflectors." *Moon*, Vol. 4, 1972.

Möller, F. and P. Sieber. "Über die Abweichung zwischen Wind und geostrophischen Wind in der freien Atmosphäre." *Ann. Hydrogr. Marit. Meteorol.*, Vol. 7, 1937.

Naval Oceanographic Offices. H.O. Publication No. 1-N, "Catalog of Nautical Charts and Publications, Introduction Part II." Washington, DC., May 1966.

Naval Oceanographic Offices, H.O. Publication No. 224, "Omega Skywave Correction Tables." Washington, D.C., April 1973.

North American Datum, National Academy of Sciences, Government Printing Office, Washington, D.C., 1971.

Norton, K. A. "Propagation of Waves," *Proc. IRE*, September 1937.

Parm, T. "A Determination of the Velocity of Light Using Laser Geodimeter." Reports of the Finnish Geodetic Institute, No. 73/3, Helsinki, 1973.

Pederson, Donald O., Jack J. Studer, and John R. Whinnery. *Introduction to Electronic Systems, Circuits, and Devices.* McGraw-Hill, 1966.

Pettersen, S. *Weather Analysis and Forecasting.* McGraw-Hill, 1940.

Pierce, J. A. "Radux." Cruft Laboratory, Harvard University Technical Report No. 17, Cambridge, Mass., July 1947.

Pierce, J. A. "The Use of Composite Frequencies at Very Low Radio Frequencies." Harvard University Technical Report No. 552, Cambridge, Mass., February 1968.

Pierce, J. A., R. H. Woodward, W. Palmer, and A. D. Watt. "Omega: A World-Wide Navigation System-System Specifications and Implementation." Pickard and Burns Electronics Pub. No. 886B, 2nd ed., Waltham, Massachusetts, May 1966.

Pierson, W. J. and E. Mehr. "The Effects of Wind, Waves, and Swell on the Ranging Accuracy of a Radar Altimeter." School of Engineering Science, New York, Technical Report, January 1970.

Piscane, V. L. et al. "Recent (1973) Improvements in the Navy Navigation Satellite System." *Navigation*, Vol. 20, No. 3, Fall 1973.

Poder, K. and O. Bedsted. "Results and Experiments of Electronic Distance Measurements 1960–1963." Paper presented at the Thirteenth General Assembly of the International Union of Geodesy and Geophysics, Berkeley, Calif., 1963.

Polhemus Navigation Sciences, Inc., "Radio Aids to Navigation for the Coastal Confluence." Technical Report No. 72-0661, June 1972.

Powell, Claud. "An Informal Review of Shore-Based Radio Position-Fixing Systems." *Proc. Soc. Underwater Technol.*, Vol. 3, No. 1, 1973.

Powell, C. "The Decca Hi-Fix/6 Position Fixing System." *Int. Hydrogr. Rev.*, Vol. 51, No. 1, 1974.

Powell, C. and A. R. Woods. "LAMBDA: A Radio Aid to Hydrographic Surveying." *J. Brit. IRE*, Vol. 25, No. 6, 1963.

Prescott, N. J. D. "Experiences with Secor Planning and Data Reduction." Symposium on Electromagnetic Distance Measurement, Oxford, September 1965. Hilger-Watts Ltd. London.

Pressey, B. E., G. E. Ashwell, and C. S. Fowler. "The Measurement of Phase Velocity of Ground-Wave Propagation at Low Frequencies over Land Path." *Proc. IRE*, Part III, No. 64, 1953.

Radar Electronic Fundamentals. Navship 900,016, Bureau of Ships, Navy Department, Washington, D.C., 1944.

Ridenour, L. *Radar System Engineering.* McGraw-Hill, 1947.

Romaniello, Charles G. "Advancing Technology in Electronic Surveying." *J. Surv. Mapping Div., Proc. ASCE*, Vol. SU 2, September 1970.

Ross, J. E. R. "Geodetic Application of Shoran." Geodetic Survey of Canada, Pub. No. 78, Ottawa, 1955.

Saastamoinen, J. "Contributions to the Theory of Atmospheric Refraction." *Bull. Geod.* No. 107, 1973.

Saastamoinen, J. "Reduction   of Electronic Length Measurements." *Can. Surv.*, Vol. 27, No. 2, March 1962.

Saltzman, H. and G. Stavis. "A Dual Beam Planar Antenna for Janus Type Doppler Navigation System." General Precision Laboratory, Technical Series, Vol. 9, Pleasantville, N.Y., 1958.

Sanderson, W. T. "Randon Phase Variation of CW Signals in the 70–100 kHz Band." *J. Brit. IRE*, Vol. 12, No. 3, 1952.

Schnelzer, G. A. "A Comparison of Atmospheric Refractive Index Adjustments for Laser Satellite-Ranging." Thesis, Department of Geology and Geophysics, University of Hawaii, Honolulu, 1972.

Schönholzer, A. "Das Statoskop." *Schweiz. Z. Vermess. Kulturtech.*, Vols. 5 and 6, Zürich, Switzerland, 1938.

Senus, W. J. and Richard S. Jones. "A Comparison of Loran-C and Loran-D." RADC-CO-TM-71-4, Rome, New York, August 1971.

Seppelin, T. O. "The Department of Defense World Geodetic System 1972." Defense Mapping Agency, Washington, D.C., May 1974.

Sodano, E. M. "A Rigorous Non-Iterative Procedure for Rapid Inverse Solution of Very Long Geodesics." *Bull. Geod.*, Nos. 47 and 48, 1958.

Stansell, T. A. "The Navy Navigation Satellite System: Description and Status." *Navigation*, Vol. 15, No. 3, Fall 1968.

Stansell, T. A. "Transit, The Navy Navigation Satellite System." *Navigation*, Vol. 18, No. 1, Spring 1971.

Stansell, T. A. "Extended Applications of the Transit Satellite Navigation System." Offshore Technology Conference Preprints, OTC 1397, Dallas, Texas, April 1971.

Stansell, T. A. "Accuracy of Geophysical Offshore Navigation Systems." Offshore Technology Conference Preprints, OTC 1789, Dallas, Texas, April–May 1973,

Stansell, T. A. "Integrated Navigation Systems." Telecommunications Equipment and Systems Technical Seminar, September 1974 (available from Magnavox Research Laboratories, Torrance, Calif.).

Strasser, Georg. "Die elektronischen Entfernungsmesser Wild Distomat DI 50 und DI 10." *Messtechnik*, Vol. 77, No. 12, Verlag Friedr. Vieweg & Sohn GmbH, Braunschweig, 1969.

Swanson, E. R. "Accuracy of the Omega System Using 1969 Skywave Corrections." *NELC TR* 1675, San Diego, Calif., December 1968.

Swanson, E. R. "Omega." *Navigation*, Vol. 18, Summer 1972.

Swanson, E. R. and R. P. Brown. "Omega Propagation Prediction Primer." *NELC* TN 2101, San Diego, Calif., 1972.

Swanson, E. R. and E. J. Hepperly. "Composite Omega." Naval Electronics Center Report No. 1657, San Diego, Calif., October 1969.

Swanson, E. R., P. H. Levine, and D. J. Adrian. "Differential Omega for the U.S. Coastal Confluence Regions." Part I and II, *NELC* TR 1905, January 1974.

Terrien, J. "News from the Bureau International des Poids et Measures." *Meteorologia*, Vol. 10, 1974, pp. 75–77. Springer-Verlag.

Thompson, Moody C., Jr. "Space Averages of Air and Water Vapor Densities by Dispersion for Refractive Correction of Electromagnetic Range Measurements." *J. Geophys. Res.*, Vol. 73, No. 10, 1968.

Thompson, M. C. and L. E. Wood. "The Use of Atmospheric Dispersion for the Refractive Index Correction of Optical Distance Measurements." *Symposium on Electron. Distance Measurement*, Oxford, 1965. Hilger-Watts Ltd. London.

Thompson, M. C. and L. E. Wood. "Radio-Optical Refractometer for Measuring Integrated Water Vapor Refractivity." U.S. Patent Office, Patent No. 3,437,821, 1969.

Tikka, M. "Electronic Methods in Geodesy" (in Finnish). Lecture Duplications, Helsinki Institute of Technology, Department of Surveying, Helsinki, 1973.

Tuori, H. "Seaman's Guides from Kochab to the Satellite" (in Finnish). Reprint from Tähtitiedettä Harrastajille IV, Helsinki, 1965.

U.S. Department of Defense National Geodetic Mapping Program. "AFCRL Geodetic Laser System." Department of Defense, Defense Mapping Agency, Washington, D.C., 1975.

U.S. Geological Survey. "Instructions for Computing UTM Coordinates from Shoran Distances." U.S. Department of the Interior, Geological Survey, Topographic Division, Washington, D.C., 1950.

U.S. Standard Atmosphere 1962. *Government Printing Office*, Washington, D.C., 1962.

Verstelle, J. T. "Use of the Decca Navigator Survey System in New Guinea for Hydrography and as a Geodetic Framework." Report, International Union of Geodesy and Geophysics, Eleventh General Assembly, Toronto, 1957.

Wadley, T. L. "The Tellurometer System of Distance Measurement." *Empire Surv. Rev.*, Vol. 14, Nos. 105 and 106, 1957.

Wadley, T. L. "Electronic Principles of the Tellurometer." *Trans. IEE*, May 1958.

Wait, J. R. *Electromagnetic Waves in Stratified Media*. Macmillan, 1962.

Wait, J. R. "Introduction to the Theory of VLF Propagation." *Proc. IRE*, Vol. 50, No. 7, 1962.

Walsh, Edward J. "Analysis of Experimental NRL Radar Altimeter." *Rad. Sci.*, Vol. 9, Nos. 8 and 9, August–September, 1974.

Weekes, K. and R. D. Stuart. "The Ionospheric Propagation of Radio Waves with Frequencies near 100 kHz over Short Distances." *Proc. IEE*, Vol. 99, No. 2, 1952.

Williams, C. "Low Frequency Radio Wave Propagation by the Ionosphere with Particular Reference to Long-Distance Navigation. *Proc. IEE*, Vol. 98, No. 52, 1951.

Wilson, A. M. "Shoran Photogrammetric Mapping Instructions." Engineering Research and Development Laboratory Report No. 1168, Fort Belvoir, Va, 1950.

Wong, R. E. "Conversion of Shoran Measurements to Geodetic Distances." Mapping and Charting Research Laboratory Technical Paper No. 62, Ohio State University, Columbus, 1949.

Wood, L. E. "A Technique for Transmission of Coherent Broadband Millimeter-Wave Signals and its Application to Propagation, Precise Distance Measuring and Remote Oscillation Synchronization." Dissertation, Department of Electrical Engineering, University of Colorado, Boulder, 1973.

Yionoulis, Steve M. and Harold D. Black. "A Two Satellite Technique for Measuring the Deflection of the Vertical (The Dovimeter)." International Symposium on Applications of Marine Geodesy. Battelle Memorial Institute, Columbus, Ohio, 1974.

NAME INDEX

Numbers in italics refer to pages where complete references are listed.

531

SUBJECT INDEX